AF361812

Challenges and Directions Forward for Dealing with the Complexity of Future Smart Cyber-Physical Systems

Challenges and Directions Forward for Dealing with the Complexity of Future Smart Cyber-Physical Systems

Editors

Martin Törngren
Didem Gürdür Broo
Elena Fersman
Harold (Bud) Lawson
Vincent Aravantinos

MDPI • Basel • Beijing • Wuhan • Barcelona • Belgrade • Manchester • Tokyo • Cluj • Tianjin

Editors

Martin Törngren
KTH Royal Institute of
Technology
Sweden

Didem Gürdür Broo
University of Cambridge
UK

Elena Fersman
KTH Royal Institute of
Technology
Sweden

Harold (Bud) Lawson
Lawson Konsult AB
Sweden

Vincent Aravantinos
Autonomous Intelligent Driving
GmbH
Germany

Editorial Office
MDPI
St. Alban-Anlage 66
4052 Basel, Switzerland

This is a reprint of articles from the Special Issue published online in the open access journal *Designs* (ISSN 2411-9660) (available at: https://www.mdpi.com/journal/designs/special_issues/ cyber_physical_systems2018).

For citation purposes, cite each article independently as indicated on the article page online and as indicated below:

LastName, A.A.; LastName, B.B.; LastName, C.C. Article Title. *Journal Name* **Year**, *Article Number*, Page Range.

ISBN 978-3-03943-020-8 (Hbk)
ISBN 978-3-03943-021-5 (PDF)

Contents

Dedication

We dedicate this book to Harold "Bud" Lawson, IEEE Computer Pioneer, who spent a large portion of his life in fighting unnecessary complexity, all the way from computer engineering and real-time systems to systems engineering.

About the Editors

Martin Törngren has served as Professor of Embedded Control Systems at the Mechatronics division of the KTH, Department of Machine Design, since 2002, at KTH Royal Institute of Technology, Sweden. He has been a pioneer in bridging the gaps between the fields of automatic control and real-time distributed systems, with research in recent years focused particularly on architectural design and safety engineering of autonomous trustworthy cyber-physical systems. He has spent several periods abroad as a visiting scholar, including at UC Berkeley and Stevens Institute of Technology. In 1994, he received the SAAB-Scania Award for qualified contributions in distributed control systems, and in 2004, the ITEA Achievement Award for contributions in the EAST-EEA project. Networking and multidisciplinary research have been characteristic throughout his career. He is the initiator and Director of the Innovative Centre for Embedded Systems, a KTH-industry competence network launched in 2008, and is moreover the Director for the recent Swedish competence center TECoSA on Trustworthy Edge Computing Systems and Applications.

Didem Gürdür Broo is an experienced researcher who is trained as a computer scientist and holds a Ph.D. degree in Mechatronics. Currently, she is a research associate at the University of Cambridge. She is associated with the Center for Smart Infrastructure, Laing O'Rourke Centre for Construction Engineering and Technology and the Center of Digital Built Britain. Dr. Didem Gurdur Broo has collaborated with universities and companies all around Europe through three EU-funded projects while conducting research on data and visual analytics for cyber-physical systems. Dr. Didem Gurdur Broo has a multidisciplinary background from arts, computer science, and mechatronics. She strongly advocates the importance of cross-disciplinary, collaborative research. She has been granted IEEE Senior Membership due to her significant performance and excellence over the last five years. She is a member of European AI Alliance and actively contributes to the discussions of all aspects of artificial intelligence development and its impact on society.

Elena Fersman is Research Director of Artificial Intelligence at Ericsson. She is responsible of a team of researchers located in Sweden, USA, India, Hungary, and Brazil. She is a docent and Adjunct Professor in Cyber-Physical Systems, specialized in Automation, at the Royal Institute of Technology in Stockholm. She holds a Ph.D. in Computer Science from Uppsala University, a Master of Science in Economics and Management from St. Petersburg Polytechnic University, and conducted a postdoctoral stay at the University Paris-Saclay. At Ericsson, she has held various positions ranging from product management to research leadership. Elena is a member of the Board of Directors of RISE Research Institutes of Sweden. Elena has co-authored over 50 patent families.

Harold (Bud) Lawson was an IEEE Life Fellow, Association for Computing Machinery Fellow, INCOSE Fellow, and recipient of the IEEE Computer Society's Computer Pioneer Award (2000), Lawson was a consultant with Lawson Konsult AB, Stockholm, Sweden.

Vincent Aravantinos was awarded his Ph.D. in Theoretical Informatics in 2010 from the University of Grenoble, France. Since then, he explored various academic domains from theory (proof theory at CNRS/TU Vienna, interactive theorem proving applied to probability theory and physics at Concordia University) to applied sciences (formal verification, requirements engineering, model-based systems engineering at fortiss, a research institute of TU Munich). Since 2017, he is working on autonomous driving, first on algorithms for perception, prediction, localization, and motion planning (at fortiss), then on safety of autonomous driving, since 2019, as a systems engineer with Autonomous Intelligent Driving—now a subsidiary of Argo AI.

Preface to "Challenges and Directions Forward for Dealing with the Complexity of Future Smart Cyber-Physical Systems"

This book represents a collection of papers stemming from the Special Issue on "Challenges and Directions Forward for Dealing with the Complexity of Future Smart Cyber-Physical Systems" of *Designs*, closed in 2018.

We would like to dedicate this book to our colleague Harold "Bud" Lawson, one of the co-editors of the Special Issue. He sadly passed away in 2019. Bud Lawson spent a large portion of his life in combatting complexity, all the way from computer engineering and real-time systems to systems engineering (see https://en.wikipedia.org/wiki/Harold_Lawson).

A key aspect of cyber-physical systems (CPS) is their potential for integrating information technologies with embedded control systems and physical systems to form new or improved functionalities. CPS thus draws upon advances in many areas. This positioning provides unprecedented opportunities for innovation, both within and across existing domains. However, at the same time, it is commonly understood that we are already stretching the limits of existing methodologies. In embarking towards CPS with such unprecedented capabilities it becomes essential to improve our understanding of CPS complexity and how we can deal with it. Complexity has many facets, including complexity of the CPS itself, of the environments in which the CPS acts, and in terms of the organizations and supporting tools that develop, operate, and maintain CPS.

This book is a result of the abovementioned journal Special Issue, with the objective of providing a forum for researchers and practitioners to exchange their latest achievements and to identify critical issues, challenges, opportunities, and future directions for how to deal with the complexity of future CPS. The contributions include 10 papers on the following topics: (I) Systems and Societal Aspects Related to CPS and Their Complexity; (II) Model-Based Development Methods for CPS; (III) CPS Resource Management and Evolving Computing Platforms; and (IV) Architectures for CPS.

Of course, many of these topics are not independent, but the above structure provides a high-level overview of the included topics and may act as a guide to readers in finding topics of their interest. A brief introduction to the 10 papers is now provided.

(I) Systems and Societal Aspects Related to CPS and Their Complexity

The first three papers in the book serve as a useful introduction by setting the scene with helicopter perspectives to complexity and "smartness" aspects of CPS.

The paper "How to Deal with the Complexity of Future Cyber-Physical Systems?" introduces a number of perspectives to complexity, including general notions as well as those specific to CPS. The paper further provides an analysis of limitations of existing methodologies for dealing with the complexity of CPS and discusses what is needed in order to address those limitations.

The second paper "Sharpening the Scythe of Technological Change: Socio-Technical Challenges of Autonomous and Adaptive Cyber-Physical Systems" relates to the first paper, and specifically treats safety challenges of future CPS considering the increasing use of artificial intelligence (AI) and machine learning (ML) as part of CPS. In addition, the paper also emphasizes the need to incorporate socio-technical aspects, including trustworthiness, responsibility, liability, as well as the ability to learn from past events.

The third paper, "A Computational Framework for Procedural Abduction Done by Smart

Cyber-Physical Systems", treats "smartness" of CPS in terms of reasoning and decision-making mechanisms within a CPS. In particular, procedural abduction as a knowledge-based computation and learning mechanism is investigated, and a computational framework is proposed as an extension and more detailed elaboration of patterns such as "monitor, analyze, plan, execute".

(II) Model-Based Development Methods for CPS

Papers 4 and 5 treat model-based development CPS as one approach to managing complexity.

Paper 4, "Testing of Complex Embedded Systems Using EAST-ADL and Energy-Aware Mutations" addresses testing in the context of embedded systems architecting with considerations of low-energy computing. The paper highlights the limits of traditional testing methods for future CPS, and advocates using early testing with architectural models, shifting some testing efforts to system models. The paper proposes a method based on fault-based testing to derive tests and executes automated tests leveraging mutation testing, statistical model-checking, and architectural models.

Paper 5, "A Full Model-Based Design Environment for the Development of Cyber-Physical Systems", addresses model-based development of CPS with emphasis on development of the application of the CPS targeting synthesis of logic controllers, i.e., the "cyber" side of a CPS. The paper describes a modeling language and a tool, and describes experiences in its design, including trade-offs, drawing upon developments of production grade systems.

(III) CPS Resource Management and Evolving Computing Platforms

Papers 6 through 9 address resource management and evolving computing platforms in the context of CPS. Paper 6, "A Lazy Bailout Approach for Dual-Criticality Systems on Uniprocessor Platforms", treats resource management and, in particular, scheduling of activities of different criticalities as part of a CPS. A mixed-criticality scheduling method is proposed along with a quality criterion for comparing mixed-criticality scheduling approaches.

Paper 7, "Fighting CPS Complexity by Component-Based Software Development of Multi-Mode Systems", addresses resource management and modeling abstractions to deal with the growing software complexity of CPS. The proposed approach combines two "partitioning approaches", dividing a system into operational modes specified at design time (switching between modes at run-time) with component-based software engineering (integration of independently developed software components). The approach addresses the reconciliation between these two partitioning strategies by introducing a hierarchical mode concept, mode mapping, and mode transformation approaches to link the bottom-up component-based approach and system-level modes.

Paper 8, on "Adaptive Time-Triggered Multi-Core Architecture", addresses time-triggered resource management for multi-core platforms, with the goal of reconciling the benefits of static resource allocation for safety critical CPS with adaptation as a key factor to deal with energy efficiency and fault recovery. Adaptive time-triggered multi-core architecture is introduced, featuring adaptation using multi-schedule graphs while preserving the key properties of time-triggered systems. The architecture is based on a network-on-a-chip with building blocks for context agreement and adaptation. An evaluation is presented using scenarios employing adaptation for improved energy efficiency.

Paper 9, "A Two-Layer Component-Based Allocation for Embedded Systems with GPUs", investigates software component design for embedded systems composed of CPU and GPUs, addressing the increasing heterogeneity—and thereby complexity—of CPS platforms. The paper proposes and evaluates and approach using a two-layer component-based architecture in order to reduce the complexity of component allocation.

(V) Architectures for CPS

Many of the above described papers address, in various ways, architectures or the architecting of CPS. Paper 10, "Developing Self-Similar Hybrid Control Architecture Based on SGAM-Based Methodology for Distributed Microgrids", explicitly addresses architectures for future CPS in the context of microgrids as a CPS domain with specific relevance to sustainable energy. As a basis for the design of the specific distributed software architecture, the characteristics of microgrids are analyzed and the requirements derived, including interoperability (for heterogenous devices and multiple communication protocols), modularity, and scalability to support various functionalities such as island mode operations, energy efficient operations, energy trading, and predictive maintenance. The architecture, which features distributed decision-making, plug-and-play capabilities, and self-similarity of software components, has been applied to a real system of residential buildings, and implementation and deployment details are discussed.

Martin Törngren, Didem Gürdür Broo, Elena Fersman, Harold (Bud) Lawson,

Vincent Aravantinos

Editors

Article

How to Deal with the Complexity of Future Cyber-Physical Systems?

Martin Törngren [1,*,†] and Paul T. Grogan [2,†]

[1] Department of Machine Design, KTH Royal Institute of Technology, 100 44 Stockholm, Sweden
[2] School of Systems and Enterprises, Stevens Institute of Technology, Hoboken, NJ 07030, USA;
    pgrogan@stevens.edu
* Correspondence: martint@kth.se; Tel.: +46-8-790-63-07
† These authors contributed equally to this work.

Received: 30 September 2018; Accepted: 19 October 2018; Published: 22 October 2018

Abstract: Cyber-Physical Systems (CPS) integrate computation, networking and physical processes to produce products that are autonomous, intelligent, connected and collaborative. Resulting Cyber-Physical Systems of Systems (CPSoS) have unprecedented capabilities but also unprecedented corresponding technological complexity. This paper aims to improve understanding, awareness and methods to deal with the increasing complexity by calling for the establishment of new foundations, knowledge and methodologies. We describe causes and effects of complexity, both in general and specific to CPS, consider the evolution of complexity, and identify limitations of current methodologies and organizations for dealing with future CPS. The lack of a systematic treatment of uncertain complex environments and "composability", i.e., to integrate components of a CPS without negative side effects, represent overarching limitations of existing methodologies. Dealing with future CPSoS requires: (i) increased awareness of complexity, its impact and best practices for how to deal with it, (ii) research to establish new knowledge, methods and tools for CPS engineering, and (iii) research into organizational approaches and processes to adopt new methodologies and permit efficient collaboration within and across large teams of humans supported by increasingly automated computer aided engineering systems.

Keywords: complexity; cyber-physical systems; systems engineering; uncertainty

---

1. Introduction

Modern technological development is fueled by simultaneous advances in software, data science and artificial intelligence (AI), communications, computation, sensors, actuators, materials, and their combinations such as 3D printing, batteries and augmented reality. Different perspectives on these advances have led to the creation of many terms—Cyber-Physical Systems (CPS), the Internet of Things (IoT), Industry 4.0, and the Swarm—to represent new classes of technologically enabled systems.

We focus on CPS as a more general notion (e.g., see [1,2]). CPS was introduced in 2006 in the U.S. to characterize "the integration of physical systems and processes with networked computing" for systems that "use computations and communication deeply embedded in and interacting with physical processes to add new capabilities to physical systems" [3]. In this context, the word *cyber* alternatively refers to the dictionary definition of "relating to, or involving computers or computer networks" [4] or more general feedback systems as in the field of *cybernetics* pioneered by Wiener [5]. We consider both interpretations of CPS to be valid and, unless otherwise noted, use the term cyber to refer to computing or software parts of a CPS. For further definitions and viewpoints of CPS, see [6] for an overview of existing agendas and roadmaps.

A CPS is thus characterized by an integration of computing and physical elements, typically including feedback loops, into various networks with both physical and cyber interfaces. We consider human-CPS to be a special (but common) category of CPS where humans are integral elements of the system together with technical parts. Figure 1 illustrates interactions between cyber (C), physical (P) and human (H) elements as well as information and physical interactions between the corresponding components. CPS are developed, produced, used and maintained by teams of people (H) and supporting tools, including computer aided engineering software (C) and hardware-in-the-loop simulation ((C) and (P))). We refer to these (people and tools) as Collaborative Information Processing Systems (CIPS). Finally, both CPS and CIPS act in specific environments. For example, the CPS environment may be a factory in which an automated vehicle operates and the CIPS environment includes related stakeholders such as component suppliers as well as applicable legislation and engineering guidelines/standards.

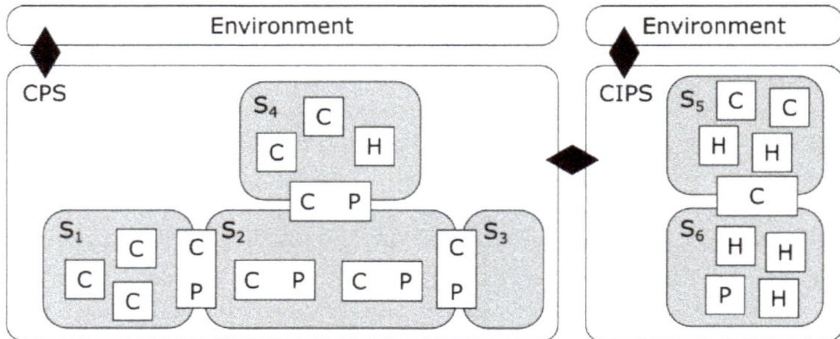

Figure 1. Conceptual view of a CPS with cyber (C), physical (P), and human (H) components arranged in multiple systems (S_i) and a CIPS responsible for design, production, and maintenance.

Cyber-physical systems play a key role in a large number of application domains including in transportation, energy, health and well-being, manufacturing, and indeed as part of smart infrastructures and cities—thus not only relating to industrial domains, see e.g., [7,8].

The combined advances of multiple technologies indicate an ongoing technological shift where new types of systems and systems-of-systems are emerging within and across domains such as those just mentioned. CPS are becoming autonomous, intelligent, connected, and collaborative, resulting in the creation of Cyber-Physical Systems of Systems (CPSoS) with unprecedented capabilities and opportunities. CPSoS are characterized by independent system evolution and the lack of single responsible system integrator (compare for instance with an intelligent transportation system), see e.g., [9].

The implications are that CPS will be widespread and a multitude of key societal functions (such as water, energy, transportation, and health-care) will rely on the proper operation of such CPSoS (henceforth we will use the term CPS to refer to both CPS and CPSoS, unless a detailed distinction is necessary).

This paper focuses on the complexity of future CPS, intuitively interpreted as a system characteristic making it difficult, and sometimes even impossible, to accurately predict behavior over time, especially in terms of understanding all relevant interactions among CPS elements and with the environment. We believe unprecedented capabilities and opportunities will be achieved by a corresponding unprecedented technological complexity, where unknown or poorly understood such interactions may lead to unexpected and undesired effects including effort overruns, poor operational performance, or even system failures.

Our work aims to improve understanding and awareness of complexity in technical artifacts and contribute to establishing new foundations, knowledge and methodologies for engineering design. Evolving approaches to manage complexity attempt to keep up the pace with technology and systems development and avoid unreasonable monetary, personal and societal risks. The full potential can only be obtained when new engineering methodologies are in place to ensure future CPS are sufficiently safe, secure, available and cost-efficient. In addressing these questions, our paper draws upon recent investigations of CPS complexity [6,10,11].

The structure of the paper follows the graphical outline in Figure 2. Section 2 establishes general perspectives on complexity to describe what general notions constitute complexity (Section 2.1), its effects (Section 2.2), and how to manage its evolution (Section 2.3). Section 3 focuses on CPS to elaborate on their specific facets of complexity (Section 3.1), the implications of future CPS on complexity (Section 3.2), and limitations of existing design or management methodologies (Section 3.3). Section 4 discusses how to improve our ability to deal with complex CPS including avenues to address the previously identified facets and limitations. Finally, Section 5 concludes with a summary including our call for action.

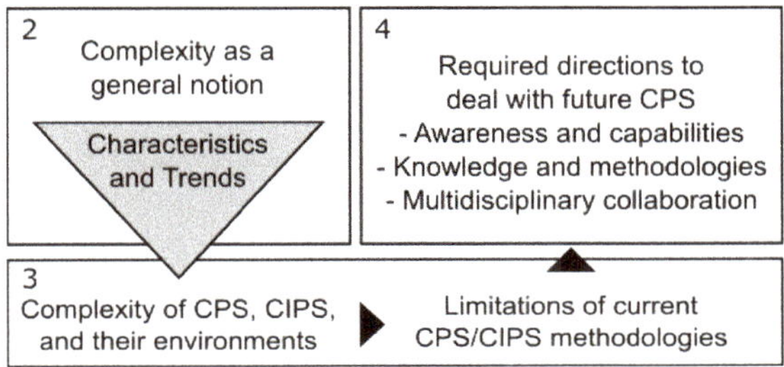

Figure 2. Graphical outline of this paper.

2. Perspectives on Complexity

2.1. Sources of Complexity

It is natural to first elaborate on what we mean by complexity because there are many interpretations and definitions. This discussion views systems design and management as a series of decisions [12] where complexity is a barrier to effective decision-making. Features such as emergence, non-linearity, temporal dynamics, memory, and interaction effects contribute challenges commonly attributed to complexity. Several frameworks argue how complex systems require a distinct set of decision-making and design methodologies.

Snowden's Cynefin framework differentiates sense-making strategies for simple, complicated, complex, and chaotic systems [13]. While simple systems present obvious fact-based solutions, complicated systems demand additional effort but still benefit from best practices available to cover known issues and phenomena including side effects. Complex systems embody unpredictability and a lack of well established practices and thus more knowledge needs to be developed. Chaotic systems omit the very existence of patterns to support decision-making. This model presents complexity as a barrier to decision-making which requires discovery of new knowledge or information to overcome.

From another perspective, Jackson and Keys classify system methodologies along two dimensions [14]. A technical dimension varies between mechanical (reducible or decomposable) and systemic (irreducible) problem types. A social dimension varies between between unitary or pluralistic decision-makers. The most complex or 'wicked' problems span systemic-pluralistic contexts where technical and social sources complicate decision-making due to an inability to separate decisions and link cause-and-effect combined with a general lack of local control over the decision-making process. This model presents complexity as a structural challenge to make decisions in alternative contexts.

Drawing from these frameworks, general sources of complexity in decision-making include large scale or scope from irreducibility of decisions, plurality of opinions or objectives, unpredictability of outcomes, and lack of causal theory or knowledge. More specific to design as a decision-making process, alternative views frame complexity as a source of increased design effort [15] or uncertainty in meeting desired goals of the designed artifact [16]. While at first these two perspectives appear similar, they highlight a fundamental difference between objective (descriptive) and subjective (perceived) sources of complexity [17].

Objective sources of complexity represent an inherent amount of work or information in a system independent of the people involved. They represent sources of effort that remain for a hypothetical omniscient designer to achieve desired system functions with perfect knowledge. For example, the number of components, types of interactions, number of paths through a logical program, and responses to external stimuli are sources of objective complexity which require explicit design effort to define, integrate, and operate. Objective complexity contributes to Cynefin's concept of complicated systems or Jackson and Key's systemic/pluralistic contexts.

Subjective sources of complexity consider the challenges to interpreting, understanding, and anticipating design as a human activity rather than an omniscient one and can be viewed as accidental or additional effort [18]. For example, limits to accumulated knowledge, working memory, and language or communication are sources of subjective complexity which contribute to uncertainty in achieving desired outcomes. Subjective sources of complexity are, by definition, always relative as advances in technology and knowledge influence human activity and relate to Cynefin's concept of complex systems which exhibit unknown unknowns (knowable but unknown to the human observers).

2.2. Effects of Complexity

Complexity contributes two competing effects on system design illustrated in Figure 3. Most first think of distinctly negative effects: uncertainty in achieving functional goals [16], difficulty in separating, analyzing, or solving sets of components [19] or more numerous tasks to achieve a function [15] resulting in increased cost, schedule, or effort [20,21]. These effects are caused by both objective (e.g., number of tasks, total information content, responding to high environmental variability) and subjective sources related to limitations of individual designers. Prior development of complexity metrics attempts to establish scaling laws to predict cost or effort as a function of observable quantities such as lines of code [22], number of logical paths through a program (similar to behavioral state–space) [23], number of components and degree of coupling [19], and architectural topology [24].

However, complexity also contributes positive effects as a key enabler for improved system performance. Well-designed objectively complex systems can afford larger capacity through parallelism, greater response to external stimuli through feedback loops, higher efficiency by closely coupling components, and increased intelligence from adaptability and learning. Under optimal control, complexity allows a system to reach greater levels of performance by more closely matching behavior to needs [25]. All of these benefits, however, critically rely on the design activity being executed *well*, subject to the limited abilities of participating designers.

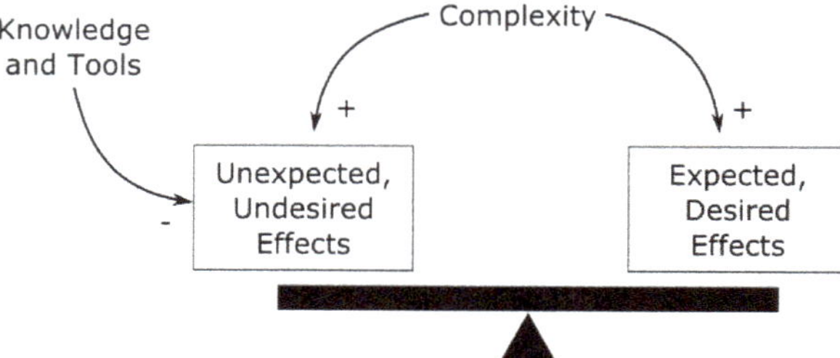

Figure 3. The balance between unexpected, undesired effects and expected, desired effects is influenced both by complexity and by knowledge and tools.

To summarize, objective complexity contributes to both expected, desired effects through essential components and to undesired effects (namely: extra effort) through accidental sources. Subjective complexity contributes only to unexpected, undesired effects due to limited human knowledge, perception, and cognitive abilities but can be directly influenced by mitigating or overcoming human limits. Thus, humans present both the source of and means to tip the balance of complexity positive by mitigating the negative downsides with improved new knowledge and tools. Knowledge establishes explanatory causal links between phenomena and helps improve anticipation and perception of negative downsides. Tools help humans to focus efforts at higher levels of abstraction, relying on automation to optimally solve well-characterized problems.

2.3. Evolution of Complexity

Technological progress is measured by the performance of an engineered system to achieve desired goals efficiently (minimize costs) and effectively (maximize value). Firms expend significant research and development effort to accumulate knowledge and refine designs for increasingly sophisticated systems. This dynamic is perceptible: artifacts transition from complex to complicated domains as a knowledge base is established and matured. Meanwhile, novel compositions may still exhibit complex features due to unknown emergent behavior. This section argues there are two paths to improve performance: knowledge-driven activities to reduce subjective sources of complexity and architecture-based activities to reduce objective sources of complexity.

Although technological revolutions are unpredictable [26], there remain consistent patterns of performance growth over long time frames. For example, Koh and Magee illustrate technological progress as long-term exponential growth of key performance measures in diverse fields such as information and energy technology [27,28]. While patterns such as Moore's law emphasize component features such as the number of transistors per square inch, Christensen argues innovation at the architectural level, not the component level, drives strategic advantage [29,30]. Figure 4 illustrates the effects of technological progress as a series of S-curves which model the initial growth and saturation of alternative product architectures.

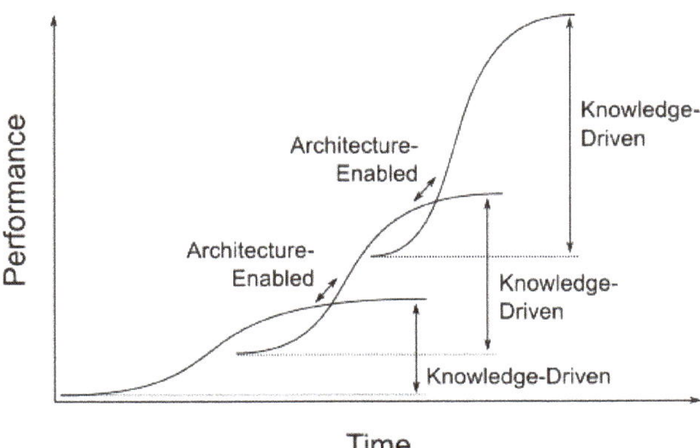

Figure 4. Notional representation of linked technology S-curves highlighting performance gains from knowledge-driven and architecture-enabling activities.

Incremental performance improvement within product architectures benefits from knowledge-driven advances, akin to mitigating the effects of subjective complexity and reducing complex systems to complicated systems. Sustained improvement over longer time frames requires architectural innovation, which resembles a paradigm change where new technologies or the integration of new technologies provide capabilities exceeding the state-of-the-art. Elegant architectures, once matured, limit negative effects of complexity such as unintended side effects without constraining desired performance (see e.g., Section 2.2.1 of [31] for a definition of composability and Section 7.4.1 of [31] for examples of such side effects for distributed real-time computer systems). This pathway deals with objective complexity and leverages design strategies such as modularity [32], layered architectures [33], and abstractions through the use of models [34]. However, some requisite level of knowledge is required to conceive of new product architectures, preventing the two modes of evolution from being separate.

3. Complexity in CPS

In Section 3.1 we elaborate complexity facets of current CPS, and then turn to discuss the evolution of such facets for future CPS (Section 3.2) and how this complexity relates to methodological limitations (Section 3.3).

3.1. Facets of Complexity

Complexity for CPS is highly multifaceted, arising from the CPS itself, its environment, its design process, and its design organization as a CIPS, recall Figure 1. These aspects or viewpoints are interrelated [35] as we will discuss further in this section.

Fundamentally, CPS complexity facets stem from:

- The environment in which a CPS acts and undertakes tasks corresponding to its functional and extra-functional requirements. Environments and tasks are intimately related to the capabilities of CPS [36].
- The cyber components, where software-defined behaviors and sophisticated electronics platforms lead to very large state-spaces with implications for understanding, maintaining and predicting system behaviors. Unintended effects may arise from behaviors and assumptions referring to other software components, resource sharing and the CPS environment. Interactions between software and the electronics platforms and reuse of legacy components (sometimes black-box or poorly documented/understood), further contribute to complexity. Interactions between components in computer systems include both local and global interactions and exhibit much faster characteristic timescales compared to physical interactions [10].
- The physical components, where a key source of complexity arises from side effects which can be the same order of magnitude as the intended behavior (e.g., friction-induced thermal effects between surfaces in contact) [37]. Component interactions and side effects are further multifold (e.g., motion, heat, electromagnetism) and are characterized by strong local effects.
- Interactions between the cyber and physical components. Combining cyber and physical components enables feedback and adaptive systems, providing cost-efficient capabilities otherwise impossible but also characterized by more complex behaviors (e.g., hybrid real-time systems) including a multitude of possible faults and failure modes. As a particular characteristic, a CPS will be characterized by a multiplicity of interfaces and interrelations that encompass both explicit and implicit dependencies among CPS components, and between the CPS and its environment [10]. Correspondingly, a change in one part of a system may affect many others, producing unexpected or undesired side effects from the close coupling and tight integration. The integration of cyber and physical components moreover requires reconciling different worldviews and traditions (see e.g., [38]). Lacking timing (real-time) abstractions for software systems poses a key limitation for predictability [39].

The CPS design process reflects these complexity facets. Establishing a number of desired properties such as performance, cost-efficiency, sustainability, upgradability and safety becomes challenging because of the strong interdependencies expressed in shared design variables. Many of these properties are of a systems nature and depend on the successful development and integration of cyber- and physical components as part of end-to-end feedback interactions, subject to the various side-effects and dependencies mentioned. This results in a situation where the properties are generally difficult to trade-off explicitly and predict during design. It is striking that several of the U.S. definitions of CPS emphasize co-design of cyber and physical parts and the importance of multi-disciplinarity because CPS requires "integration" and composition across a number of aspects and layers (see e.g., [3,10,40]).

Interestingly, integration of cyber and physical components also has a partially limiting effect on complexity. Integration requires each side (cyber and physical) to deal with the limitations and constraints of the other and, specifically, physical constraints limit cyber actions. For example, inertia and energy constraints or cost and safety considerations limit potential actions of cyber components, often forcing designs to reduce various facets of complexity in comparison to general purpose software systems with fewer such constraints.

The over time evolving complexity of CPS is manifested by the introduction of computer control systems and networked control systems that are integrated with, and control, physical systems to provide novel capabilities. This trend has led to needs for a corresponding evolution of organizational responsibilities and processes. For example, introducing local control systems in cars required new responsibilities. Connecting these control systems over Controller Area Networks (CAN) networks (adding the networking dimension to cyber) introduced distributed functions for coordinating subsystems (various physical actuators and sensory subsystems). An example category of such functions are those that provide active safety, for example vehicle stability control, further illustrating the growing scope of responsibilities (e.g., for the network, for the coordination among local controllers and sensors), see e.g., [41].

Facets of complexity for a CPS are intricately tied to its CIPS and vice-versa, linking the CPS, CIPS, and their respective environments in Figure 1. Theories alternately known as the mirroring hypothesis or Conway's law posit a product architecture parallels its parent organization structure [35,42]. Facets of organizational complexity include overcoming communication barriers across organizational units to resolve technical dependencies and aligning multiple viewpoints with the overall system objectives.

Today, thousands of engineers develop what we consider to be advanced machines such as vehicles and aircraft. No single person is able to grasp the complete design and there is no complete model capturing all dependencies in the system. As a remedy to deal with complexity, CPS designers have since long introduced computer support tools (Computer Aided Engineering, CAE) to help to manage complexity in terms of extending memory and reasoning capabilities of individual designers and to support collaboration among large teams of engineers. It only natural that a minimum amount of (objective) CIPS complexity will be required to deal with complex CPS acting in complex environments. Research has confirmed the important role of integration mechanisms, such as organizational structure, work procedures and methods, training, social systems, and computer-aided engineering, to promote multidisciplinary collaboration. In addition, strategic management in relation to the organization is essential because the organizational power base is likely to evolve (e.g., from physical/electronics to software and data) and since future CPS will require new (complementary) competences, roles and responsibilities [43]. Related to the mirroring hypothesis, the CIPS has to evolve along with the changing content of the CPS product/services.

In summary, CPS are characterized by strong coupling and interdependency between cyber and physical components as illustrated in Figure 1. They exhibit facets of complexity such as heterogeneity, size, computability, uncertainty, dynamic behavior and structural dependencies [10,11,44]. We believe CPS represents a fundamental shift from other existing engineering products similar to architecture-enabled innovation in Figure 4 and, thus, carries potential benefits of higher objective complexity with related challenges of overcoming associated subjective complexity.

3.2. Evolution of Future CPS

Overall, CPS are increasingly being fitted with new functions, deployed in new domains, and as larger systems. The implications are that future CPS are facing an increasing envelope of functional and extra-functional requirements. Examples of functional requirements include for instance typical use-cases such as predictive maintenance and increasing levels of CPS automation. Examples of more stricter (and newer) extra-functional requirements include cyber-security (due to increasing connectivity and collaboration), safety (due to extended types of tasks in more complex environments) and environmental sustainability (e.g., due to stricter legislation). The cross-domain nature of many new CPS applications, where for example robotics technologies are adopted by automated cars, and where telecom networks may find their use for smart machines, poses both large opportunities and challenges.

In the following we elaborate the following aspect for evolving CPS: (1) systems integration and scope, (2) intelligence and level of automation, and (3) advanced tasks and complex environments—and then provide conclusions based on the trends in these areas.

As introduced in Section 1, technological advances enable entirely new types of integration and connectivity in CPS across:

- Technological areas such as physical, embedded, networked and information (e.g., cloud and edge computing) systems,
- Standalone systems, for example integrating vehicles and infrastructure to form intelligent transportation systems, and
- Life-cycle stages, in particular making data available throughout the life-cycle and enabling software upgrades. These concepts are closely related to so-called DevOps (Development-Operations integration) associated with continuous software development, integration, and deployment with feedback from operational systems.

At the same time, increasing levels of automation and intelligence including data analytics are being introduced at a rapid pace in CPS. These capabilities rely on a number of Artificial Intelligence (AI) techniques including machine learning. The prospects of AI and data analytics are seen as game-changers, highlighted by a report by the National Science and Technology Council in the US [45] and Section 0.2 of the European strategic research agenda on electronic components and systems [8]. Context awareness is essential for new types of AI-based CPS, including among other things the ability to understand what entities are currently part of the near environment and to infer what their intentions are. Further maturation of AI technologies is likely to drive increasing levels of automation in a number of application domains.

Because of their potential to solve societal challenges and generate revenue, future CPS will also be tasked with increasingly difficult tasks in open environments. Automated driving on public roads provides a useful example where highly advanced technologies will be deployed and spread across society as opposed to traditional manufacturing applications where advanced technologies are enclosed and used by few. Similar trends are seen in manufacturing with robot-human collaboration systems and in many other application domains, see e.g., [8].

Such smart CPS must deal with tasks in dynamic and changing environments (e.g., highly varying traffic environments which also evolve over time with changing infrastructures and behaviors of humans and CPS). In general, this implies that all operational conditions will not be known a priori (at the time of system development) and, correspondingly, system adaptation and evolution after deployment, including learning from field data, will be essential, see e.g., [10]. The broader implications indicate open CPS will face a variety of existing and new types of uncertainties from partially unknown environments, security vulnerabilities and attacks (e.g., with the difficulty of anticipating attacks by providing appropriate "attacker models"), and changing properties of the CPS itself (e.g., due to partial failures, upgrades or learning). Uncertainty may relate to aspects for all life-cycle stages and parts of a CPS (e.g., in terms of environment perception, embedded knowledge and actual physical capabilities), as well as to the CPS environment. A systematic treatment of uncertainty thus becomes important, taking various sources of uncertainty into account, see e.g., [46].

With this evolution, new types of CPS and CPSoS will be characterized by intelligence, automation, flexibility, connectivity, and data- and service-related business models, see e.g., [1,9,47,48]. Integration with information technologies, most notably cloud and edge computing, will provide platforms to enable new functionalities, on top of which new services can be built , see e.g., [49]. These developments will further stimulate technological development and inventions, continuing the drive towards more sophisticated, large-scale, intelligent and adaptive CPS likely to continue on a path of complexity growth. Using the Cynefin terminology, it is in our understanding relevant to characterize current advanced CPS as representing a mixture of complicated and complex systems while future CPS will fall into the class of complex systems until requisite knowledge can be obtained.

In summary, the continued evolution of CPS will have a drastic impact on complexity, clearly increasing the objective complexity by orders of magnitude, and also introducing new phenomena in terms of smart evolving automated CPS, raising the subjective complexity. The new types of CPS will face more of unknowns, dependencies, hidden assumptions and accidental complexity (due to systems integration, legacy, etc.). This in turn leads to concerns whether we are really prepared to embrace such new levels of complexity.

3.3. Limitations of Existing Methodologies

This section reviews and discusses the key limitations of existing methodologies to address complexity of future CPS.

As discussed by [10], current methodologies use a number of techniques to deal with CPS complexity, including, (1) process-based approaches (e.g., as checklists for all aspects that needs to be considered, and for aligning the different life-cycles of software and hardware), (2) model-based and computer aided engineering—which support complexity management through multiple abstractions and views where the challenge then becomes the integration/reconciliation among these views), (3) design and architecting measures including layered architectures and principles to promote composability, and finally, (4) through people/organizational approaches that enhance skills, motivation and collaboration.

While these measures work for current systems, it is generally considered that they do not scale to the next generation CPS. Consequently, there are also multiple calls for new methodologies, see e.g., [8,10,50–52]. As an example, the Electronic Components and Systems for European Leadership (ECSEL) roadmap, which was developed in collaboration between three industrial-academic associations representing hundreds of organizations active in electronics, embedded systems and CPS [8]. highlights five essential areas to manage future CPS across application domains:

- Systems and components: architecture, design and integration,
- Connectivity and interoperability,
- Safety, security and reliability,
- Computing and storage,
- Electronics process technology, equipment, materials and manufacturing.

Many of these topics relate to CPS complexity facets and also consider increased levels of automation, intelligence, and evolvability, (see e.g., Chapters 6 and 8 in [8]). Roadmaps call for technical measures and improvements on a broad scale including engineering education, processes and organizations for CPS, and legislation see e.g., [7,10,51].

Dealing with the increasingly complex environments, uncertainty, and the inherent dependencies and side effects in CPS, represents core aspects that must be addressed to improve the state of methodologies. As described in Section 3.2, future CPS are likely to be tasked with increasingly challenging tasks in open, dynamic and changing environments. Describing these varying environments and systematically dealing with uncertainty represents a key challenge.

Dependencies related to a CPS, as described in Section 3, are closely related to the concept of composability, to be discussed in the following.

The success of Very Large Scale Integration (VLSI) for integrated circuits and design methods for software can be traced to high-level design abstractions and synthesis methods, and, in particular for VLSI, synthesis methods yielding correctness by design. Similar goals have been targeted for CPS; however, while we agree that correctness by design is a highly worthwhile goal and that unnecessary complexity should be eliminated where ever possible (see e.g., [31]), correctness by construction represents a challenge for CPS. Abstraction and decomposition, as traditional means to deal with complexity, assume side effects can be eliminated or managed. As investigated by [37], the success for digital systems design in VLSI arises because their characteristics enable abstractions and eliminate or robustly deal with side effects. Similar success has not been achieved for mechanical engineering because the same characteristics are not valid; for example, platform-based design is much more complicated with multi-fold side effects [10]. It follows that the endeavor of correct-by-design is even further challenging for CPS.

The essence of CPS design aims to combine and integrate elements (e.g., C, P and H) and components (e.g., C-P components) to achieve the desired functionality and extra-functional properties (e.g., performance, reliability and flexibility), within - and in interaction with - varying partly unknown dynamic environments. This integration may take place during design (off-line) or run-time for adaptive systems. We see the lack of a systematic treatment of "composability" (integration) as one overarching limitation of existing methodologies. Composability has been defined in Section 2.2.1 of [31] as "An architecture is said to be composable with respect to a specified property if the system integration will not invalidate this property, once the property has been established at the subsystem level. ... Examples of such properties are timeliness or testability. In a composable architecture, the system properties follow from the subsystem properties".

The likely characteristics of future CPS (recall Section 3.2) will clearly require novel ways for reasoning about system level properties and in achieving composability. In the following we elaborate the multidimensional topic of CPS composability and identify what we see as as specialized composability issues including dealing with human-CPS, CPS incorporating AI, trustworthiness and CPSoS.

- **Composability across components, disciplines and aspects.** When composing CPS components, multiple technologies needs to be integrated (electrical, mechanical, hydraulic, software, etc.) and multiple aspects needs to be reconciled — referring to for example properties such as cost, safety and availability. When reasoning about such compositions, multiple theories will apply, each focusing on one or more aspects of integration (e.g., logic-interface automata, composing transfer functions, scheduling, overall reliability, etc.). Real and successful composition must consider all these aspects, including characteristic side effects and dependencies of CPS. For this there is no existing comprehensive theory or methodology. Current best practices can manage systems of today's complexity, albeit using heuristics and with questionable cost-efficiency. Beyond composition at design or production time, CPS will be increasingly adaptive and able to reconfigure during run-time. There are thus needs to improve interoperability and reason about composability over multiple aspects also at run-time. In a similar vein, with the use of DevOps and learning systems with CPS, the software behaviors of parts of the CPS will change during operation, requiring the ability to monitor and ensure proper operations, see e.g., [1,8].

- **Human-CPS integration.** Regardless of the system type and level of automation, humans acting as developers, operators, users, and maintainers will increasingly have to deal with and interact with CPS. Increasing levels of automation poses challenges for human-machine interaction, e.g., for cases where humans (e.g., pilots in an aircraft) may still be responsible to act in emergency situations. We are currently transitioning to increasing levels of automation and smarter systems and the lessons learned in the automation history remain highly relevant to improve support for humans interacting with highly capable CPS [53]. As an example of a related challenge in automated driving, leading companies have abandoned the so-called "level 3" of automated driving, moving directly to "level 4" where the automation system is also responsible for fall-back maneuvers, see e.g., [54]. A key notion for human CPS is that of intent, i.e., the understanding of what drives the action or inaction of an agent. There is also a need to identify deviations from normal behaviour of such (human) agents, in particular behaviour that leads to decisions/actions of significance for the functioning of CPS [8] (e.g., Chapter 5.3.3).

- **AI within CPS.** While AI and machine learning technologies enable entirely new types of applications, the use of in particular machine learning in terms of neural networks raises concerns about how to deal with transparency (black-box behavior), robustness and predictability (e.g., when data is outside of a training set), and how to cost-efficiently verify, validate and assure such systems, see for example [55,56].

- **Trustworthiness.** Trust involves properties such as privacy, cybersecurity, safety and availability, which affect each other, requiring new holistic methodologies. Security risks already exist for current CPS and will increase as CPS become even more widely adopted, connected and with an increasing use of open source software. Absolute safety and security will not be possible so on-line measures are necessary to deal with security breaches and safety related anomalies. Moreover, the increasing deployment and use of CPS increases the importance of their availability, implying traditional fail-safe solutions are not an option. Future systems must be fault-tolerant while balancing the complexity increase due to the introduction of redundancy, adaptability and fall-back measures. Finally, issues related to liability, ethics and assurance are recognized as essential. Who if any will/should take responsibility when a complex CPS fails, what are the ethical considerations of decisions made by highly automated systems, and what is required by an assurance case for a future complex CPS? [8]
- **CPSoS.** Future CPS are likely to form part of CPSoS. Such systems, may also because of their novelty and scale, relate to multiple domains, and require consideration a larger set of stakeholders, jurisdictions, regulations and standards [9].

Existing development methodologies are usually centered in either physical systems, software, or data management. Methodologies are also not integrated when it comes to providing a holistic view of complex CPS, CIPS and environments. Each entity itself is complex but also strongly interrelated. For example, the corresponding CIPS encompasses an increasing quantity of tools, data, information and knowledge. Structuring, managing and ensuring interoperability among such assets provides a great challenge on its own. Information capture and formalization requires extra work efforts including the maintenance of relations and validity of this information. Methodologies are also not aligned across the life-cycle stages. This is, for example, visible with the increasing difficulty of CPS maintenance, requiring people and tools with additional skills/capabilities. Future methodologies will need to embrace and integrate all these perspectives.

4. Addressing Limitations to Complexity

Given our analysis and review of CPS complexity and state of the art, we conclude that progress is required to deal with future CPS. This section addresses how we should act to overcome the identified limitations and the required means and strategies required. Research agendas discussed in previous sections propose a variety of measures to deal with future CPS including strategic research areas identified by ECSEL [8] and a mix of socio-technical priorities. We contrast these perspectives by relating to our discussion on the evolution of complexity in general (Section 2.3) and for CPS (Section 3.2).

With the human creativity and ingenuity, we envision a future better prepared to deal with unprecedented complexity, corresponding to our ability to make a paradigm-changing architectural transition as indicated in Figure 4. This future development scenario implies that organizations will have access to existing and new knowledge and methods that explicitly address and manage uncertainty, dependencies and composability. Future tools and automation will assist humans in CPS design to: (i) improve efficiency and effectiveness of inter-human communication, (ii) better accommodate (design) tasks for which our memories and processing capabilities are insufficient, i.e., automating complex tasks such as design space exploration, change management and verification, and (iii) provide adequate collaboration between humans and supporting AI systems. Organizations and management will be strategically prepared by establishing greater awareness of complexity and adopting life-long learning. Awareness relates to creating an understanding of both the benefits and risks associated with complexity.

To reach such a setting/scenario, and to be able to deal with the transition towards such a mature state, there is a need to promote and drive efforts that address:

- Complexity awareness by people and organizations, as well as improving training, leveraging existing knowledge and best practices,
- Research towards establishing CPS foundations that addresses the identified limitations of current CPS methodologies,
- Multi-disciplinary collaboration for the above mentioned educational and research efforts, and also in terms of exchange of best practices across industrial domains and between academia and industry.

In order to address complexity there is a need to create and raise awareness for various CPS stakeholders. This represents an important first step, implying that organizations would be better prepared to assess costs and risks for both projects and products/services. Training and spreading knowledge and best practices play important roles and go beyond engineers. There is a need to consider and include a broader set of stakeholders including policy-makers and the general public to raise awareness of CPS implications (opportunities, risks and challenges) [7].

Complexity metrics have potential to bring awareness to developing organizations and decision-makers. Relevant metrics developed with respect to objective and subjective complexity could give insights and guide decisions. Developing representative metrics has, however, shown to be very difficult. While many have been proposed, they tend to emphasize objective complexity and focus on a few aspects [57]. Few have been adopted in practice, and those that have (e.g., lines of code) are very coarse grain. Thus, while it would be valuable to have operational complexity metrics, other measures will be needed for raising awareness, including through training and guidelines with attention to the multiple aspects of CPS complexity. In addition, it is necessary to develop new ways of organizing life-long learning as a way to systematically increase organizational capabilities through use of new knowledge and tools.

There are several challenges for education and training in relation to future CPS, see e.g., [51,58]. Fundamentally these refer to the trade-off between breadth and depth and identifying appropriate knowledge profiles, preparing students for continued learning (self-learning), the fact that teachers as well as universities have disciplinary backgrounds/profiles with corresponding difficulties in teaching topics cutting across the aspects of a CPS, and finally, the fact that teaching, in many cases, has a low status compared to research. In the short term, curricula need renewal to consider the needs for the "cross-cutting" aspects of CPS, such as developing new courses that address safety and security, safety and AI for CPS, etc. There is clearly room for new schemes for continued education and strong needs to raise the importance and status of teaching and training in this area.

As emphasized by several research agendas, see e.g. [1,7], there is a need for new foundations of CPS engineering to address the issues elaborated in Section 3.3. We envision that such new foundations would encompass new abstractions,(uncertainty) modeling frameworks, composable multi-view models and composable architectures. Based on new knowledge in these areas, novel methods and tools can be developed, eventually yielding new methodologies.

In parallel, there is a need to drive research into organizational approaches and processes that can adopt such methodologies while catering for efficient collaboration within and across large teams of humans supported by increasingly automated computer aided engineering systems. There is a need for alignment and tailorability among the product, organization and process architectures as discussed earlier. This structuring should align with the corresponding structure of teams and their interactions and processes with specific consideration for synchronization and alignment and software and hardware processes. Automation and assistance by CAE systems will be needed to support the above and in particular effective communication between people and teams (currently requiring huge efforts as part of development). Strategic management efforts will be essential for all these endeavours.

Many of the above efforts will require multidisciplinary collaboration because of the nature of CPS. Collaboration between industry and academia will also be essential. While we hope to make much progress in dealing with future CPS, they will continue to evolve, demanding longer-term efforts that continuously work towards spreading existing knowledge and for gaining new knowledge. There will correspondingly be a need to advance formal methods and techniques (for dealing with complicated CPS) as well as engineering heuristics (for dealing with truly complex CPS). Correspondingly, it will be important to leverage cross-domain collaboration encompassing the exchange of best practices across industrial domains and between academia and industry.

5. Discussion and Conclusions

This paper argues CPS, as close integrations of physical, cyber, and human elements, represent an architectural innovation over existing engineered systems carrying both unprecedented potential to improve performance but also unprecedented challenges. From an objective perspective, CPS complexity arises from technical sources such as number and heterogeneity of components, degree of interdependency and response to stimuli in an open environment. While some some aspects of complexity are desired for performance, from a subjective perspective the existing knowledge base on CPS is limited, generating difficulty for engineers to anticipate and predict their behavior. We argue new design methods, tools and, ultimately, knowledge will mature CPS.

Limitations to dealing with future CPS arise from the CIPS responsible for designing and operating CPS. Design of future CPS must overcome current limitations for describing highly varying environments, and in dealing with uncertainty and composability. Composability covers both the integration of cyber and physical components and the integration of technical systems with human counterparts. Issues such as effective collaboration between humans and intelligent systems, establishing trustworthiness, and considering a broader system of systems perspective must be addressed.

Our call to action can be summarized in three points. First, for people and organizations involved with CPS, efforts are needed to increase the awareness of the sources of complexity and approaches to mitigate the unexpected and undesired side effects arising from an insufficient or immature knowledge base. Second, new research should specifically address the limitations to current CPS methodologies including identifying approaches to better understand uncertainty and composability. Finally, improved collaboration across disciplines and between industry and academia is necessary to implement the educational and research efforts described above.

Responding to this call will be important to reach the projected societal-scale advantages of CPS while managing the new risks introduced by such CPS.

Author Contributions: Both authors contributed equally to this article. M.T. led efforts on Sections 1, 3, and 4. P.T.G. led efforts on Sections 2 and 5 and implemented the four figures.

Funding: This research was partially funded for Martin Törngren by the Platforms4CPS project, supported by the European Commission H2020 program, Grant Agreement No. 731599, and a KTH sabbatical grant.

Conflicts of Interest: The authors declare no conflict of interest.

References

1. Damm, W.; Sztipanovits, J.; Baras, J.S.; Beetz, K.; Bensalem, S.; Broy, M.; Grosu, R.; Krogh, B.H.; Lee, I.; Ruess, H.; et al. Towards a Cross-Cutting Science of Cyber-Physical Systems for Mastering all-Important Engineering Challenges. 2016. Available online: https://cps-vo.org/node/27006 (accessed on 4 August 2018).
2. Lee, E.A.; Seshia, S.A. *Introduction to Embedded Systems: A Cyber-Physical Systems Approach*, 2nd ed.; MIT Press: Cambridge, MA, USA, 2017.
3. Cengarle, M.V.; Bensalem, S.; McDermid, J.; Passerone, R.; Sangiovanni-Vincentelli, A.; Törngren, M. Characteristics, Capabilities, Potential Applications of Cyber-Physical Systems: A Preliminary Analysis. 2013. Available online: http://www.cyphers.eu/sites/default/files/D2.1.pdf (accessed on 11 July 2018).
4. Cyber. *Merriam-Webster Dictionary*; Merriam-Webster: Springfield, MA, USA, 2018.

5. Wiener, N. Cybernetics. *Sci. Am.* **1948**, *179*, 14–19. [CrossRef] [PubMed]

6. Platforms4CPS. Foundations of CPS—Related Work. 2017. Available online: https://platforum.proj.kth.se/tiki-index.php?page=Foundations+of+CPS+-+Related+Work (accessed on 11 July 2018).

7. Schätz, B.; Törngren, M.; Bensalem, S.; Cengarle, M.V.; Pfeifer, H.; McDermid, J.; Passerone, R.; Sangiovanni-Vincentelli, A. Research Agenda and Recommendations for Action. 2015. Available online: http://cyphers.eu/sites/default/files/d6.1+2-report.pdf (accessed on 20 October 2018).

8. AENEAS; ARTEMIS Industry Association; EPoSS. Strategic Research Agenda for Electronic Components and Systems. 2018. Available online: https://efecs.eu/publication/download/ecs-sra-2018.pdf (accessed on 20 October 2018).

9. Engells, S. European Research Agenda for Cyber-Physical Systems of Systems and Their Engineering Needs. 2015. Available online: http://www.cpsos.eu/wp-content/uploads/2016/06/CPSoS-D3.2-Policy-Proposal-European-Research-Agenda-for-CPSoS-and-their-engineering-needs.pdf (accessed on 10 August 2018).

10. Törngren, M.; Sellgren, U. Complexity Challenges in Development of Cyber-Physical Systems. In *Principles of Modeling*; Lohstroh, M., Derler, P., Sirjani, M., Eds.; Lecture Notes in Computer Science; Springer: Cham, Switzerland, 2018; Volume 10760, doi:10.1007/978-3-319-95246-8_27.

11. Grogan, P.T.; de Weck, O.L. Collaboration and Complexity: An experiment on the effect of multi-actor coupled design. *Res. Eng. Des.* **2016**, *27*, 221–235, doi:10.1007/s00163-016-0214-7. [CrossRef]

12. Ullman, D.G. Robust Decision-making for Engineering Design. *J. Eng. Des.* **2001**, *12*, 3–13, doi:10.1080/09544820010031580. [CrossRef]

13. Snowden, D.J.; Boone, M.E. A Leader's Framework for Decision Making. *Harv. Bus. Rev.* **2007**, *85*, 68–76. [PubMed]

14. Jackson, M.; Keys, P. Towards a System of Systems Methodologies. *J. Oper. Res. Soc.* **1984**, *35*, 473–486. doi:10.1057/jors.1984.101. [CrossRef]

15. Bashir, H.A.; Thomson, V. Models for estimating design effort and time. *Des. Stud.* **2001**, *22*, 141–155, doi:10.1016/S0142-694X(00)00014-4. [CrossRef]

16. Suh, N.P. A theory of complexity, periodicity and the design axioms. *Res. Eng. Des.* **1999**, *11*, 116–132, doi:10.1007/PL00003883. [CrossRef]

17. Schlindwein, S.L.; Ison, R. Human knowing and perceived complexity: Implications for systems practice. *Emerg. Complex. Organ.* **2004**, *6*, 27–32.

18. Brooks, F.P. No Silver Bullet: Essence and Accidents of Software Engineering. *IEEE Comput.* **1987**, *20*, 10–19. [CrossRef]

19. Summers, J.D.; Shah, J.J. Mechanical engineering design complexity metrics: Size, coupling, and solvability. *J. Mech. Des.* **2010**, *132*, 021004, doi:10.1115/1.4000759. [CrossRef]

20. Braha, D.; Maimon, O. *A Mathematical Theory of Design: Foundations, Algorithms and Applications*; Springer Science & Business Media: Dordrecht, The Netherlands, 1998. doi:10.1007/978-1-4757-2872-9.

21. Arena, M.V.; Younossi, O.; Brancato, K.; Blickstein, I.; Grammich, C.A. *Why Has the Cost of Fixed-Wing Aircraft Risen? A Macroscopic Examination of the Trends in U.S. Military Aircraft Costs over the Past Several Decades*; Monograph MG-696-NAVY/AF; RAND Corporation: Santa Monica, CA, USA, 2008.

22. Albrecht, A.J.; Gaffney, J.E. Software Function, Source Lines of Code, and Development Effort Prediction: A Software Science Validation. *IEEE Trans. Softw. Eng.* **1983**, *SE-9*, 639–648, doi:10.1109/TSE.1983.235271. [CrossRef]

23. McCabe, T.J. A Complexity Measure. *IEEE Trans. Softw. Eng.* **1976**, *4*, 308–320, doi:10.1109/TSE.1976.233837. [CrossRef]

24. Sinha, K.; de Weck, O.L. Empirical Validation of Structural Complexity Metric and Complexity Management for Engineering Systems. *Syst. Eng.* **2016**, *19*, 193–206, doi:10.1002/sys.21356. [CrossRef]

25. Deshmukh, A.V.; Talavage, J.J.; Barash, M.M. Complexity in Manufacturing Systems. Part 1: Analysis of Static Complexity. *IIE Trans.* **1998**, *30*, 645–655, doi:10.1023/A:1007542328011. [CrossRef]

26. Simon, H.A. The Steam Engine and the Computer: What Makes Technology Revolutionary. *Comput. People* **1987**, *36*, 7–11.

27. Koh, H.; Magee, C.L. A functional approach for studying technological progress: Application to information technology. *Technol. Forecast. Soc. Chang.* **2006**, *73*, 1061–1083, doi:10.1016/j.techfore.2006.06.001. [CrossRef]

28. Koh, H.; Magee, C.L. A functional approach for studying technological progress: Extension to energy technology. *Technol. Forecast. Soc. Chang.* **2008**, *75*, 735–758, doi:10.1016/j.techfore.2007.05.007. [CrossRef]

29. Christensen, C.M. Exploring the Limits of the Technology S-Curve. Part I: Component Technologies. *Prod. Oper. Manag.* **1992**, *1*, 334–357, doi:10.1111/j.1937-5956.1992.tb00001.x. [CrossRef]

30. Christensen, C.M. Exploring the Limits of the Technology S-Curve. Part II: Architectural Technologies. *Prod. Oper. Manag.* **1992**, *1*, 358–366, doi:10.1111/j.1937-5956.1992.tb00002.x. [CrossRef]

31. Kopetz, H. *Real-Time Systems: Design Principles for Distributed Embedded Applications*, 1st ed.; Kluwer Academic Press: Dordrecht, The Netherlands, 1997; doi:10.1007/978-1-4419-8237-7.

32. Baldwin, C.Y.; Clark, K.B. *Design Rules: The Power of Modularity*; MIT Press: Cambridge, MA, USA, 2000.

33. Doyle, J.C.; Csete, M. Architecture, constraints, and behavior. *Proc. Natl. Acad. Sci. USA* **2011**, *108*, 15624–15630, doi:10.1073/pnas.1103557108. [CrossRef] [PubMed]

34. Maier, M.W. *The Art of Systems Architecting*, 3rd ed.; CRC Press: Boca Raton, FL, USA, 2009; doi:10.1002/inst.200912364.

35. Eppinger, S.D.; Salminen, V. Patterns of Product Development Interactions. In Proceedings of the International Conference on Engineering Design, Glasgow, UK, 21–23 August 2001.

36. Simon, H. *The Sciences of the Artificial*, 3rd ed.; MIT Press: Cambridge, MA, USA, 1996.

37. Whitney, D.E. Why mechanical design cannot be like VLSI design. *Res. Eng. Des.* **1996**, *8*, 125–138, doi:10.1007/BF01608348. [CrossRef]

38. Derler, P.; Lee, E.A.; Torngren, M.; Tripakis, S. Cyber-Physical System Design Contracts. In Proceedings of the ICCPS '13: ACM/IEEE 4th International Conference on Cyber-Physical Systems, Philadelphia, PA, USA, 8–11 April 2013.

39. Lee, E.A. Computing Needs Time. *Commun. ACM* **2009**, *52*, 70–79, doi:10.1145/1506409.1506426. [CrossRef]

40. National Institute of Standards and Technology, Cyber Physical Systems Public Working Group. Framework for Cyber-Physical Systems—Release 1.0. 2016. Available online: https://pages.nist.gov/cpspwg/ (accessed on 4 August 2018).

41. Johansson, K.H.; Törngren, M.; Nielsen, L. Vehicle Applications of Controller Area Network. In *Handbook of Networked and Embedded Control Systems*; Birkhäuser Boston: Boston, MA, UAS, 2005; pp. 741–765, doi:10.1007/0-8176-4404-0_32. [CrossRef]

42. MacCormack, A.; Baldwin, C.; Rusnak, J. Exploring the duality between product and organizational architecture: A test of the "mirroring" hypothesis. *Res. Policy* **2012**, *41*, 1309–1324, doi:10.1016/j.respol.2012.04.011. [CrossRef]

43. Adamsson, N. Interdisciplinary Integration in Complex Product Development: Managerial Implications of Embedding Software in Manufactured Goods. Ph.D. Thesis, Department of Machine Design, KTH Royal Institute of Technology, Stockholm, Sweden, 2007.

44. Sillitto, H. *Architecting Systems: Concepts, Principles and Practice. Volume 6: Systems*; College Publications: London, UK, 2014.

45. Networking and Information Technology Research and Development Subcommittee. *The National Artificial Intelligence Research and Development Strategic Plan*; Executive Office of the President or the United States: Washington, DC, USA, 2016.

46. Zhang, M.; Selic, B.; Ali, S.; Yue, T.; Okariz, O.; Norgren, R. Understanding Uncertainty in Cyber-Physical Systems: A Conceptual Model. In Proceedings of the 12th European Conference on Modelling Foundations and Applications, Vienna, Austria, 4–8 July 2016; Springer-Verlag: Berlin/Heidelberg, Germany, 2016; Volume 9764, pp. 247–264, doi:10.1007/978-3-319-42061-5_16. [CrossRef]

47. Rajkumar, R.; Lee, I.; Sha, L.; Stankovic, J. Cyber-physical systems: The next computing revolution. In Proceedings of the 47th Design Automation Conference, Anaheim, CA, USA, 13–18 June 2010; IEEE: Anaheim, CA, USA, 2010; doi:10.1145/1837274.1837461. [CrossRef]

48. Horváth, I.; Rusák, Z.; Li, Y. Order beyond chaos: Introducing the notion of generation to characterize the continuously evolving implementations of cyber-physical systems. In Proceedings of the ASME 2017 International Design Engineering Technical Conferences and Computers and Information in Engineering Conference, Cleveland, OH, USA, 6–9 August 2017; ASME: Cleveland, OH, USA, 2017; doi:10.1115/DETC2017-67082. [CrossRef]

49. Satyanarayanan, M. The Emergence of Edge Computing. *IEEE Comput.* **2017**, *50*, 30–39. [CrossRef]

50. Jacobson, I.; Lawson, H. Software and systems. In *Software Engineering in the Systems Context: Addressing Frontiers, Practice and Education*; College Publications, 2015; Volume 6, Chapter 1.

51. National Academies of Sciences, Engineering, and Medicine. *A 21st Century Cyber-Physical Systems Education*; The National Academies Press: Washington, DC, USA, 2016; doi:10.17226/23686. [CrossRef]
52. Acatech National Academy of Science and Engineering. Living in a Networked World. Integrated Research Agenda Cyber-Physical Systems (agendaCPS). 2015. Available online: http://www.cyphers.eu/sites/default/files/acatech_STUDIE_agendaCPS_eng_ANSICHT.pdf (accessed on 4 August 2018).
53. Bainbridge, L. Ironies of automation. *Automatica* **1983**, *19*, 775–779, doi:10.1016/B978-0-08-029348-6.50026-9. [CrossRef]
54. Waymo. Waymo Safety Report: On the road to Fully Self-Driving. 2018. Available online: https://storage.googleapis.com/sdc-prod/v1/safety-report/Safety%20Report%202018.pdf (accessed on 30 August 2018).
55. Wagner, M.; Koopman, P. A Philosophy for Developing Trust in Self-driving cars. In *Road Vehicle Automation 2*; Meyer, G., Beiker, S., Eds.; Lecture Notes in Mobility; Springer: Cham, Switzerland, 2015; pp. 163–171, doi:10.1007/978-3-319-19078-5_14.
56. Amodei, D.; Olah, C.; Steinhardt, J.; Christiano, P.; Schulman, J.; Mané, D. Concrete Problems in AI Safety. *arXiv* **2016**, arXiv:1606.06565.
57. Sheard, S.A.; Mostashari, A. A Complexity Typology for Systems Engineering. In Proceedings of the INCOSE International Symposium, Chicago, IL, USA, 12–15 July 2010; doi:10.1002/j.2334-5837.2010.tb01115.x.
58. Törngren, M.; Grimheden, M.E.; Gustafsson, J.; Birk, W. Strategies and considerations in shaping cyber-physical systems education. *ACM SIGBED Rev.* **2016**, *14*, 53–60, doi:10.1145/3036686.3036693. [CrossRef]

Article

Sharpening the Scythe of Technological Change: Socio-Technical Challenges of Autonomous and Adaptive Cyber-Physical Systems

Daniela Cancila [1,*], **Jean-Louis Gerstenmayer** [2,†], **Huascar Espinoza** [1] **and Roberto Passerone** [3]

[1]   CEA, LIST, CEA Saclay, PC172, 91191 Gif-sur-Yvette, France; huascar.espinoza@cea.fr
[2]   MEFi-DGE French Ministry of Economy, 94201 Ivry-sur-Seine, France; jean-louis.gerstenmayer@cea.fr or jean-louis.gerstenmayer@finances.gouv.fr
[3]   Dipartimento di Ingegneria e Scienza dell'Informazione, University of Trento, 38123 Trento, Italy; roberto.passerone@unitn.it
*   Correspondence: daniela.cancila@cea.fr; Tel.: +33-(0)1-6908-0107
†   Researcher CEA at present in position of project manager and policy advisor at the French Ministry of Economy. Ideas and opinion in this paper are these of the author, they are not representative of the French Ministry of Economy opinions.

Received: 28 September 2018; Accepted: 21 November 2018; Published: 28 November 2018

Abstract: Autonomous and Adaptative Cyber-Physical Systems (ACPS) represent a new knowledge frontier of converging "nano-bio-info-cogno" technologies and applications. ACPS have the ability to integrate new 'mutagenic' technologies, i.e., technologies able to cause mutations in the society. Emerging approaches, such as artificial intelligence techniques and deep learning, enable exponential speedups for supporting increasingly higher levels of autonomy and self-adaptation. In spite of this disruptive landscape, however, deployment and broader adoption of ACPS in safety-critical scenarios remains challenging. In this paper, we address some challenges that are stretching the limits of ACPS safety engineering, including tightly related aspects such as ethics and resilience. We argue that a paradigm change is needed that includes the entire socio-technical aspects, including trustworthiness, responsibility, liability, as well as the ACPS ability to learn from past events, anticipate long-term threads and recover from unexpected behaviors.

Keywords: autonomous cyber-physical systems; resilience; ethics; nano-bio-info-cogno technologies

1. Introduction

In 2014, the Cyber-Physical European Roadmap and Strategy (CyPhERS) [1,2] pointed out artificial intelligence (AI) as a distinctive characteristic of cyber-physical systems (CPS). In 2018, after two years of research, the Platform4CPS European project [3] introduced a list of recommendations together with the main scientific topics and business opportunity for markets [4]. Among the main topics, the project highlights the importance of *autonomous Cyber-Physical Systems* (especially those including AI) and their impact on the incoming period with respect to several aspects and disciplines. For example, the introduction of AI into autonomous Cyber-Physical Systems demands new design and development technologies able to consider the learning phase, adaptability and requirement's trace. Among the main concerns, the project highlights safety, resilience, security and confidence on these systems before their production and exploitation. Finally, the impact on social, legal and ethics issues is inherently involved by these systems. To clearly address the features of autonomy and environment adaptability of CPS [5], the European Commission refers to Autonomous Cyber-Physical Systems with the acronym ACPS. In this paper, we adopt then the term ACPS.

The design and development of ACPS requires the convergence of the cyber-side (computing and networking) with the physical side [6] and AI. More generally, tremendous progress becomes possible through converging technologies stimulated by advances in four core fields: *Nanotechnology, Biotechnology, Cognitive and Information technologies* [7,8]. Such convergence focuses on both the brain and the ambient socio-cultural environment. ACPS represent an example of that convergence which embraces not only engineering and technological products, but also legal (regulations) and ethical perspectives.

More specifically, the ACPS technologies are expected to bring large-scale improvements through new products and services across a myriad of applications ranging from healthcare to logistics through manufacturing, transport and more. The convergence of the "nano-bio-cogno-info" fields in ACPS could significantly improve the quality of our lives. However, the empowerment of sensitive intelligent components together with their increasing interaction with humans in shared spaces, e.g., personal care and collaborative industrial robots, raises pressing concerns about safety.

The technical foundations and assumptions on which traditional safety engineering principles are based worked well for human-in-the-loop systems where the human was in control, but are inadequate for ACPS, where autonomy and AI are progressively more active in this control loop. Incremental improvements in traditional safety engineering approaches over time have not converged to a suitable solution to engineer ACPS, having increased levels of autonomy and adaptation. In addition to physical integrity, safety in ACPS is tightly related to ethical and legal issues such as trustworthiness, responsibility, liability and privacy. A paradigm change is needed in the way we engineer and operate this kind of systems.

The main contribution of this paper is twofold: an analysis of the "nano-bio-cogno-info" convergence and a discussion on the aspects that are currently stretching the limits of ACPS safety engineering, including:

(i)   reduced ability to assess safety risks issues due to their analytical nature that relies on accident causality models,
(ii) inherent inscrutability of Machine Learning algorithms due to our inability to collect an etymologically sufficient quantity of empirical data to ensure correctness,
(iii) impact on certification issues, and
(iv) complex and changing nature of the interaction with humans and the environment.

The results of both analysis highlight the importance of the notion of *resilience*. This notion emerges in the "nano-bio-cogno-info" convergence analysis as a means to the survival of the 'competitors' after an irreversible damage. Moreover, our analysis shows that traditional safety engineering cannot be applied to ACPS as it is. The European commission is aware of this topic and introduced it as high rpiority R&D&I aerea in the Strategic Research Agenda for Electronic Components and Systems [9]. Like the "nano-bio-cogno-info" convergence study, the emerging notion of resilience in the literature seems more appropriate than safety when we deal with ACPS. In [10,11], for example, the author highlights two ways to ensure safety: "avoiding that things go wrong" (called Safety I) and "ensuring that things go right" (named Safety II), i.e., the ability for a system to accomplish its mission under acceptable outcomes. In this regards -the author continues- "resilience does not argue for a replacement of Safety-I by Safety-II, but rather proposes a combination of the two ways of thinking". But, more work should be done in this direction to establish safety engineering techniques and tools tailored to safety for ACPS.

The paper is organized as follows. Section 2 presents an overview of the nano-bio-cogno-info convergence. The section introduces the resilience issue and concludes with two open questions related to scientific and socio-ethics issues (Section 2.3) which are discussed in Section 3 and Section 4, respectively. More in particular, Section 3 discusses some scientific challenges to assess ACPS safety and Section 4 focuses on social and ethics issues, including privacy in ACPS. Finally, Section 5 shares our conclusions.

2. An Historical View of the Convergence Paradigm

The accelerating convergence of scientific and technological developments for the spheres of nano-systems industry, information and communication systems, as well as spheres of biology and medical applications, is an emerging phenomenon, based on our recently acquired capacity to measure, manipulate and organize the matter at the nanometer scale. A capital example of that convergence is the introduction of nano-material technologies tailored to medical applications. This is the case of intelligent microchip implants for humans (included in the brain) to control (lost) functionality. The design and development of that system demands microchips compatible with the human body, safety embedded software, prevention of potential attacks to the systems, and medical science. Quantum computing is another example: it combines neuromorphic architectures with computer science and physics. Finally, AlphaGo is able to automatically learn go, integrating artificial intelligence and automatic learning into embedded software and hardware [12].

Life sciences present the characteristics to become the keystone of that convergence. Nevertheless, as a consequence of maturing of artificial intelligence, more and more areas of convergence will be born 'in silico' before any 'material' development (e.g., scientists are building databases of thousands of compounds so that algorithms predict which ones combine to create new materials and design).

Historical Remarks: earlier digital technologies convergence and the development of interdisciplinary synergies are the precursors of a global (almost really 'big') convergence, known in the literature as Natotechnologies (N), Biotechnologies (B), Information Technologies (I) and Cognitive sciences (C) (NBIC) convergence [7] or as Bio Nano & Converging Technologies (BNCT) [8]. In 1998, E. O. Wilson (sociobiologist) writes a book on the emerging harmony among the sciences [13] meaning the age of the concern of the knowledge unity ('universe' comes from 'unus vs. multa' (lat.)). However, the first consistent 'Nano-Bio-Info-Cogno convergence' paradigm was worldwide diffused only in 2005 by the very influential book of Mihail Roco and William Sims Bainbridge [7]. The authors prophesied a future overlapping of disciplines with extensive transdisciplinary fecundity. The figure represents the OCDE's mapping, first introduced in [14], and reveals how distinct academic and technological domains can create an area of overlapping and increasing convergence (see Figure 1).

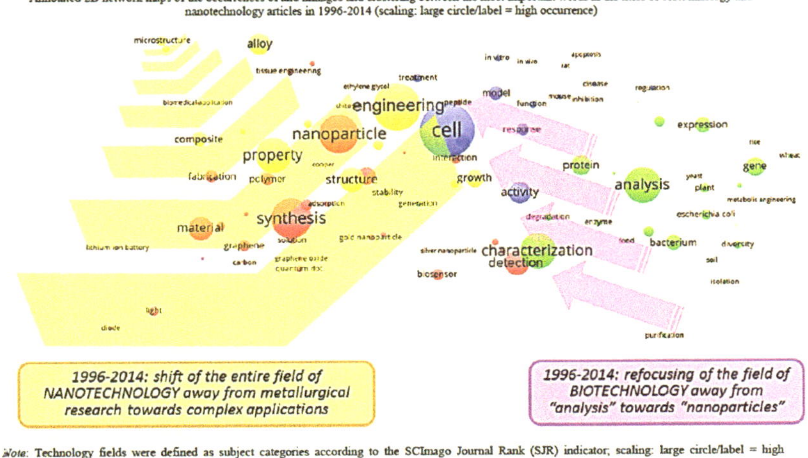

Note: Technology fields were defined as subject categories according to the SCImago Journal Rank (SJR) indicator; scaling: large circle/label = high occurrence.
Source: OECD calculations based on Scopus Customs Data, Elsevier, Version 12.2015, October 2016; text-mining: VOSviewer, version 6.1.3.

Figure 1. OCDE's mapping on the convergence. The Figure is extracted from [14] with permission.

2.1. Several Underlying Realities in the Convergence

Several factors, scientific research, technological development, market adoption of diversified products and regulation, are contributing, differently, to convergence as a whole. A bird view may mask the distinction between the different aspects of the convergence having different motivations and distinct ethical concerns. The knowledge construction starts from non-resolved questions or theoretical dissonances. As achievement of our natural and permanent search for meaning and for efficiency, a theoretical convergence is first, regarding the theoretical sciences, the result of our natural need to know—when we synthesize powerful concepts unifying (anytime with genius) partial or multiple empirical knowledge in a global vision. As examples, when Einstein wrote, in 1905, his paper "Zur Elektrodynamik bewegter Körper" (On the Electrodynamics of Moving Bodies) known as 'Einstein's special theory of relativity', he succeeded in the unification between Maxwell's equations for electricity and magnetism with the laws of mechanics, which were formerly mutually inconsistent. In a second time, a technological background convergence results of our need for efficiency—when we build more complex patterns and combine means in order to fulfill new needs. The technological construction background starts from the lack of common functionality that has to be satisfied. We should not forget the recursive growth of sciences and technology: a progressive integration of previously independent disciplinary sciences leads to a larger disciplinary perimeter, increasing de facto the resource reservoir in use by engineers and inventors. Simultaneously, a cumulative growth of available technological solutions leads increasingly to a wider number of technological inventions potentially combinable. As a result, we obtain a combinatorial law of growth for the creative activity (obviously limited by a logistic function regarding the exchange fluxes or the management of stocks, on a longer period). Researchers propose diverse models to predict technological convergence and singularities [15]. Moreover, progress regarding the logistic limitations on the fluxes are also contributing to reinforce the acceleration and enhance the maximum level (e.g., digitalization, big data, human–machine interfacing, artificial intelligence) and the offer of capital determined by the level of trust. Truth is largely determined by governance and transparency issues. The judgment of rightness or wrongness upon the intellectual activity, regarding the knowledge construction, is the epistemology. The judgment on the goodness or the badness regarding an action, of individuals or of groups, having an impact on persons or on the whole society (equivalent to the Greek word ethos), is in the field of morals. As a whole, disruption, growth and convergence are largely driven by human appetite of knowledge and by trust. Trust is the fuel of business when science is the reliable rock where the imagination becomes rational. The concept of the technological convergence is becoming of greater importance because convergence is natural and its "road-map" highlights the decision making for more massive amount of funding needed by the industrial technology development. Then, enlarging the coherence on the scientific and technological basis ensures the investments and enlarges the opportunity of development. First benefit: to drive the convergence of micro-nano-bio technologies towards integrated biomedical systems. Imagination, initiative and boldness to launch new business are energies for good, but they always come with some indetermination and even with risks. Technological discoveries and innovations have always triggered fears and rejection, inspiring imagination for catastrophic and even apocalyptic scenarios. Recently, nanomaterials and nanotechnology incarnated the latest darkest fears related to technological development. Today, artificial intelligence with its increasing speed of development (relayed to the public for instance by the AlphaGo performance) is challenging the first place. The increasing speed of technological evolution, the enlarging perimeter of knowledge, the decreasing confidence in the expert, lead to a kind of cognitive dissonance due to the time required for constructing a coherent mental representation and a stable vision of our social ecosystem, thus, we both like and fear the technology, and the society is in tension: are we able to secure our trajectory of development? How could we proceed? This speed of change is often compared to a 'shock', a technological shock. Shock implies irreversibility, memory losses, change faster than the (social) speed of assimilation. The response to a change, to the indetermination, to a shock or to a probability of shock has to be discussed.

2.2. Resilience in the Convergence Paradigm

Innovation ability characterizes human societies and, anytime, fast developing social innovation or disruptive technological innovation feels as a shock for more vulnerable people. Physical sciences characterize a shock by a propagation speed larger than the speed of information (e.g., the speed of sound). Similarly, we can define a societal shock as a social transformation propagating faster than the 'speed of acceptance' along the social network (according to individual acceptance time and communication efficiency). A shock is also characterized by irreversibility, i.e., a partial 'memory loss' (explicitly coded -symbolic- information/or inherent -implicit- structural information (Structural information relative to a system, similarly to the philosophical notion of 'essentia', is from a bird view the 'information needed to *build* it', according to one of these two modalities of 'being': in knowledge (symbolic information) or with matter (inherent information) [16]) in the case of complex systems, and by a disruption of the physical state variables (temperature, pressure, speed of sound . . .). Resilience is the capability of a strained body to recover its size and shape after deformation caused especially by compressive stress [17]. In the field of societal issues, resilience is defined as "the ability to prepare and plan for, absorb, recover from, and more successfully adapt to adverse events" [18]. In the case of a learning autonomous-cyber-physical-system based on artificial intelligence, widely diffusing in the civil society [19], the capacity of *resilience* [from *resilire* (lat.), to back jump] seems to be a solution to restore this big 'information system', similar to—but more complex than—a software backup capacity. Nevertheless, it is not so simple when there are expensive infrastructures or humans in the loop and humans impacted. Hereafter, the 'resilient' subject, of interest, is holistically the whole interrelated system and some of its essential elements, including ACPS, humans (society), critical things and critical links. When is the resiliency most useful? According to Grime's theory [20], ecological succession theory teaches us that plants adopt one survival strategy amongst three: (C) 'Competitors' maximize growth if resource is abundant and stable; (S) 'Stress-tolerators' maximize (individual) survival; (R) Ruderals maximize the survival of the species by means of a large fecundity. Two strategies are adapted to bad or very bad conditions of life, (S) to constantly unfavorable and (R) to largely fluctuating environment. Inversely, competitors (C) are optimized for growth in a context of resource abundance and are the least prepared for a (rare) shock (see Figure 2).

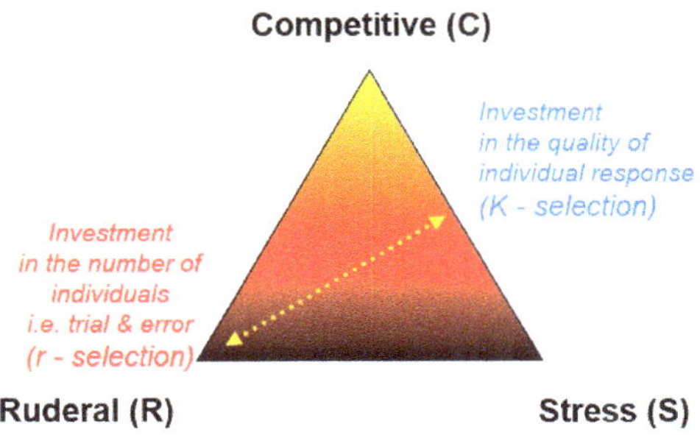

Figure 2. Illustration of the Philip Grime's CSR theory.

As a conclusion, resilience is essential for the survival of the competitors after an irreversible damage leading from harmony to chaos (because of partial erasing of work and structural memory). Thus, what does resilience need? To be able to rebuild, resilience capacity needs, to be operational, an access to a preserved or a reconstructed memory (e.g., memory distributed in the social network) and an access to sources of efficiency (energy, tools, materials, money, willingness, etc.). Then the ability Resilience needs an efficient distribution of resources and of information about the system which can be stored outside of the system (strong importance of the weak links). In case of incomplete information, it may be difficult to rebuild the initial situation and the degraded resilience operation may drive the system towards emerging new solutions of survival (evolution) worse or better adapted to the new conditions.

The case of France: with other countries, France participates in a conferences about "social impact" and Awareness-Readiness-Response [21]. The Joint Research Centre (JRC), as European Commission's science and knowledge service, supports EU policies with independent scientific evidence throughout the whole policy cycle and develops innovative tools for policy makers challenges (including ethics, privacy, liability, etc.). In this context, JRC works on the social resilience too, defining a program of work (see Table 1). Basically, resilience or resiliency is the recovering of the initial state (resilire (lat.) = back jump) and 'resistance' is 'remaining at the same place'.

Table 1. Information extracted from the table in [22].

What Could Be Tested?	How Should Be Tested?
Resilient Design	e.g., tests of the resiliency of a component or a system based on real failure data
(Inter)Dependency	Computer-based modeling and simulation (inter)dependency
Redundancy (including interoperability, adaptive capacity)	e.g., passive and active redundancy testing
Restoring capacity	e.g., measuring the ration of the lost performance

2.3. Open Questions

The technological convergence involves the notion of resilience, as described in the above section, and leaves two wide open questions:

- *A scientific issue* The restoration of a system at a previous state implies the use of memorized information. Thus, the resilience capacity is closely related to the accessibility to information on the oldest state to be restored, explicit and structural information, and to the accessibility to an energy source (physical energy, financial energy, social energy), leading to the obligation to preserve a good financial ratio between the costs of these functionalities (it may not be an economically sustainable option).
- *Socio-ethical questions* According to the point of view of the observer, the edge of a system is relative; it is simultaneously an ecosystem for smaller parts, a subsystem for a larger point of view, or the system itself for the designer. What part of the system must be protected: e.g., the 'autonomous car', its passengers, the dog in the street?

Life does not use invariably the resilience but often use the adaptability or the evolution, recovering a new equilibrium, different, even better than the oldest. Is it justifiable, on the economic or political view, to favor a back jump towards a past reference compared with other solutions more opened on an adaptive evolution?

In the next sections, we study these questions further.

3. Scientific Challenges of Safety-Critical ACPS

Autonomy in ACPS intrinsically involves an automatic decision-making and, then, more extensively, embraces Artificial Intelligence, as clearly stated in [1,2]. The introduction of AI in ACPS is revolutionizing safety techniques as traditionally used. Still a few year ago, some distinguished scientists wrote *Transcending Complacency on Superintelligent Machines*, acting as a gadfly to the community: "So, facing possible futures of incalculable [AI] benefits and [AI] risks, the experts are surely doing everything possible to ensure the best outcome, right? Wrong. [...] some of us—not only scientists, industrialists and generals—should ask ourselves what can we do now to improve the chances of reaping the [AI] benefits and avoiding the risks." [23].

After that, in 2015, S. Russell et al. [24], in a letter signed by many scientists, point out three challenges related to AI safety:

- *verification* (how to prove that a system satisfies safety-related properties);
- *validation* (how to ensure that a system meets its formal requirements and does not have unwanted behaviors); and
- *control* (how to enable meaningful human control over an AI system after it begins to operate).

In 2016, the problems raised by safety and AI have been further analyzed in [25], where the authors focus on accidents in the machine learning systems. At the same time, the Future of Life Institute is becoming a leading actor for AI Safety Research [26].

ACPS are expected to bring a technological revolution that will influence the lives of a large part of the world population and will have a deep impact on nearly all market sectors. Increasing levels of AI will allow ACPS to drive on our roads, fly over our heads, move alongside us in our daily lives and work in our factories, offices and shops, soon. In spite of this disruptive landscape, deployment and broader adoption of ACPS in safety-critical scenarios remains challenging. Broadly speaking, the technical foundations and assumptions on which traditional safety engineering principles are based worked well for human-in-the-loop systems but are inadequate for autonomous systems where human beings are progressively ruled out from this control loop. The following sections discuss aspects that are currently stretching the limits of ACPS safety engineering. In other terms, they cannot be applied to ACPS as they are.

3.1. Ability to Appraise Safety Issues and to Learn from Experience

If we consider ACPS applications as a whole (e.g., drones, robots, autonomous vehicles), we discover that safety-related analysis, as applied in the traditional application domains such as railway or nuclear energy, suffers from given limits when we focus on ACPS. Of the extensive topic, we only discuss few examples that are, however, sufficiently expressive to show some safety-related limits.

Some of the most important assumptions in traditional safety engineering lie in models of how the world works. In this regards, current approaches include probabilistic risk assessment such as Preliminary Hazard Analysis, Fault Tree Analysis (FTA) and Failure Modes and Effects Analysis (FMEA), working on the assumption that, once the system is deployed, it does not learn and evolve anymore. Another strong hypothesis of the traditional safety approach is based on the constructor's responsibility. In traditional domains, specialized teams use the final product and are responsible of their maintenance. In the railway, for example, we have specialized train-drivers, teams specialized in the train maintenance, teams specialized in the physical infrastructure, team specialized in the electrical infrastructures, etc. The responsibility of an accident is totally in charge of the (train) constructors and/or the (railway) company. Rarely, a passenger is fully responsible to mitigate a possible accident. This situation does not happen in some applications of ACPS. Compare the railway organization with the case of a drone. The drone driver is not a safety engineer and/or a specialized drone pilot. Often, s/he has another work and uses the product, for example, to control agriculture or to film video. In case of an accident, the pilot's responsibility is analyzed. For example, in 2015, at the Pride Parade in Seattle, a drone's pilot has been judged guilty of a drone's accident and he was sentenced to 30 days in jail [27]. The role of a user of a (semi-autonomous) drone could be similar to the driver of a vehicle. Among the main differences, however, is the use of redundancy as a means to reduce the risk of an accident. In a vehicle, to ensure the precision of a measure, (a given set of) redundant and heterogeneous software and hardware mechanisms and architectures can be considered to be deployed. Of course, a vehicle is strongly limited in space and cost with respect to a train or a nuclear plant, but we can introduce some redundant mechanisms and architectures (e.g., more sensors to detect an obstacle). This technique, however, is strongly limited in the case of a drone due to space constraints. Preventing accidents in ACPS requires using models that include the entire socio-technical aspects and treat safety as a dynamic control problem. Future intelligent autonomous systems need to be able to appraise safety issues in their environment, self-learn from experience and interactions with humans, and adapt and regulate their behavior appropriately. Furthermore, a higher level of autonomy in uncertain, unstructured, and dynamic environments involves many open systems science and systems engineering challenges. Autonomous systems interact in open-ended environments, making decisions and taking actions based on information from other systems and facing contingencies and events not known at design time. In recent years, new advances in technology have provided systems with the ability to anticipate, resist, recover from, and reconfigure after failures and disturbances. Safety assurance approaches have not kept up with the rapid pace of this kind of technological innovation.

Towards Ubiquitous Approaches for Resilience in ACPS

Recently, some scientists promoted methods and tools for *robust control* as a possible solution [28] to achieve the control of ACPS functionality even in the presence of uncertainty. Its cornerstone combines safety and performance over multi-sensor and multi-actuator heterogeneous networking architecture and—continue the authors in [28]—Model-Based Design represents a means to achieve a robust control design. Albeit robust control does not fix the AI-safety issues yet, it could be interesting to investigate in this direction further. First of all, robust control is closely associated with the concept of resilience [28]. We define two different types of resilience. *Exogenous* and *Endogenous* Resilience (see the side bar An Illustrative use case). Both types of resilience could represent a possible answer to ensure safety in ACPS. Hence, the paradigm changes: from safety, as traditionally studied in embedded systems, to resilience. Exogenous and Endogenous Resilience are pretty new in ACPS and, in our knowledge, established methodologies and tools for ACPS, including AI, do not exist yet.

An illustrative use case: drones are an illustrative example of ACPS and, extensively, of the technological convergence. Figure 3 shows the interaction between a fully-autonomous drone and a ground control station in the case of an accident. *Endogenous resilience* is related to the drone ability to accomplish its mission under physical constraints (e.g., battery level), error-detection (raising from embedded software, software and hardware integration, sensor failures, as well as malicious attack on the net), and safety-related measures to decrease the risk of an accident [29]. *Exogenous resilience* is related to the drone ability to accomplish its mission in presence of dynamic obstacles (e.g., birds) [29]. The Joint Authorities for Rulemaking of Unmanned Systems (JARUS) suggests some guidelines on Specific Operations Risk Assessment (SORA) for the category of specific drones [30], which do not require certification [31,32]. By the European regulation [31,32], only the category of fully automated drones ought to be certified. The blank hypothesis of the safety and certification process is that a drone does not change its internal behavior, for example, with respect to the number of flights (see Sections 3.1 and 3.3). This hypothesis is broken if AI is embedded in the software of the drone, for example by allowing a learning phase. Although specific drones do not require certification, the assessment of safety critical properties could be problematic by the introduction of AI and, in particular, of the machine learning algorithms, as discussed in [33].

Figure 3. Interaction between a fully-autonomous drone and a ground control station in the case of an accident. The figure is extracted from Emine Laarouchi. *An approach of safety analysis of CPS-IoT*. PhD on-going work. Supervisors: Daniela Cancila and Hakima Chaouchi.

3.2. Inherent Inscrutability of Autonomy Algorithms

Adaptation to the environment in autonomous systems is being achieved by AI techniques such as Machine Learning (ML) methods rather than more traditional engineering approaches. Recently, certain ML methods are proving themselves especially promising, such as deep learning, reinforcement learning and their combination. However, the inscrutability or opaqueness of the statistical models underlying ML algorithms (e.g., accidental correlations in training data or sensitivity to noise in perception and decision-making) pose yet another challenge. In traditional control systems, deductive inference logically links basic safety principles to implementation. Early cyber-physical systems tried to use deductive reasoning to construct sophisticated rule systems that fully defined behavior. However, cognitive systems have made spectacular gains using inductive inference based on ML, which may not produce semantically understandable rules, but rather find correlations and classification rules within training data. Inductive inference can yield excellent performance, in nominal conditions, but validating inductive learning is tough, due to our inability to collect an epistemologically sufficient quantity of empirical data to ensure correctness. The combination of autonomy and inscrutability in these inductive-based autonomous systems is particularly challenging for traditional safety techniques as their assurance is becoming intellectually unmanageable.

Towards Robustness Validation Approach for Autonomy

In the literature, one potential approach to deal with the inscrutability of AI-based systems is to identify the key elements of safety-related requirements on perception, see e.g., [34]. In that work, the authors specify the sources of perceptual uncertainty and suggest a reasoning on the system (for example, Conceptual uncertainty, Development situation and scenario coverage, Situation or scenario uncertainty, etc.). Their idea is to control these factors to ensure that the system meets a threshold of acceptability.

Another potential approach to deal with the inscrutability of AI-based systems is robustness testing, which is a relatively mature technology for assessing the performance of a system under exceptional conditions. Fault Injection (FI) is a robustness testing technique that has been recognized as a potentially powerful technique for the safety assessment and corner-case validation of fault-tolerance mechanisms in autonomous systems [35]. The major aim of performing FI is not to validate functionality, but rather to probe how robust the vehicle is—or their components are—to arbitrary faults under unforeseen circumstances. The potential benefits of using FI into design phases of autonomous systems range from providing early opportunities for integration of inductive technologies—e.g., machine learning algorithms that use training sets to derive models of camera lens—to reducing costs and risks associated to autonomy functions. Such techniques have already been used successfully to find and characterize defects on autonomous vehicles [36].

3.3. Certification

The certification of a system happens after the design and development and the integration of the subsystems. A certified system can include subsystems having different level of criticality, i.e., different levels of certification [37].

Warning. Because of the description of the certification process in all details is extremely complex, we here provide only a simplistic overview of the process and we refer the interested reader to the related norms and literature.

Broadly speaking, a system is first specified by requirements, then designed, developed and, finally, (the components are) integrated. Safety analyses follow the design and development phases by ensuring compliance of the system under development to the referred safety-related standards (CENELEC for railway, ISO26262 for automotive, IEC 60880 for the nuclear, and so on). For example, in the railway application domain, safety teams assess the safety level (SIL) of the components during the initial phases of the design of the system, by proving the risk assessment analysis. During the system development, safety analyses exploit specific techniques to ensure that a component implements the SIL level, via, for example, the computation of the MTBF (mean time between failures) parameter and the implementation/justification of the underlying redundant system architecture (e.g., 2oo4). Once the system is integrated and analyzed, safety teams provide the arguments for the certification. A third certification entity first analyzes them and, then, discusses them via audit. We highlight that a system is always certified with respect to a particular use and related norms. For example, a system having a Technical Standard Order (TSO) authorization cannot be installed and used in an aircraft without passing the avionics certification [38].

Although the benefits of certified systems are marked and outstanding, the cost of the certification remains expensive and, in many case, prohibitive. In avionics, the cost to ensure the most critical level (level A) is estimated to be more than 55% with respect to the minimum level (level E) [37]. In the drone application domain, only one category requires certification ('Certified' Category for operations with an associated higher risk) [31,32].

Towards Certification Approaches for ACPS

The European commission has financially supported research projects in the last decade with the aim to reduce the cost of the certification. In this regard, the OPENCOSS European project integrates the certification's arguments already in the preliminary phases of the design [39]; Goal Structuring Notation (GSN) is a graphical formalism to safety arguments [40].

Today, certification studies mechanisms to ensure modularity and mixed-criticality on many and multi-core. Mixed-criticality is the ability of a system to execute functionality, having different levels of criticality, in the same hardware by guaranteeing the safety level associated to each function.

The PROXIMA European project [41] is part of the mixed-criticality European cluster, together with CONTREX [42] and DREAMS [43] European projects, financially supported by the European commission. The PROXIMA project analyzes "software timing using probabilistic analysis for many and multi-core critical real-time embedded systems and will enable cost-effective verification of software timing analysis including worst case execution time" [41]. The result is compliant with DO-178B (Software Considerations in Airborne Systems and Equipment Certification). A modular certification aims to study mechanisms to restrain certification from the certified system as a unique whole to the certified component, which is modified or substituted. These mechanisms primarily address contract-based system interfaces [44], and impact analysis of the component modification/substitution in the reaming system. One main difficulty of the certification process does not concern correctness of the automatically generated code with respect to a given model, but the assurance that the system, which is executed, meets the system requirements and only implements the specified functionality. In other terms, engineers have to avoid to automatically generate correct code with respect to a wrong model specification. As in the case of safety, certification is based on the hypothesis that the system to be certified will have the same behavior (it always implements only the specified functionality). In other words, no learning phase is allowed. For example, in the avionic application domain, a program does not change its behavior with the increasing number of performed flights. Similarly, in the railway application domain, the control command to automatically open/close the doors of an automatic metro will have the same behavior forever. However, the AI introduction and, in particular, the (on-line or off-line) learning phase overstretch the fundamental hypothesis on which the traditional certification process is based (i.e., the learning phases overstretch reproducibility of proofs that ensure the same system behavior under the same inputs). This situation is also problematic for the category of specific drones (which do not require certification [31,32]) to assess safety-related requirements [30] as discussed in [33]).

In addition, if AI is coupled with a full or a high level of system autonomy (i.e., human does not have control of the system), then they unwind civilian responsibility: who is responsible of an accident of ACPS having AI? How can we protect the ACPS and humans? Is it possible to certify an ACPS having AI? And how?

In this regards, an interesting study is *Certification Considerations for Adaptive Systems* by NASA [45]. In the document, the authors address the impact on adaptive and AI system in the avionics certification. The authors conclude with a number of recommendations and a road-map to improve certification approaches. One issue is to relax some strict assumptions and requirements to allow a more streamlined certification process.

3.4. New Forms of Interaction between Humans and Autonomous Systems

There are unique challenges in moving from human-machine interaction in automation, where machines are essentially used as tools, to the complex interactions between humans and autonomous systems or agents. As autonomous systems progressively substitute cognitive human tasks, some kind of issues become more critical, such as the loss of situation awareness, or the overconfidence in highly-automated machines. The Tesla accident occurred in 2016 is a clear illustration of the loss of situation awareness in semi-autonomous vehicles, as stated by the National Transportation Safety Board (NTSB): "the cause of the crash was overreliance on automation, lack of engagement by the driver, and inattention to the roadway" [46]. One of the main causes of this problem is a reduced level of cognitive engagement when the human becomes a passive processor rather than an active processor of information [47].

In [48], the author addresses the "Ironies of Automation", especially for the control process in industries. The main paradox is that any automated process needs (human) supervisors. In the case of an accident or a problem in industry, the human supervisor may not be sufficiently prepared or reactive to solve it, because automation hides and directly manages ex-manual operations. Therefore, the supervisor could be less prepared in autonomous industrial control systems. Of course, this situation could appear in industry and not in traditional critical systems. In a nuclear plant, for example, engineers in the control rooms receive an expensive and strong (theoretical and experimental via simulation) training, required after the TMI-2 accident [49,50].

While the potential loss of situation awareness is particularly relevant in autonomous systems needing successful intervention of humans, there is a number of more general situations where risks in human-machine interaction must be better understood and mitigated:

- Collaborative missions that need unambiguous communication (including timescale for action) to manage self-initiative to start or transfer tasks.
- Safety-critical situations in which earning and maintaining trust is essential at operational phases (situations that cannot be validated in advance). If humans determine the system might be incapable of performing a dangerous job, they would take control of the system.
- Cooperative human-machine decision tasks where understanding machine decisions are crucial to validate autonomous actions. This kind of scenario implies providing autonomous agents with transparent and explainable cognitive capabilities.

Towards Trusted and Safe Human-Machine Relationships

Several authors [51–54] argue that interactions with autonomous agents must be considered as "human relationships", as we are delegating cognitive tasks to these entities. This perspective opens the door to the application of existing fundamental knowledge from the social sciences (psychology, cognitive modelling, neuropsychology, among others), to develop trustable and safe interactions with autonomous systems. For instance, the authors of [51] propose to encode the human ability of building and repairing trust into the algorithms of autonomous systems. They consider trust repair a form of social resilience where an intelligent agent recognises its own faults and establishes a regulation act or corrective action to avoid dangerous situations. Unlike other approaches where machine errors remain unacknowledged, this approach builds on creating stronger collaboration relationships between humans and machines to rapidly adjust any potential unintended situation.

In 2017, some influential leaders in AI established the Asilomar AI Principles aimed at ensuring that AI remains beneficial to humans [55]. One of these principles is value alignment, which demands that autonomous systems "should be designed so that their goals and behaviours can be assured to align with human values throughout their operation". In this context, some researchers such as Sarma et al. [53] argue that autonomous agents should infer human values by emulating our social behaviour, instead of embedding these values into their algorithms. This approach would also apply to the way human interact and it would be the basis to create learning mechanisms emphasizing trust and safe behaviours, including abilities such as intentionality, joint attention, and behaviour regulation. While these ideas look more appealing with the rise of AI and machine learning, the concept of learning "safe" behaviours in intelligent agents was already explored in 2002 by Bryson et al. [54]. Nevertheless, the problem of finding meaningful safety mechanisms for human-machine interaction inspired from human social skills remains largely open, because of the complex and intricate nature of human behaviour and the need of a provably-safe framework to understand and deploy artificial relationships.

4. Socio-Ethical Challenges of Safety-Critical ACPS

Social trust is a major challenge for the future of ACPS. The recent catastrophic accidents involving autonomous systems (e.g., Tesla fatal car accident), show that sole engineering progress in the technology is not enough to guarantee a safe and productive partnership between a human and an autonomous system. The immediate technical research questions that come to mind are how to quantify social trust and how to model its evolution? Another direction that is key to understanding and formalizing of social trust is how to design a logic that allows the expression of specifications involving social trust? It is immediately followed by the questions of how to verify (reason about) such specifications in the context of a given human–machine collaborative context or, even more prominent, how to synthesize (design) an autonomous system such that, in a collaborative context with a human, these specifications are guaranteed?

4.1. Ethics and Liability in ACPS

Ethics in safety-critical autonomous systems is closely related to the question of risk transfer. If any safety risk is transferred from some people to others then the risk transfer must be explicitly justified, even where the overall risk is reduced. Indeed, while it may be possible to argue that the introduction of ACPS in certain situations (e.g., automated cars) reduces the overall harm, from an ethical point of view this may not be sufficient. Ethical issues can be regarded in terms of the trade-off associated with reducing one risk posed by an ACPS at the potential cost of increasing another risk. This is essential to understand ethics principles in terms of safety [56]. In the literature, Andreas Matthias introduced the notion of *responsibility gaps* to identify the situation in which "nobody has enough control over the machine's actions to assume the responsibility for them" [57]. This notion has been developed further to cover two dimensions: the control on the "what" and on the "how" of the system behaviors [58].

One fundamental problem when dealing with ACPS is the liability for accidents involving autonomous systems. As a general rule, the more a machine can learn, the less control the manufacturer has, which is important for product liability. On the 16 of February 2017, following the suggestion made by the Legal Affairs Committee (LAC), the European Parliament (EP) made a resolution in favor of a robust European legal framework to ensure that autonomous systems are and will remain in the service of humans. Regarding liability, the EP suggests to explore, analyze, and consider the implications of all possible legal solutions, such as "establishing a compulsory insurance scheme", "compensation funds" or even "creating a specific legal status for autonomous systems in the long run". Other studies suggest addressing ACPS-specific risks in comparison with the risks of traditional technologies (without learning or adaptive abilities) used for the same purpose. This may facilitate the analysis of liability cases under existing law or be used as elements in new legal rules.

An important issue is related to the implementation of ethical and other legal rules in the inherent behavior of autonomous systems. In this area we need:

- Support the modeling, verification and transformation of safety and ethical rules into machine-usable representations (safety/ethics constraint set) and reasoning capabilities for run-time monitoring and adaptation. This includes the integration of ethics and safety in the whole system engineering life-cycle.
- Embody building blocks (regulators, adviser, adaptors) to ensure that the design of intelligent behaviors only provide responses within rigorously defined safety and ethical boundaries (safety/ethics constraint set). This includes the definition of architectural patterns to systematically reduce the development complexity -including integration- and modular assurance of AI-based systems.
- Support mechanisms to permit the monitoring and adaptation of a safety constraint set and underlying behavioral control parameters. Human–machine interaction abilities to warn human operators and users about safety issues preventing the autonomous system from acting, to manage the override of a constraint, and to inform about any change in the behavioral (safety/ethical) constraint set.

4.2. Privacy

Privacy is a live issue in the society and crosses several levels. In [4] privacy is identified as a societal and legal grand challenge. It is especially emphasized whenever AI functionality will be widely developed in the ACPS systems. In [24], for example, the authors reveal how "the ability of AI systems to interpret the data obtained from surveillance cameras, phone lines, emails, and so on, can interact with the right to privacy". In this section we investigate the issue via two examples: one related to drones and one to autonomous vehicles. These examples show how the privacy right has acquired importance in our society and could be weakened by technical solutions on ACPS. We share the Platform4CPS vision and recommendation [4] which states that privacy will be a major topic in the future, reinforced by the AI introduction and AI expected exploitation on ACPS products.

4.2.1. Privacy in the Use of Drones

In 2015, just after an extremely violent earthquake, Nepal decided to use drone technology. These later have proved to be extremely useful in catastrophic scenarios. Drones are expected to increase the chance of survival to 80% because they are more reactive and, then, more efficient than the traditional surgery means [59]. For example, drones have been used after the Irma Hurricane in Florida (2017) to understand the situation and target the first-aide, and in Paris (Spring 2018) to control the dangerous increasing level of Senna and prevent catastrophic consequences for humans and the city (Several videos and resources are available on the net. For example, Paris https://www.francetvinfo.fr/meteo/inondations/video-la-crue-de-la-seine-filmee-par-un-drone-a-paris_2580526.html and Florida https://unmanned-aerial.com/drones-artificial-intelligence-impacted-hurricane-irma-claims-response). However, in Nepal drones have been banned after their use in the earthquake, because of a privacy conflict due to having captured images on heritage sites [60]. Privacy issue for drone applications is an extremely sensitive issue, as analyzed by the works of Silvana Pedrozo and Francisco Klauser (sociologists). The authors are (also) known to having made a *sondage* on the Swiss population on the acceptability or not to drones in 2015. The results have been diffused and discussed via several means (radio, scientific journals, etc.) [61,62]. In [61], the authors state that: "whilst the majority of respondents are supportive of the use of unarmed military and police drones (65 and 72% respectively), relative numbers of approval decrease to 23 and 32% when it comes to commercial and hobby drones". The underlying reason of such acceptability is not based on the guarantee of safety level of that systems. For example, in 2015, a drone injured a woman at the Pride Parade in Seattle [27] and the news has had a very broad resonance [63]. Instead, it is based on privacy concern and individual freedom [62]. This challenge is so relevant that the European regulation for drones addresses privacy issue with the support of a lawyer European team [64] (see, for instance, Article 29 Data Protection Working Party). Similarly, the Platform4CPS project [3] suggests to "enforce General Data Protection Regulation (GDPR), mandate in-built security mechanisms for key applications and clarify liability law for new products and services". The project identifies the following potential implementation: "Put in place enforcement measures for GDPR, enforce built in security for European products and put in place appropriate legislation for products and services" [4].

4.2.2. Privacy in Autonomous Vehicles Communications

Privacy covers more aspects in autonomous vehicles: from embedded software (e.g., how we deal with the personal data of the driver) to vehicle to vehicle (V2V) communication. Of the extensive issue, we only focus here on the latter.

The main problem concerning privacy in the V2V communications regards the full traceability of a vehicle. Vehicles should communicate with each other, for example to signal an accident or know if another vehicle is incoming to the same crossroad. The simplest technical solution to manage this scenario is to provide a unique identification to each vehicle. Problem solved. However, this technical solution collides with privacy: everyone can trace the trajectory of sensible targets. Examples of sensible targets are people having a protection, transfer of jailed people, bank transfer. All these targets can be easily traced ad then, potentially, attacked. Moreover, this social phenomenon could have a wider impact: a wife/husband can trace the partner, an employer traces the employees, etc. To avoid a scenario, as described in the book 1984 by George Orwell [65], the technical solution is a pseudonym for each vehicle. However, the social impact is similar to the previous one: it is technically a simple game to understand the association pseudonym/target, trace it and attack it. Another technical solution consists in periodically changing the pseudonym. However, if this change is too slow, Bob can easily trace Alice. On the contrary, if it is too fast, an autonomous vehicle receives the information that ten vehicles are incoming in the same crossroad whilst, in fact, only one is engaging the crossroad (the vehicle has automatically changed the pseudonym nine times in a short elapsed time).

This example clearly shows how technical solutions impact our society and could limit some rights, as privacy. This technical and society challenge is extremely sensible in the community: a special working group in ETSI TC ITS is addressing the question [66,67], the European commission is devoting its financial effort to support several projects on that direction (e.g., see [68]). Finally, in the literature we can find interesting surveys regarding privacy and V2V communications (see for example [69]), technical investigations and promoted solutions (see e.g., [70]).

5. Conclusions

We are witnessing a convergence in Nanotechnologies, Biotechnologies, Information technologies and Cognitive sciences [7]. In this regard, our analysis showed the importance of the notion of resilience (Section 2.2) and opens the door to two open questions on scientific and socio-ethical issues (Section 2.3). The rest of the paper developed both questions further. To do this, we focus on ACPS which are a key example of the convergence in Nanotechnologies, Biotechnologies, Information technologies and Cognitive sciences convergence and we mainly addressed four topics: (i) reduced ability to assess safety risks, (ii) inherent inscrutability of Machine Learning algorithms, (iii) impact on certification issues and (iv) complex and changing nature of the interaction with humans and the environment.

(i) and (iii) ACPS having AI and a high level of autonomy are showing the limits of safety and certification processes when applied to these systems. Traditional safety techniques and related methods cannot be applied as they are. (ii) The core concern is the unpredictability due to machine learning, which is, on one hand, a means to reduce and manage complexity inherently involved by ACPS, but, on the other hand, there still are no widely accepted techniques and processes to manage the impact of machine learning in the assessment of safety-related properties and certification. Hereafter, the main response to this kind of issue are collected, where "responsibility sharing" is the key concept in addition to other levels of response (resilience, autopoiesis, insurance, legal solution, etc.). As a generic example, AI inside ACPS leads to new issues due to unpredictability of the comportment emerging from unpredicted competences and behavior that machine learning will develop in different contexts of use. This unpredictability may be caused by the impossibility for the designers and safety teams to specify the multiplicity of situations and contexts, or by the practical impossibility to study a behavior that will be learned during the life of the system, or by the practical impossibility to compute them, or by the practical impossibility to collect all return experience data of interest. Hence, there are too many cases or these cases are difficult to be defined precisely, so that it is difficult to build an exhaustive base of examples dedicated to constructing exhaustive incidental cases virtual simulations as a base of reliability. (iv) In addition, an inadequate communication between humans and machines is becoming an increasingly important factor in accidents, which can create significant new challenges in the areas of human–machine interaction and mixed initiative control.

One interesting solution, proposed by NASA [45], is to limit the range of possibilities and relax some hypotheses by adopting a more streamlined (safety and certification) process. A similar problem appears in the education of a young human: some of them may become incompetent or immoral. Nevertheless, despite this individual human unpredictability, a human society is globally stable on the long term because people have individually and collectively the ability to learn a safer behavior (cautiousness) and to become 'responsible'. What can we do in the case of a hybrid society including learning machines? It is possible to transpose, adapt or divide the notion of 'responsibility' (technical and legal point of view), alone or with the complement of other dispositions as insurance and resilience? These are the way of progress to be explored.

Author Contributions: J.-L.G. contributed to Sections 2, 2.1, 2.2, 2.3 and 5. H.E. contributed to Sections 1, 3, 3.1, 3.2, 3.4 and 4.1. D.C. contributed to Sections 1, 3, 3.1, 3.3, 4.2 and 5, LaTex Editing and team coordination. R.P. has reviewed all the work.

Acknowledgments: We thank Francoise Roure of Ministère de l'Economie et des Finances—France for her studies and suggestions on social resilience. We thank Steffi Friedrichs for the figure "OECD's mapping". We thank Emine Laarouchi for the figure with drones and for his help with the reviews integration. We thank the journal editorial board and the reviewers for their comments and valuable suggestions.

Conflicts of Interest: The authors declare no conflict of interest.

Abbreviations

The following abbreviations are used in this manuscript:

AI	Artificial Intelligence
ACPS	Autonoumous and Adaptative Cyber-Physical Systems
BNCT	Bio Nano & Converging Technologies
EP	European Parliament
ETSI TC ITS	ETSI Technical Committee (TC) for Intelligent Transport Systems (ITS)
FI	Fault Injection
ML	Machine Learning
NBIC	Nano-Bio-Info-Cogno (technologies)
V&V	vehicle to vehicle

References

1. CyPhERS FP7 Project. Cyber-Physical European Roadmap and Strategy. Available online: http://cyphers.eu/ (accessed on 26 November 2018).
2. Törngren, M.; Asplund, F.; Bensalem, S.; McDermid, J.; Passerone, R.; Pfeifer, H.; Sangiovanni-Vincentelli, A.; Schätz, B. Characterization, Analysis, and Recommendations for Exploiting the Opportunities of Cyber-Physical Systems. In *Cyber-Physical Systems*; Song, H., Rawat, D.B., Jeschke, S., Brecher, C., Eds.; Intelligent Data Centric Systems; Academic Press, Elsevier: Cambridge, MA, USA, 2016; Chapter 1, pp. 3–14.
3. Platform4CPS European Project. Available online: https://www.platforms4cps.eu/ (accessed on 26 November 2018).
4. Thompson, H.; Reimann, M.; Ramos-Hernandez, D.; Bageritz, S.; Brunet, A.; Robinson, C.; Sautter, B.; Linzbach, J.; Pfeifer, H.; Aravantinos, V.; et al. *Platforms4CPS: Key Outcomes and Recommendations*; Steinbeis: Stuttgart, Germany, 2018.
5. D'Elia, S. *CPS in EU Programmes*; European Commission DG CONNECT: Brussels, Belgium, 2017.
6. Sangiovanni-Vincentelli, A.; Damm, W.; Passerone, R. Taming Dr. Frankenstein: Contract-Based Design for Cyber-Physical Systems. *Eur. J. Control* **2012**, *18*, 217–238. [CrossRef]
7. Bainbridge, W.S.; Roco, M.C. (Eds.) *Managing Nano-Bio-Info-Cogno Innovations: Coverging Technologies in Society*; Springer: New York, NY, USA, 2005.
8. Winickoff, D. *Working Party on Biotechnology, Nanotechnology and Converging Technologies: BNCT Project Updates*; Technical Report DSTI/STP/BNCT(2015)6; OECD: Paris, France, 2015.
9. Aeneas. *Strategic Research Agenda for Electronic Components and Systems (ECS-SRA)*; EpoSS: Berlin, Germany, 2018.
10. Woods, D.D.; Hollnagel, E. *Resilience Engineering: Concepts and Precepts*; CRC Press: Boca Raton, FL, USA, 2006.
11. Hollnagel, E. *Safety-I and Safety-II: The Past and Future of Safety Management*; CRC Press: Boca Raton, FL, USA, 2014.
12. Silver, D.; Schrittwieser, J.; Simonyan, K.; Antonoglou, I.; Huang, A.; Guez, A.; Hubert, T.; Baker, L.; Lai, M.; Bolton, A.; et al. Mastering the game of Go without human knowledge. *Nature* **2017**, *550*, 354–359. [CrossRef] [PubMed]
13. Wilson, E.O. *Consilience: The Unit of Knowledge*; Alfred A. Knopf: NewYork, NY, USA, 1998.
14. Friedrichs, S. *Report on Statistics and Indicators of Biotechnology and Nanotechnology. Documents de Travail de L'OCDE sur la Science, la Technologie et L'Industrie*; Technical Report 06; OECD: Paris, France, 2018.
15. Sandberg, A. An Overview of Models of Technological Singularity. In *The Transhumanist Reader*; Wiley-Blackwell: Hoboken, NJ, USA, 2013; Chapter 36, pp. 376–394.

16. Burgin, M.; Feistel, R. Structural and Symbolic Information in the Context of the General Theory of Information. *Information* **2017**, *8*, 139. [CrossRef]

17. Merriam-Webster. Dictionary. Available online: https://www.merriam-webster.com/dictionary/resilience (accessed on 26 November 2018).

18. The National Academies Press. *Disaster Resilience: A National Imperative*; The National Academies Press: Washington, DC, USA, 2012.

19. Villani, C. Donner un Sens à l'Intelligence Artificielle. Rapport au Premier Ministre, Mission Confiée par le Premier Ministre Edouard Philippe, Mission Parlementaire du 8 Septembre 2017 au 8 Mars 2018. Available online: https://www.ladocumentationfrancaise.fr/var/storage/rapports-publics/184000159.pdf (accessed on 26 November 2018).

20. Silvertown, J.; Franco, M.; McConway, K. A Demographic Interpretation of Grime's Triangle. *Funct. Ecol.* **1992**, *6*, 130–136. [CrossRef]

21. Roure, F. Nanotechnology: 10 Years of French Public Policy towards a Responsible Development. Available online: https://www.youtube.com/watch?v=W4QzHaok_Xo (accessed on 26 November 2018).

22. Pursiainen, C.; Gattinesi, P. *Towards Testing Critical Infrastructure Resilience*; Joint Research Centre (JRC) Scientific and Policy Reports; JRC: Ispra, Italy, 2014.

23. Huffington Post. Stephen Hawking and Max Tegmark and Stuart Russell and Frank Wilczek. In *Transcending Complacency on Superintelligent Machines*; Huffington Post: New York, NY, USA, 2014.

24. Russell, S.J.; Dewey, D.; Tegmark, M. Research Priorities for Robust and Beneficial Artificial Intelligence. *AI Mag.* **2015**, *36*, 105–114. [CrossRef]

25. Amodei, D.; Olah, C.; Steinhardt, J.; Christiano, P.F.; Schulman, J.; Mané, D. Concrete Problems in AI Safety. *arXiv* **2016**, arXiv:1606.06565.

26. Future of Life. AI Safety Research. Available online: https://futureoflife.org/ai-safety-research/ (accessed on 26 November 2018).

27. Steve Miletich. Pilot of Drone That Struck Woman at Pride Parade Gets 30 Days In Jail. 2017. Available online: https://www.seattletimes.com/seattle-news/crime/pilot-of-drone-that-struck-woman-at-pride-parade-sentenced-to-30-days-in-jail/ (accessed on 26 November 2018).

28. Amin, S.; Schwartz, G.; Sastry, S.S. *Challenges for Control Research: Resilient Cyber-Physical Systems*; Technical Report; IEEE CSS: New York, NY, USA, 2014.

29. Laarouchi, E.; Mouelhi, S.; Cancila, D.; Chaouchi, H. Robust Control Predictive Design for Resilient Cyber-Physical Systems. *Ada Eur.* **2019**, submitted.

30. Joint Authorities for Rulemaking of Unmanned Systems (JARUS). *JARUS Guidelines on Specific Operations Risk Assessment (SORA)*; Technical Report JAR-DEL-WG6-D.04; Swiss Federal Office of Civil Aviation (OFAC): Ittigen, Switzerland, 2017.

31. European Aviation Safety Agency. *Advance Notice of Proposed Amendment (A-NPA) 2015-10—Introduction of a Regulatory Framework for the Operation of Drones*; EASA: Brussels, Belgium, 2015.

32. European Aviation Safety Agency. *'Prototype' Commission Regulation on Unmanned Aircraft Operations*; EASA: Brussels, Belgium, 2016.

33. Schirmer, S.; Torens, C.; Nikodem, F.; Dauer, J. Considerations of Artificial Intelligence Safety Engineering for Unmanned Aircraft. In Proceedings of the First International Workshop on Artificial Intelligence Safety Engineering (WAISE), Vasteras, Sweden, 18 September 2018.

34. Czarnecki, K.; Salay, R. Towards a Framework to Manage Perceptual Uncertainty for Safe Automated Driving. In Proceedings of the First International Workshop on Artificial Intelligence Safety Engineering (WAISE), Vasteras, Sweden, 18 September 2018.

35. Juez, G.; Amparan, E.; Ruiz, A.; Perez, J.; Lattarulo, R.; Espinoza, H. Early Safety Assessment of Automotive Systems Using Sabotage Simulation-Based Fault Injection Framework. In *Computer Safety, Reliability, and Security, Proceedings of the International Conference on Computer Safety, Reliability, and Security, Trento, Italy, 12–15 September 2017*; Springer: Cham, Switzerland, 2017.

36. Vernaza, P.; Guttendorf, D.; Wagner, M.; Koopman, P. Learning Product Set Models of Fault Triggers in High-Dimensional Software Interfaces. In Proceedings of the 2015 IEEE/RSJ International Conference on Intelligent Robots and Systems (IROS), Hamburg, Germany, 28 September–2 October 2015.

37. Hilderman, V.; Baghai, T. *Avionics Certification: A Complete Guide to DO-178 (Software) DO-254 (Hardware)*; Avionics Communicaitons Inc.: Leesburg, VA, USA, 2008.

38. FAA. Aircraft Certification Design Approvals Technical Standard Order (TSO). Available online: https://www.faa.gov/aircraft/air_cert/design_approvals/tso/ (accessed on 26 November 2018).

39. OPENCOSS Open Platform for EvolutioNary Certification of Safety-Critical Systems (OPENCOSS), FP7 European Project. Available online: http://www.opencoss-project.eu/ (accessed on 26 November 2018).

40. The GSN Working Group Online. Goal Structuring Notation (GSN). Available online: http://www.goalstructuringnotation.info/documents/GSN_Standard.pdf (accessed on 26 November 2018).

41. PROXIMA FP7 European Project. Probabilistic Real-Time Control of Mixed-Criticality Multicore and Manycore Systems (PROXIMA). Available online: https://cordis.europa.eu/project/rcn/109947_fr.html (accessed on 26 November 2018).

42. CONTREX FP7 European Project. Design of Embedded Mixed-Criticality CONTRol Systems under Consideration of EXtra-Functional Properties (CONTREX). Available online: https://contrex.offis.de/home/ (accessed on 26 November 2018).

43. DREAMS FP7 European Project. Distributed REal-Time Architecture for Mixed Criticality Systems (DREAMS), FP7 European Project. Available online: https://contrex.offis.de/home/ (accessed on 26 November 2018).

44. Benveniste, A.; Caillaud, B.; Nickovic, D.; Passerone, R.; Raclet, J.B.; Reinkemeier, P.; Sangiovanni-Vincentelli, A.L.; Damm, W.; Henzinger, T.A.; Larsen, K.G. Contracts for system design. *Found. Trends Electron. Des. Autom.* **2018**, *12*, 124–400. [CrossRef]

45. Bhattacharyya, S.; Cofer, D.; Musliner, D.J.; Mueller, J.; Engstrom, E. *Certification Considerations for Adaptive Systems*; Technical Report NASA/CR-2015-218702; NASA: Washington, DC, USA, 2015.

46. National Transportation Safety Board. *Collision between a Car Operating with Automated Vehicle Control Systems and a Tractor-Semitrailor Truck near Williston, Florida, 7 May 2016*; Accident Report NTSB/HAR-17/02-PB2017-102600; National Transportation Safety Board: Washington, DC, USA, 2017; p. 42.

47. Endsley, M.R. Situation Awareness in Future Autonomous Vehicles: Beware of the Unexpected. In *Advances in Intelligent Systems and Computing, Proceedings of the 20th Congress of the International Ergonomics Association (IEA 2018), Florence, Italy, 26–30 August 2018*; Bagnara, S., Tartaglia, R., Albolino, S., Alexander, T., Fujita, Y., Eds.; Springer: Cham, Switzerland, 2018; Volume 824.

48. Bainbridge, L. Ironies of Automation. *Sci. Direct* **1983**, *19*, 775–779.

49. Clément, B.; Jacquemain, D. *Nuclear Power Reactor Core Melt Accidents*; Chapter Lessons Learned from the Three Mile Island and Chernobyl Accidents and from the Phebus FP Research Programme—Chapter 7; IRSN: Paris, France, 2015.

50. Walker, S. *Three Mile Island a Nuclear Crisis in Historical Perspective*; University of California Press: Berkeley, CA, USA, 2006.

51. de Visser, E.J.; Pak, R.; Shaw, T.H. From automation to autonomy: The importance of trust repair in human-machine interaction. *J. Ergon.* **2018**. [CrossRef] [PubMed]

52. Kohn, S.C.; Quinn, D.; Pak, R.; de Visser, E.J.; Shaw, T.H. *Trust Repair Strategies with Self-Driving Vehicles: An Exploratory Study*; SAGE Publications: Thousand Oaks, CA, USA, 2018; Volume 62, pp. 1108–1112.

53. Sarma, G.P.; Hay, N.J.; Safron, A. AI Safety and Reproducibility: Establishing Robust Foundations for the Neuropsychology of Human Values. In *Computer Safety, Reliability, and Security*; Gallina, B., Skavhaug, A., Schoitsch, E., Bitsch, F., Eds.; Springer: Berlin, Germany, 2018; pp. 507–512.

54. Bryson, J.J.; Hauser, M.D. What Monkeys See and Don't Do: Agent Models of Safe Learning in Primates. In Proceedings of the AAAI Symposium on Safe Learning Agents, Palo Alto, CA, USA, 25–27 March 2002.

55. The Future of Life Institute. Asimolar AI Principles. Available online: https://futureoflife.org/ai-principles/ (accessed on 1 November 2018).

56. Menon, C.; Alexander, R. Ethics and the safety of autonomous systems. In Proceedings of the Safety-Critical Systems Symposium, York, UK, 6–8 February 2018.

57. Matthias, A. The responsibility gap: Ascribing responsibility for the actions of learning automata. *Ethics Inf. Technol.* **2004**, *6*, 175–183. [CrossRef]

58. Porter, Z.; Habli, I.; Monkhouse, H.; Bragg, J. The Moral Responsibility Gap and the Increasing Autonomy of Systems. In Proceedings of the First International Workshop on Artificial Intelligence Safety Engineering (WAISE), Vasteras, Sweden, 18 September 2018.

59. Scott, J.E.; Scott, C.H. Drone Delivery Models for Healthcare. In Proceedings of the Hawaii International Conference on System Sciences (HICSS), Honolulu, HI, USA, 4–7 January 2017.

60. Choi-Fitzpatrick, A.; Chavarria, D.; Cychosz, E.; Dingens, J.P.; Duffey, M.; Koebel, K.; Siriphanh, S.; Yurika Tulen, M.; Watanabe, H.; Juskauskas, T.; et al. *Up in the Air: A Global Estimate of Non-Violent Drone Use 2009–2015*; Technical Report; University of San Diego Digital USD: San Diego, CA, USA, 2016.

61. Klauser, F.; Pedrozo, S. Big data from the sky: Popular perceptions of private drones in Switzerland. *Geogr. Helv.* **2017**, *72*, 231–239. [CrossRef]

62. Silvana Pedrozo and Francisco Klauser. Drones Policiers: Une Acceptabilité Controversée. 2018. Available online: https://www.espacestemps.net/articles/drones-policiers-une-acceptabilite-controversee/ (accessed on 26 November 2018).

63. BCC. Seattle's Ferris Wheel Hit by Drone. Available online: https://www.bbc.co.uk/news/technology-34797182 (accessed on 26 November 2018).

64. European Aviation Safety Agency. *Advance Notice of Proposed Amendment 2015-10*; EASA: Brussels, Belgium, 2015.

65. Orwell, G. *Nineteen Eighty-Four (Also Published as 1984)*; Martin Secker and Warburg Ltd.: London, UK, 1949.

66. Lonc, B.; Cincilla, P. Cooperative ITS security framework: Standards and implementations progress in Europe. In Proceedings of the IEEE 17th International Symposium on A World of Wireless, Mobile and Multimedia Networks (WoWMoM), Coimbra, Portugal, 21–24 June 2016.

67. ETSI TC ITS. Automotive Intelligent Transport Systems. C-ITS Security. Available online: https://www.etsi.org/technologies-clusters/technologies/automotive-intelligent-transport (accessed on 26 November 2018).

68. PRESERVE FP7 Project. Preparing Secure Vehicle-to-X Communication Systems. Available online: https://www.preserve-project.eu/ (accessed on 26 November 2018).

69. Petit, J.; Schaub, F.; Feiri, M.; Kargl, F. Pseudonym Schemes in Vehicular Networks: A Survey. *IEEE Commun. Surv. Tutor.* **2015**, *17*, 228–255. [CrossRef]

70. Gisdakis, S.; Laganà, M.; Giannetsos, T.; Papadimitratos, P. SEROSA: SERvice Oriented Security Architecture for Vehicular Communications. In Proceedings of the IEEE Vehicular Networking Conference (VNC), Boston, MA, USA, 16–18 December 2013.

Article

A Computational Framework for Procedural Abduction Done by Smart Cyber-Physical Systems

Imre Horváth

Cyber-Physical Systems Design Section, Department of Design Engineering, Faculty of Industrial Design Engineering, Delft University of Technology, Landbergstraat 15, 2628 CD Delft, The Netherlands; i.horvath@tudelft.nl; Tel.: +31-15-278-3520

Received: 11 October 2018; Accepted: 19 December 2018; Published: 25 December 2018

Abstract: To be able to provide appropriate services in social and human application contexts, smart cyber-physical systems (S-CPSs) need ampliative reasoning and decision-making (ARDM) mechanisms. As one option, procedural abduction (PA) is suggested for self-managing S-CPSs. PA is a knowledge-based computation and learning mechanism. The objective of this article is to provide a comprehensive description of the computational framework proposed for PA. Towards this end, first the essence of smart cyber-physical systems is discussed. Then, the main recent research results related to computational abduction and ampliative reasoning are discussed. PA facilitates beliefs-driven contemplation of the momentary performance of S-CPSs, including a 'best option'-based setting of the servicing objective and realization of any demanded adaptation. The computational framework of PA includes eight clusters of computational activities: (i) run-time extraction of signals and data by sensing, (ii) recognition of events, (iii) inferring about existing situations, (iv) building awareness of the state and circumstances of operation, (v) devising alternative performance enhancement strategies, (vi) deciding on the best system adaptation, (vii) devising and scheduling the implied interventions, and (viii) actuating effectors and controls. Several cognitive algorithms and computational actions are used to implement PA in a compositional manner. PA necessitates not only a synergic interoperation of the algorithms, but also an objective-dependent fusion of the pre-programmed and the run time acquired chunks of knowledge. A fully fledged implementation of PA is underway, which will make verification and validation possible in the context of various smart CPSs.

Keywords: smart cyber-physical systems; self-generated intelligence; ampliative reasoning mechanism; procedural abduction; data-driven system control; run-time acquired data; computational functions; self-adaptation capability; human/socially-centered applications

---

1. Introduction

1.1. An Evolutionary View on Cyber-Physical Systems

The aspiration for having cyber-physical systems (CPSs) emerged more than 60 years ago [1]. However, the principles of their practical manifestation were formulated just a decade ago [2] and the enabling technologies are becoming available nowadays [3]. At the time of inception, CPSs were seen as large-scale, distributed, self-controlling, software integrated application systems, which combine the functionalities and technological enablers of embedded control systems, advanced robotic systems, networked distributed systems, real-time control systems, and collaborative software agents systems [4]. Their operations were supposed to be coordinated, controlled, and monitored by a digital computing, communication and control core [5]. Interestingly, this novel system paradigm has been going through changes from both theoretical and practical points of view in the last years [6]. And this process will go on [7]. Some envisage even a kind of metamorphosis of the paradigm due to new technological affordances provided by, for instance, massive data networking technologies, cloud computing, artificial system intelligence, and cognitive system engineering technologies [8–10].

The almost infinite number of networked sources of sensor data, the databases residing on servers in the cloud, the growing availability of massive data flows, the development of smart data analytics approaches, the emergence of run time adapting system controls, and the other technologies shown in Figure 1, are the main enablers of the radical paradigm shift that is already observable in the academic research [11,12]. This shift encourages us to look at smart CPSs as the transformative systems of the near future. Nevertheless, it is worth pointing out that the traditional definitions and interpretations of CPSs are not able to capture the essence of the above circumscribed changes [13]. The main reason is that they place the emphasis on having a predefined and deterministic tight coupling between the physical world and the cyber-world, rather than on achieving synergism between run-time acquired data and dynamic operational objectives [14]. By doing so, they actually restrict the paradigmatic evolution of this family of engineered systems. This is a vital issue since S-CPSs can offer novel functional affordances and services that cannot be provided by other systems [15].

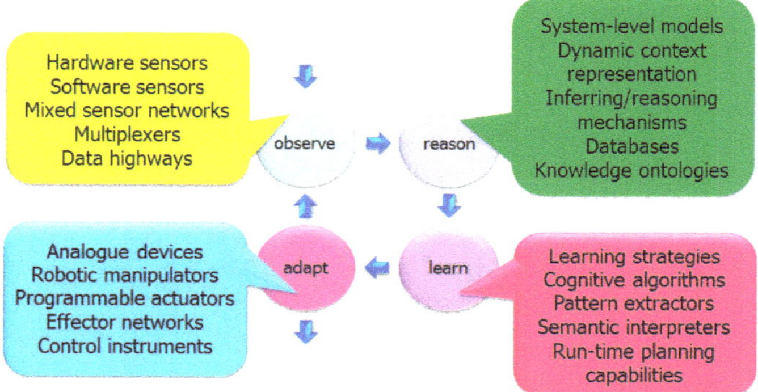

Figure 1. Four clusters of enablers for smart cyber-physical systems.

The understanding of the fundamentals and affordances of CPSs is more nuanced today than it was a decade ago. Therefore, new interpretations and definitions have been proposed. For instance, smartness has been put into a historical dimension and interpreted as a given stage of progression with regards to having cognitive competences and operationalization of a body of knowledge (without considering possible socialized collective intelligence of a system of systems yet). It has been identified as a distinguishing paradigmatic feature of next generation CPSs [8]. Equipping systems with cognitive capabilities, i.e., the process of intellectualization, is the major concern of cognitive engineering of CPSs. However, this domain of interest is still in a premature stage. The guiding assumption is that smart systems should be able to: (i) operate according to dynamically varying, even undefined, circumstances and control regimes, (ii) build awareness of the operational state of their entirety, components, and embedding environment, and (iii) adapt themselves in order to achieve the best possible operational objectives and performance. In other words, smart operation and servicing of CPSs require the capabilities of self-awareness, contextualized reasoning and learning, and functional and architectural morphing [16].

Owing to context-dependent reasoning and system-level adaptation, smart CPSs can be applied in many non-traditional human- and socially-centered applications. However it has also been comprehended that smartness may be utilized in rather different forms in different systems, such as smart cities, homes, transportation, clouds, manufacturing, production, and smart service systems [17], and even smart everything [18]. These systems will eventually show largely diverse levels of intellectualization. Evidently, implementation of smartness is inseparable from the realization of cyber-physical computing (CPC) that intends to go beyond the classical (predefined explicit algorithms-based) computation that is underpinned by the von Neumann theory of computing [19]. CPC complements or replaces predefined flows of computation with run-time devised flows, and appends run-time constructed or acquired (hidden, but learnable) implicit algorithms to the (preprogrammed) explicit algorithms. The implicit algorithms may manifest as patterned data structures, situated computing strategies, and context-driven reasoning mechanisms.

Obviously, an intense foundational research and a lot of system prototypes-based engineering studies are needed to fully explore and exploit smart system behaviors and affordances [20]. In particular, for the reason that even higher level of intellectualization as smartness is supposed to be present in third- and fourth-generation CPSs. Namely, second-generation CPSs self-generate awareness and perform some-level of functional/architectural self-adaptation under varying conditions [21], whereas third generation CPSs are supposed to have self-cognizance (i.e., awareness with semantic understanding) and non-biological self-evolution capability. As ultimate realizations, fourth-generation CPSs are assumed to have human-resembling system intelligence, which may involve computational consciousness, dependable reasoning, decisional autonomy, and non-genetic self-reproduction [8]. Current systems science and artificial general intelligence (AGI) research are still unable to describe and explain the above-mentioned 'beyond-smartness' types of system behaviors and to inform system engineering about feasible and efficient implementations. At the same time, many recent results validate the concept of second-generation CPSs. The feasibility and utility of specific implementations powered by context management, machine learning and other reasoning mechanisms have been demonstrated.

1.2. On the Background Research and the Research Issue Addressed in This Article

Apprehension of the above-mentioned novel system features calls for a progressive definition, which puts CPSs into the position of self-managing engineered systems. Our definition claims that smart CPSs: (i) employ the principles of cyber-physical computing, (ii) closely interact with the hosting environment(s), (iii) deeply penetrate into physical, biological, social, cognitive, etc. processes, (iv) act as a purposefully arranged set of adaptable actors in these contexts, and (v) possess the capability of non-organic complexification and self-organization. According to this definition, not only the software (middleware) establishes a tight relationship between the physical realm and the cyber realm, but also the mentalware, which is jointly possessed by the concerned system and its stakeholders. This was the starting point and motivation for our background research.

The reported work was conducted in the framework of the portfolio research of the Cyber-Physical Systems Design Section of the Faculty of Industrial Design Engineering at the Delft University of Technology. The shared theme of all related research projects is cognitive engineering of smart cyber-physical systems. The more than ten interrelated projects of staff members and PhD students addressed issues such as: (i) capturing and inferring from dynamic contexts representations, (ii) building system awareness real-time, (iii) computational mechanisms for reasoning, (iv) strategies for functional/architectural adaptation, (v) dynamic system messaging, and (vi) specific applications of smart CPSs. The main research questions for the portfolio research were:

- Based on what theoretical and methodological foundations can system-level smartness be implemented in next generation cyber-physical systems?
- What way can dynamic context information processing be extended to provide semantically enriched awareness representation?
- In what forms can procedural abstraction be implemented as a system-level reasoning mechanism of smart CPSs?
- Based on what knowledge can an active engineering framework support transdisciplinary development of compositional CPSs?
- What do human/system-in-the-loop and supervisory/operative control mean in the context of smart CPSs?

On the one hand, the completed research involved a comprehensive analysis of the state of the art as well as investigation of the pioneering research initiatives. On the other hand, it included a purposeful synthesis of underpinning theories and computational methodologies, and development of testable prototypes.

1.3. Content and Structure of the Article

The most important scientific concerns of the background research were: (i) development of a new reasoning model (a conceptual advancement model) concerning the generations of CPSs based on their level of self-intelligence and self-organization, (ii) implementation of multiple demonstrative prototype (sub)systems in various applications, and (iii) conceptualization of a robust computational framework for procedural abduction (PA) as an ampliative system-level reasoning and adaptation mechanism for S-CPSs. The content of this article contributes to the last concern. PA comprises a contemplation part (that helps realize some level of system self-awareness) and an alteration part (that supports functional and architectural self-adaptation under varying conditions). From an implementation point of view, PA consists of a large set of computational algorithms, which are run-time activated and integrated for reasoning and strategizing.

Though the presented results blend many research aspects of cognitive engineering of cyber-physical systems, only the functional and architectural specification of the PA framework is discussed in the rest of the article. The Section 2 analyses the current state of the art, considering five perspectives: (i) theoretical understanding of mind-like behavior of artefactual systems, (ii) potential enabling technologies for smart CPSs, (iii) achieving system-level holism in reasoning and decision making, (iv) the logical basis of system level reasoning, and (v) recent efforts to exploit abduction as an ampliative computational mechanism. The Section 3 provides a brief overview of the four prototype systems that hinted at the necessary constituents of procedural abduction as a system level reasoning mechanism. It discusses the experimental implementation of the systems for: (i) system-level features composition and compositional system modeling, (ii) identification and forecasting failures in a resilient cyber-physical greenhouse, (iii) representation and reasoning with dynamic context knowledge in a fire evacuation aiding system, and (iv) reasoning and adaptation in an engagement monitoring and enhancing system for stroke rehabilitation. The issues of synthesizing the constituents of the PA mechanism are also considered. The Section 4 presents the computational framework of procedural abduction. It provides a formal specification of the general workflow and the constituents, the underpinning knowledge, and the required functionality of the enabling algorithms of PA. The Section 5: (i) reflects on the completed work, (ii) states some vital propositions, and (iii) makes an inventory of the immediate and future research opportunities.

2. Current State of the Art

2.1. Theoretical Understanding of Mind-Like Behavior of Artefactual Systems

Thinking about transferring human-like intelligence to artificial structures has a long tradition. As known, this process commenced with computerization and, through informatization and cyberization, reached the stage of intellectualization of engineered systems. The development of software science/engineering and information/knowledge engineering proved to be the major driver of the progress. Using the terms and the argumentation of D.M. MacKay, the essence is in the move from 'slave-machines' to 'actor-machines' [1]. As actor-machines, intellectualized artefactual systems are not alternatives of current digital computing systems, but something else and more. Mentioned by the above author, the most fundamental question is "Can an artefact be made to show the behavioral characteristics of an organism?" MacKay argued that this kind of systems should be able to perform important functions such as (i) receiving, selecting, storing, and sending information, (ii) reacting to changes in their 'universe', including data on their own state, (iii) reasoning deductively from premises which are results of previous deductions and data on the different courses, (iv) observing and controlling their own activities, or otherwise, so as to further some goal, and (v) changing their own pattern of behavior so as to develop quite complex and superficial characteristics capable of rational description in purposive terms.

If we systematically replace the above phrases with our modern terms such as (i) networked sensing, (ii) awareness building, (iii) situated reasoning, (iv) self-organization, and (v) adaptation in context, then we can conclude that D.M. MacKay eventually described the most important paradigmatic features of smart cyber-physical systems. It is a surprising fact since the description happened at the very beginning of 1950s. In addition to stating that future engineered systems are subjects of a reasonably high level of intellectualization and socialization, feature self-planned and goal-directed activities, and potent to be self-sustaining, MacKay also called the attention to the need of sophisticated system-level reasoning mechanisms. He proposed a probabilistic reasoning-mechanism as a possible implementation of it. In addition, like many others, he also emphasized the importance and necessity of the capability of abstraction (by an intellectualized system) [22].

Fifty years later, Russell, S. and Norvig, P. proposed the following taxonomy of systems with non-natural intelligence: (i) systems that think like humans (e.g., cognitive architectures and neural networks), (ii) systems that act like humans (e.g., natural language processing, knowledge representation, learning, and automated reasoning systems), (iii) systems that think rationally (e.g., logic solvers, inference, and optimization systems), and (iv) systems that act rationally (e.g., software agents and embodied robots with perception, planning, reasoning, learning, communicating, and decision-making capabilities) [23]. The above taxonomy indicates the versatility of current systems with non-natural intelligence. They typically require a correct and complete model of the problem as well as of the application domain. However, in the case of a large problem or domain, construction of such models is either rather difficult or not possible at all. Furthermore, if the intellectualized system or its environment, or the set of tasks and the contexts of problem solving frequently and dynamically change, then a steady problem model and application domain model may become inadequate or even inappropriate [24]. In this case, run-time adaptation or complementation of the system model is needed.

2.2. Potential Enabling Technologies for Smart CPSs

In the last sixty years, many efforts were made to develop formal reasoning and learning mechanisms and to implement AI-based systems. Concerning the set objectives, we differentiated: (i) ambitious (striving for ultimate intelligence and autonomy), (ii) realist (constructively exploiting technologies), and (iii) modest (underestimating the affordances and potentials) visions of AI research. Domingos, P. identified five approaches to creating AI-based systems, namely: (i) "symbolist", implementing logical reasoning based on abstract symbols, (ii) "connectionist", building structures inspired by the human brain, (iii) "evolutionist", using methods inspired by Darwinian evolution, (iv) "Bayesians", using probabilistic inference, and (v) "analogizer", extrapolating from similar cases seen previously [25]. Within each approach, a large number of different realizations can be found. For instance, symbolist/analyst approaches include (i) standard learning algorithm for monomials, (ii) rule-based inductive algorithms, (iii) instance/pattern/evidence-based algorithms, (iv) search space/decision tree-based algorithms, and (v) probabilistic/fuzzy/non-monotonic algorithms. The group of sub-symbolic approaches includes, among other, (i) genetic algorithms, (ii) neural network algorithms, (iii) support vector machines, (iv) naive Bayesian-classifier algorithms, (v) Bayesian-learning algorithms, and (iv) extreme learning machines. As complements of purely symbolic and sub-symbolic approaches, composite approaches are also often considered. One of the main issues is reasoning with incomplete knowledge [26]. In addition, as discussed by Reed, S.K. and Pease, A.; practically all systems confront obstacles when reasoning needs to be done based on imperfect knowledge (consisting of ambiguous, conditional, contradictory, fragmented, inert, misclassified, or uncertain parts) [27].

Several papers have recently been published on the implementation of various levels of human intellectual/cognitive behavior in consumer durable-type of products, as well as in cyber-physical systems [28,29]. Though some 'wicked-problems' have been considered, mainstream AI research focuses on problem solving means for reasonably well-defined problems. Based on this, it became known that two-valued logical reasoning techniques are not sufficient in the case of systems working with uncertainty or multiple choices [30]. However, the major issue is that integration of problem solving means is complicated since they usually rely on different information/knowledge representation schemata and are implemented by non-compatible algorithms. Among the limited number of related integrative works, the effort of Lees, B. to combine a rule-based reasoning mechanism with a case-based reasoning mechanism is worth paying attention, since it casts light on the challenge of co-evolution of the different bodies of knowledge and the mechanisms themselves [31]. Lumer, C. and Dove, I.J. pointed at the fact that some schemes of reasoning proposed in the literature, including abduction, has been found defective due to the lack of an epistemological backing and, in most cases, the inability to differentiate various degrees of uncertain justification [32].

Though the latest developments of computational learning (CL), machine learning (ML) and deep learning (DL) gave an impetus for smart systems development, the need for explanatory learning has also been identified [33]. This is still hardly supported by robust algorithms of near-zero learning time. ML is mainly concerned with deriving rules, patterns or procedures that explain a body of data or predict future data typically based on statistical processes [34]. The approaches are typically sorted into three categories: (i) supervised learning (based on decision trees, support vector machines, production rules inference, artificial neural network, genetic algorithms, game theory driven approaches, neural network, etc.), (ii) unsupervised learning (generalized additive statistical methods, tree-based methods Bayesian non-parametric approaches, semi-supervised clustering, etc.), and (iii) reinforcement learning (model-free reinforcement learning algorithms, genetic algorithms/programming, feudal Q-learning, adaptive heuristic critic, transfer learning methods, multi-agent reinforcement learning, real-time dynamic programming, etc.). DL typically uses structures of large number of processing layers loosely inspired by the human brain to explore practically the same [35]. Many of these learning technologies have reached a high level of sophistication and increased the potentials of context-independent problem solving [36]. For instance, extreme learning machines combine conventional artificial learning techniques and biological learning mechanism into a suite of machine learning techniques for feedforward neural networks with single and multiple hidden layers, in which hidden neurons need not be tuned [37]. These and other emergent learning techniques aim at computational learning approaches for advanced industrial systems, such as cyber-physical production systems.

2.3. Achieving System-Level Holism in Reasoning and Decision Making

System scientists claim that smartness, like safety, reliability, awareness, adaptability, security, etc., is an overall paradigmatic feature of the systems as a whole, and is not a behavioral characteristic of their specific components, though their functionalities are needed for the realization. This means, smartness is not a reductionist property and cannot be realized by a simple composition of the operations of some smartness-enabling, interface-able components. Furthermore, system smartness cannot be regarded just as a momentary operation of the intellectual mechanisms of a given system, but it should be considered as an always present (ubiquitous and lasting) set of capabilities (with which the system has been equipped with by its developers, or what the system itself develops based on the preprogrammed or the learnt/aggregated knowledge). Owing to learning, system smartness may evolve over the life of a system. Thus, system learning is to be seen as a sustained and combined cognitive process of knowledge acquisition and behavior adaptation.

According to Finocchiaro, M.A., reasoning is the activity of the human mind that consists of reaching conclusions on the basis of reasons, giving reasons for conclusions, and/or drawing consequences from premises [38]. The arguments created for reasoning by human mind are linguistically expressed and arranged based on others. A quasi-automated generation and representation of arguments or decision points are seen as the major challenge for computational reasoning. In addition, recognizing the sufficiency of argumentation and maintenance of the validity of the arguments are additional challenges [39]. In particular generating informal proofs that are not directly driven by logical rules or physical causalities is problematic [40]. Informal arguments typically depend on their logical form as well as on their content and contexts. Informal proves cannot be expressed in a general logical language (i.e., by explicitly defined well-formed formulae), and cannot be applied successively on explicitly specified logical inference rules or axioms [41].

In the context of system-level reasoning, holism can be addressed properly only if the reasoning strategies, reasoning mechanisms, and reasoning algorithms are simultaneously considered. Reasoning strategies establish the logical and/semantic framework of reasoning. Reasoning mechanisms are implementations of the strategies as computational processes. Reasoning algorithms are active elements used in the holistic reasoning process. Håkansson, A. et al. identified eight strategies of reasoning as dominant ones for smart CPSs, namely: (i) deductive, (ii) inductive, (iii) abductive, (iv) analogical, (v) common sense, (vi) non-monotonic, (vii) case-based, and (viii) probabilistic reasoning strategies [42]. The three enablers (strategies, mechanisms, and algorithms) may be: (i) provided for, (ii) modified by, and/or (iii) developed by a smart CPS. If one strategy is not sufficient enough, then a purposeful mix of different strategies can be considered. This raises the issue of process, algorithm and data integration. The majority of the algorithms known from AI research is typically not interchangeable and cannot be integrated directly into comprehensive reasoning mechanisms. Towards this end, the standard reasoning algorithms included in a complex reasoning mechanism must be adapted to the other related algorithms in the design time. Like customization of the mechanisms to applications, adaptation of algorithms run time by the system itself is a complicated task.

2.4. Possible Logical Bases of System-Level Ampliative Reasoning Mechanisms

A starting point of our mechanism synthesis was the differentiation between formal and informal reasoning. Formal reasoning (often also called deductive reasoning or logic) manipulates the statements of evidence by evaluating them in virtue of their sentential structure and content. Deductive inference supports only explicative inference, where the conclusion is explicitly or implicitly included in the premises. Thus, deductive inference is enumerative since it spells out information that is already contained in the given premises [43]. Deductive arguments provide grounds for making their conclusions inescapable. If the premises are true, then the argument is valid for the reason of syllogism. If all premises are true and the argument is valid, then it is sound. Informal reasoning arrives at a conclusion by means of informed guess-work, relying on amplified evidences. Also called inductive reasoning or logic, informal reasoning interprets the evidences in their correlations or context to arrive at a conclusion. As the principle of reasoning, inductive arguments attempt making a conclusion likely or probable by delivering evidences for it. Provided the premises are true and an inductive argument succeeds in this attempt, the argument is strong. If the argument is both strong and has all true premises, then it is regarded as cogent.

In the context of computational reasoning, the term 'ampliative' is used in the meaning of 'extending' or 'adding to what is already known'. It refers to the fact that the conclusion of such argument goes beyond, or amplifies upon, the premises. Actually, this type of inferences is called ampliative and the reasoning mechanisms providing these are referred to as ampliative reasoning engines [44]. Ampliative reasoning may produce conclusions that contain genuinely new information. While deductive inference is enumerative, inductive inference is ampliative in the sense that it goes beyond merely spelling out the information already contained in its supporting evidential premises. While merely explicative (analytic) logical reasoning typically does not add anything to the content of cognition, ampliative (synthetic) logical reasoning increases the given cognition. A characteristic of ampliative reasoning is that the conclusions it yields may be mistaken. Thus, an ampliative argument is not deductively valid or invalid.

Though the idea of ampliative reasoning is well known in mathematical logics and artificial intelligence research, the idea of ampliative system-level reasoning mechanisms is new. Both deductive and inductive reasoning strive after rendering the best judgement in a holistic way, i.e., considering all influencing matters. If this is not the case, then the reasoning is restricted to the available evidences and inferring can target only the best explanation. This process of hypothesis formation and inferring has been described as abduction. As the third kind of logical inferential reasoning, abduction proceeds from observational data or events to a set of hypotheses, which best explains or accounts for the data. Abduction resembles induction in that it involves a reasoning process for providing hypotheses that explain the given facts, while induction is used to derive general rules from specific facts [45]. Peirce, C.S. argued that in spite of some resemblance, abduction may not be regarded merely as a variant of induction, because the mental processes involved are sufficiently distinctive [46]. Abduction involves coming up with plausible explanations for existing data, with the possibility of predicting the existence of additional data which, if subsequently discovered in accordance with its predictions, would tend to confirm the validity of the original hypothesis [47]. Some researchers argued that the psychology of abduction is somewhat mysterious, since it requires creative thought and imagination. In other words, it supposes the ability to imagine possible factual scenarios and to concentrate on just those having the greatest salience for the task in hand. On the other hand, abduction has a strongly ampliative character, which comes from hypothesizing. It has to be mentioned that reproduction of human creative thought and innate imagination by a computational reasoning process is a huge challenge that is not coped successfully by current artificial intelligence research.

In the context of smart CPSs, synthetic judgments are supposed to be system-derived elements of reasoning. A synthetic judgment can be ampliative, if its predicate adds something new to its subject. Ampliative reasoning is heuristic in the sense that it involves obtaining new pieces of beliefs, which are not entailed in the given premises (not already implied by what is known by the system and captured in the system operation/servicing model). It depends on the number and relationships of the finite set of evidences available in the process of reasoning, and is limited by the sophistication of the inferring capabilities of the computational reasoning mechanism concerned. Meheus, J. et al. refer to abduction as ampliative adaptive logics [48]. In their view, abduction is non-monotonic (i.e., conclusions derived at some stage of the reasoning process may be modified or even rejected at a later stage). Their research effort concentrated on a proof theory that warrants that the conclusions derived at a given stage are justified in view of the insight in the premises at that stage. This logic-based theory is supposed to support justified propositions even when the premises imply some sort of undecidable conclusions. Treating abduction as a form of forward reasoning was the logical solution for ending up with ampliative adaptive logics [49].

The above analysis suggests that ampliative reasoning cannot be else but a risk-taking strategy when implemented by smart CPSs. Thus, an objective of the development of system-level reasoning and adaptation mechanism is to reduce the risk of educated guess, which always accompanies decision making by systems. Usually, there are no exact criteria, conditions or measures of when the outcome of reasoning is approximately correct or good enough. It may well be that further evidence, which does not affect the truth of the premises, renders the outcome of reasoning false. This is the reason of why some traditional computational reasoning mechanisms of computer science and artificial intelligence, such as non-monotonic logics, probabilistic logics, reasoning by circumscription, and default reasoning, are not considered as fully-fledged implementations of system-level reasoning and adaptation.

2.5. Recent Efforts to Exploit Abduction as an Ampliative Computational Mechanism

Abduction has multiple interpretations and views [45]. Peirce, S.C. described it as a mode of reasoning that justifies beliefs about the probable truth of theories [46]. It is also seen as a recipe for generating new theoretical discoveries as well as a mode of reasoning that justifies beliefs about the probable truth of theories [47]. Other authors understood it as the process of forming an explanatory supposition [48] or a speculative reasoning strategy [49]. It is the only logical operation which introduces any new idea and a type of hypothesis formation and logical inference akin to guessing [50]. Abduction may have its suppositions in logic or in knowledge [51] [52]. Hermann, M. and Pichler, R. formally described logic-based abduction as follows: Given a logical theory T formalizing an application, a set M of manifestations, and a set H of hypotheses, find an explanation S for M, i.e., a suitable set $S \subseteq H$ such that $T \cup S$ is consistent and logically entails M [53]. In addition to reasoning in science, abduction has also been considered as a reasoning strategy in engineering and design contexts. As early as 1993, Kean, A.C. described a comprehensive framework for a domain independent abductive reasoning system and proposed to separate inference from domain dependent problem solvers in a computational reasoning framework [54]. This leads to domain independent inference engines, which are portable and applicable to many application domains, and reduce the repetitive efforts to build inference engines.

Computational implementation of abduction has been part of artificial intelligence research. Eiter, T. et al. defined a general abduction model for logic programming to allow the user to define the inference operator (i.e., the programming semantics to be applied on programs) [55]. Gottlob et al. showed that identifying explanations for a given set of observations algorithmically is intractable in the case of logically-based abduction [56]. Therefore, they suggested achieving tractability by reducing the underlying clausal theory so as to have a bounded width for the search tree. However, they also found that turning the theoretical principles of tractability into practically efficient algorithms is very problematic. Though several criteria for selection of explanations have been proposed in the literature, as Poole and Rowen discussed in the context of medical reasoning, several of these criteria are conflicting [57]. Though the efforts to consider uncertainty in computational abductive reasoning proved to be useful, the various approaches were restricted in handling complex, multi-component reasoning problems and in terms of the efficiency of knowledge representation [58]. For example, when used in a smart CPS, one limitation of the Bayesian belief network is that it requires the generation of a network and instantiation of the nodes to be able to explore the posterior probability by some methods.

With regards to computability of abduction, Psillos, S. derived two conjectures: (i) the reasoning process underlying abduction has a certain logical, though not algorithmic, structure, and (ii) the more conceptually adequate a model of abduction becomes, the less tractable it is computationally [59]. The latter finding puts conceptual richness and computational tractability into juxtapositions. He also argued that a rich conceptual model of abduction cannot be adequately programmed. Some major attempts to provide computational models of abductive reasoning are as follows: Pagnucco, M. and Foo, N. proposed an approach to computational abduction based on clausal conceptual graphs, and pointed at some limitations originating, e.g., in the influences of the syntactic restrictions of the graphs and the lack of criteria for selecting the best abduction from among those derived [60]. The abduction model presented by Boutilier, C. and Becher, V. non-monotonically generates explanations that predict an observation, but require some deductive relationship between explanation and observation [61]. Kakas, A.C. et al. used abductive logic programming to develop an abductive reasoning system, called A-System, for declarative problem solving. This involves a hybrid computational model that implements the abductive search in (i) the process of reducing the high level logical representation to a lower-level constraint store, and (ii) a lower-level constraint solving process [62]. Verdoolaege, S. et al. proposed a framework for consideration of temporal information in abductive reasoning in natural language processing, which cannot however be applied directly in the context of CPSs [63].

2.6. Synthesis of the Major Findings

Several theories and technologies such as logic, probability, complexity, physical, biological, cognitive, social theories and technologies, and their various combinations have been developed to realize smart system operations. Direct integration of a bunch of technologies and algorithms does not seem to offer a solution [64]. Since smart systems are supposed to make decisions about their operations and services under varying conditions, they should be able to make conditional inference. Though probability calculus offers formulas for binomial conditional deduction, they are restricted to dealing with the measure of logical probability. They do not capture the meaning of conditioning and the interplay of multiple conditions [65]. However, conditional reasoning offers a principle for it [66]. Technically, conditional reasoning expresses conditional relationships between parent and child propositions, and then combines those conditionals with evidence about the parent propositions in order to conclude about the child propositions.

Abduction was claimed as a powerful ampliative computational mechanism by many researchers [67]. Abduction was also considered as a logical model of designing. What is to be explained in design is the properness of the overall design objective or goal, and the assumptions are the building blocks of the designs (artefacts) to be synthesized. The explanation is the design (artefact), which is based on the background professional knowledge as well as on the dynamic knowledge associated with design analysis and synthesis. Consistency of reasoning means that designing was possible and that the design (artefact) provably achieved the design goal [68]. Computational complexity was discussed as a challenging issue of abduction [69].

3. Pilot Systems Hinting at Necessary Constituents of Procedural Abduction

3.1. Forerunning Projects and their Outcomes

The idea of procedural abduction was stimulated by the results of forerunning PhD research projects, which investigated various issues of using smart functionalities in second-generation CPSs. They developed different testable pilot implementations for real life scenarios. Figure 2 shows the relationship and contribution of these promotion projects. Their foci were: (i) system-level feature-based conceptualization and integrated operational and architectural modeling of first generation cyber-physical systems [70], (ii) exploring the role of system-initiated changes of operation modes in failure analysis in the context of the above family of CPSs [71], (iii) developing computational mechanisms for dynamic context information representation and inferring, context-based strategy and action planning, and situated messaging in a second generation CPS [72], and (iv) development of a smart reasoning mechanism, and adaptation and intervention planning for a second generation stroke rehabilitation monitoring and enhancing cyber-physical system [73].

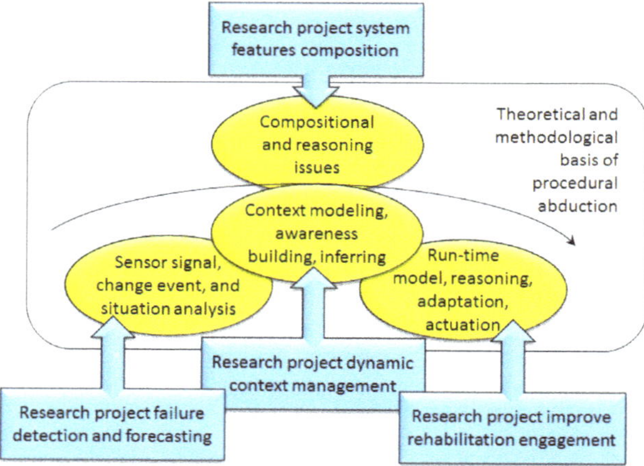

Figure 2. The interplay of the completed PhD research projects in the inception of the concept of procedural abduction.

A common assumption of the above studies is that S-CPSs partially self-determine their operational objectives and system-level adaptation based on run-time collected data, using built-in or acquired smart learning and reasoning algorithms. Relying on multiple streams of input data, the prototype systems build dynamic operational models, which in turn allow them to alter their operational objectives and to adapt their functionality and (software) architecture accordingly. In the order of mention, these studies cast light on the elements (computational activities) of a complex system-level reasoning process, which reflects the logic and characteristics of abductive reasoning. Consequently, this reasoning process has been called 'procedural abduction' (PA) [74]. The proposed theoretical foundations of PA are being challenged currently through analytic investigations (critical systems thinking) and computational implementation in other running PhD [75] and staff research projects [76]. The relevant work will be concisely summarized below.

3.2. System-Level Feature-Based Conceptualization and Modeling of First Generation Cyber-Physical Systems

The objective of this research project was to develop a software toolbox (SMF-TB) for pre-embodiment design of first-generation CPSs. The novelty of this toolbox is that it (i) tries to cope with the inherent heterogeneity of CPSs (caused by interrelated hardware, software and cyberware constituents) on system-level, (ii) combines operational and architectural modeling through (transdisciplinary) system-level features (SLFs), and (iii) provides effective methodological support for creating SLFs by using genotypes, phenotypes and instances [77]. The two main parts and the respective components of the SMF-TB are shown in Figure 3. Likewise the workflow of modeling using SMFs, the chunks of information needed for defining genotypes and phenotypes of SMFs have been clarified and organized into information schema constructs (ISCs). The proposed ICSs connect the chunks of operation and architecture information in the relational data tables of the warehouse databases. The computational algorithms and the overall system modeling mechanisms, including the feature instantiation-based composition mechanism, have been worked out.

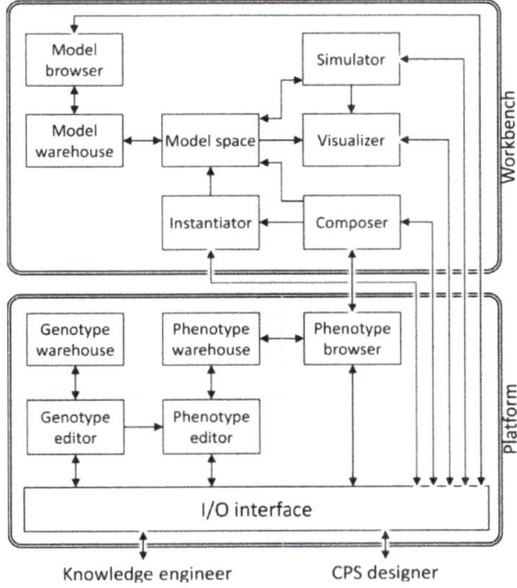

Figure 3. The parts and components of the SMF-TB. (I/O: input/output; CPS: cyber-physical system)

The proposed SMF-TB imposes a strictly physical view, i.e., it models system constituents in the 3D Euclidian space and captures their relationships as mereotopological relations (rather than just representing them by abstract modeling surrogates). The ISCs for: (i) representation of operation and architecture relations, (ii) assigning values to parameter variables, (iii) managing the meta-level knowledge-base of the model warehouse, (iv) recording the composition and parametrization history, and (v) recording the state, event and stream unification history of the instantiation and composition of system constituents are discussed in [78]. The overall process of instantiation involves three cycles, on genotype, phenotype and instance-type level. Though the proposed information schema constructs, tool-box functionality, system-level modeling methodology, and procedural and computational schemata provide effective support for SMFs-based modeling of first generation CPSs, the completed investigations cast light on various limitations and/or deficiencies with a view to context capturing and self-awareness building, self-reasoning and self-learning, and self-adaptation and self-organization capabilities of smart systems.

The proposed resources proved to be suitable for designing composable systems with definitive interfaces that fulfill input assumption and output guarantee specifications. These conventions however cannot be applied to run-time organized smart systems, which construct their reasoning capabilities during operation and acquire knowledge from run-time processed data streams. Consequently, this project revealed that reasoning mechanisms cannot be synthesized in a components-based manner, that is, by directly combining existing AI/cognitive algorithms. Only compositional synthesis, relying on a holistic knowledge processing framework, can guarantee that the constituents of an ampliative reasoning mechanism provide a synergistic body of knowledge for achieving the objectives of system operation under dynamically varying circumstances. On the other hand, using highly adaptable or self-adaptive system-level (or mechanism-level) features is a reusable idea to address composability challenges of reasoning mechanisms.

3.3. Identification and Forecasting Failures in a Resilient Cyber-Physical Greenhouse Testbed System

This research started out of the assumption that future CPSs will be characterized by growing self-intelligence and self-organization, and considered that these abilities enable them to maintain their operations according to the set objectives under the effects of various influential factors, such as internal failures or external environmental disturbances. Their control regime will set their operation modes, as well as the values of the operational parameters, to achieve the relative best output even under irregular conditions. However, due to the run-time adjustments, they hinder an early recognition of emergent failures. In other words, the transitions self-enabled by a resilient CPS cause uncertainty and hamper the use of the conventional failure analysis methods.

It was hypothesized that the intensity and the trend of the changes of system operation modes (SOMs) can be used as the basis of the diagnosis, recognition and forecasting of failures in resilient CPSs To test this hypothesis empirically, and to contribute to the theoretical understanding of the failure recognition and forecasting problem, a self-regulatory and self-tuning testbed greenhouse system was developed [79]. The roles of system initiated compensatory actions and operational mode changes were systematically investigated from the aspects of emergence and proliferation of technical failures in this first generation CPS. However, the influence of functional and structural self-adaptation on failure recognition and forecasting was not studied, since it would assume the implementation of an even more sophisticatedly controlled (i.e., smartly behaving) testbed system.

The novelty of this study was that it focused on system-level behavior, rather than on the behavior of the individual components. The input and output signals/data were interpreted on system level (Figure 4), which made it possible to generate a set of indicators that could inform about the SOMs that were triggered by emerging failures. For instance, certain failures pushed the testbed system into an 'abnormal' operation mode, which involved a combination of component operation modes not typical under regular circumstances. On the other hand, certain SOMs did not occur due to the effect of failures, or specific SOMs emerged that did not occur during regular system operation. Consequently, the main contribution of this work was providing novel means, such as the concept of failure induced operation modes, for monitoring changes of states, events and situations of a quasi-dynamic system [80]. The changes were captured through the observed variations in the frequency and the duration of SOMs—that is, in the system dynamics rather than only by signal deviations. This lent itself to a shift from timed system model-based reasoning about the system operation to a data-driven, run-time conditions-based reasoning. Our investigations disclosed that the above concepts could be reused for event and situation diagnoses in other first-generation CPSs and in feeding reasoning mechanisms with information higher than component input/output data. We also understood that it still needs further elaboration in the case of self-adaptive second-generation systems.

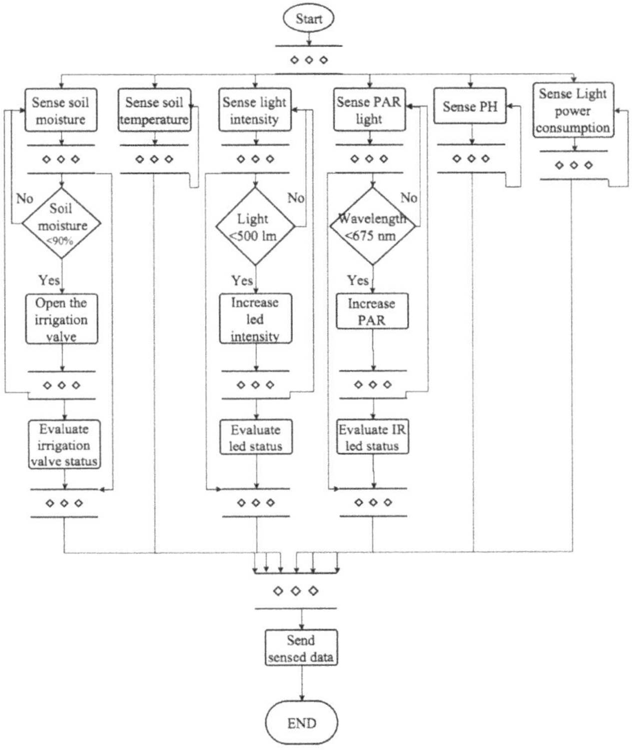

Figure 4. Example of processing sensor information in the testbed greenhouse system. (PAR: photosynthetic active radiation; IR: infrared; PH: scale of acidity).

3.4. Representation and Reasoning with Dynamic Context Knowledge in a Fire Evacuation Aiding System

Smart CPSs often works under highly dynamic circumstances and need to make decisions in dynamic contexts. Typical application examples are such as home care servicing, traffic management on roads, and aiding fire evacuation of buildings. Therefore, it is necessary to provide fast mechanisms for dynamic context computation (DCC) and building awareness by the system that enables it to interpret context changes and to infer about their implications on physical processes. This PhD research project developed a conceptual/logical framework for DCC and implemented computational algorithms for building system awareness. The DCC mechanism was developed based on a layered (context) knowledge model, which included (i) the layer of states (of entities and their spatial, attributive, temporal and semantic (SATC) relationships), (ii) the layer of events (changes in the SATC relationships), (iii) the layer of situations (logical/semantic relations of events and states), and (iv) the layer of scenes (logical/semantic relations of interplaying situations).

The knowledge model assumes information integration on each layer, and information abstraction to support semantic transitions between the subsequent layers. The semantic abstraction over the integrated information constructs provides opportunity to computationally infer not-explicitly described states, events, situations and arrangements in various forms. The data describing dynamic contexts were arranged in a specific computational scheme called context information reference cube (CIR-cube), which proved to be a dexterous computational means for handling spatial, attributive and temporal data in a cohesive manner (Figure 5) [81]. An inference mechanism was developed that updates and (re)computes the contents of the matrices included in the CIR-cube and eventually builds an awareness model of the time-varying process at hand. The CIR-cube is able to capture both physical time and computational time. The procedure of awareness building includes the following main computation steps: (i) determining the state of the concerned entities, (ii) recognition of events, (iii) identification possible situations, (iv) judging the relevance of situations to the concerned entities, (v) revealing the interplays of the relevant situations, and (vi) interpreting the implications of the interplaying situations [82]. The functional scheme of the inferring mechanism for building awareness is shown in Figure 6.

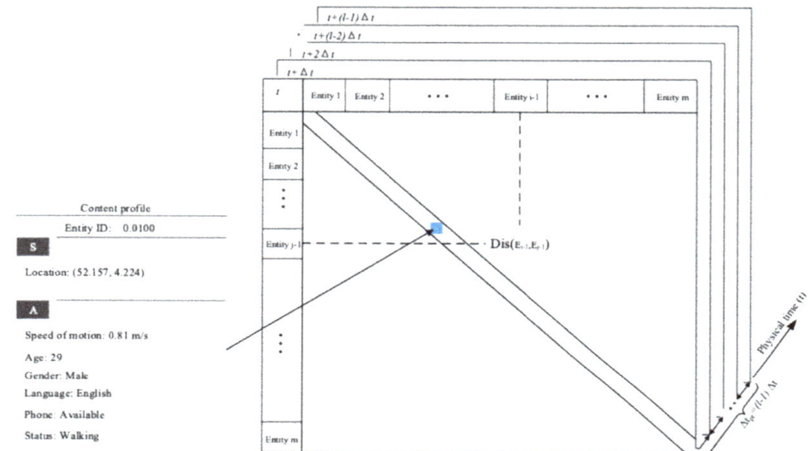

Figure 5. The context information reference cube (CIR-cube).

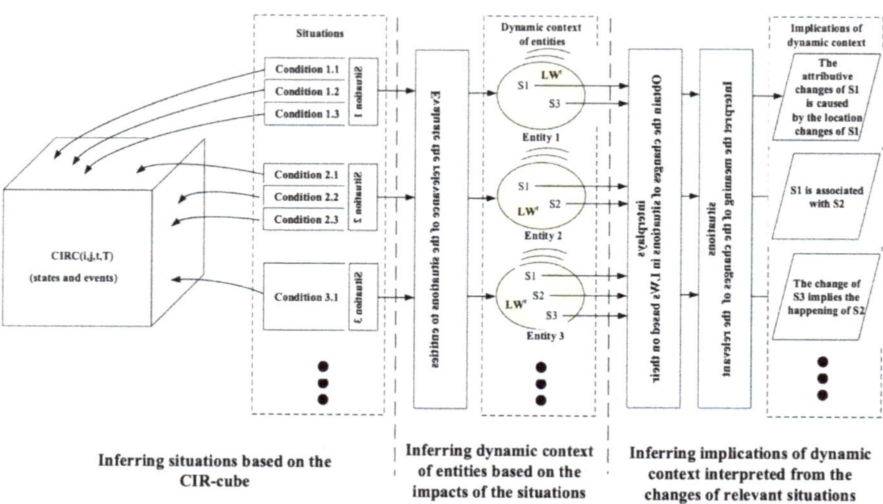

Inferring situations based on the CIR-cube

Inferring dynamic context of entities based on the impacts of the situations

Inferring implications of dynamic context interpreted from the changes of relevant situations

Figure 6. The functional scheme of the inferring mechanism for building awareness. (LW: local world; S: situation).

The developed dynamic context computation mechanism is complemented with a reasoning mechanism for operational strategy synthesis. Ultimately, this mechanism generates personalized action plans that make it possible to achieve the specific operational/servicing objective of a smart CPS. Together with the implemented additional computational mechanisms for message generation and distribution, the dynamic context computation and action plan generation mechanisms were tested in an indoor fire evacuation guiding application. The application case was investigated based on a high-fidelity simulation of presumed real-life fire propagation and of the behaviors of the human, artefactual, and natural stakeholders. The experimental results proved the efficiency of the interoperating computational mechanisms and algorithms. They also confirmed our hypothesis that the proposed dynamic context computation mechanism was able to provide descriptive knowledge about emergent situations as well as about the implications of the interplaying situations on the concerned entities.

The contribution of this research to the conceptualization of the main constituents of procedural abduction is as follows: The proposed solution uses sensor-provided and/or preprogrammed spatial, attributive and temporal data to describe all involved entities and to characterize their varying states in their local worlds, as a system of reference. A state was defined as a static representation of the momentarily characteristics of an entity, whereas an event as the change of the states of an entity at two subsequent points in time (in the computational realm). A situation is generated by the aggregation of a series of operationally non-independent events and/or states appearing at a given point in time. In addition to computationally combining/integrating logically related events and states, the mechanisms were also able to extract specific meaning of combined and/or integrated events and states by abstraction towards a higher level comprehension. The descriptive data-based context representation is supplemented with various derived elements of semantic intelligence, which are generated by the inferring algorithms of the proposed dynamic context management mechanism.

3.5. Reasoning and Adaptation in an Engagement Monitoring and Enhancing System for Stroke Rehabilitation

Task-oriented training exercises need to be practiced in upper limb rehabilitation after stroke. The hypothesis of this work was that a smart cyber-physical stroke rehabilitation system (CP-SRS), which is able to increase the motor, perceptive, cognitive, and emotional involvement of patients during rehabilitation exercises can be a promising solution. Thus, the objective of this PhD research project was to augment a robot-assisted upper limb rehabilitation subsystem with a cyber-physical computation-based engagement management subsystem. Since there is no opportunity for preprogramming in this specific application case, realization of engagement management raised the need for run time reasoning capabilities. At the beginning of the project it was not completely understood which factors (e.g., game difficulty, personal interest, game design, and immersiveness of environment) are the most influential on the engagement of patients [83]. Furthermore, no quantitative method was found to evaluate momentarily engagement. Based on the outcome of the explorative research, the main functions and architectural elements of the CP-SRS have been defined as: (i) a multi-modal sensor network, which monitors the states of patients, (ii) real time information processing, which interprets the actual signals and generates engagement models, (iii) reasoning and decision making, which provides personalized stimulation plans for different patients, and (iv) situated learning that facilitates run time generation of a reasoning model concerning the system state/objectives and the necessary/possible adaptation. These functionalities and system components were conceptualized and implemented in the CP-SRS at a testable prototype level [84].

Architecturally, the CP-SRS included five subsystems: (i) an assistive robotic subsystem, (ii) a gamification subsystem, (iii) an engagement monitoring subsystem, (iv) a smart learning mechanism (SLM), and (v) an engagement enhancement subsystem (EES) The indicators for: (i) motor engagement (ME), (ii) perceptive engagement (PE), (iii) cognitive engagement (CE), and (iv) emotional engagement (EE) were monitored using: (i) MYO$^{(TM)}$ Armband with electromyography sensors, (ii) Eyetribe eye tracking device, (iii) Emotiv$^{(TM)}$ Epoc headset (14 channel wireless electroencephalography (EEG) device), and (iv) a web camera, and Insight device, respectively. The data from these sensors were streamed to MATLAB via TCP/IP computer network transmission control protocol, where the engagement levels in the four aspects were interpreted. The workflow of information processing related to the smart learning mechanism is shown in Figure 7. Basically, when the patient's engagement level decreased, the system introduced interventions. Through the interventions it was able to re-engage the patients and to maintain their high level engagement. The interventions meant stimulations in motor, perceptive, cognitive, and emotional aspects, depending on the actual state of the patients. Stimulation strategies were created as a combination of stimulations in multiple aspects.

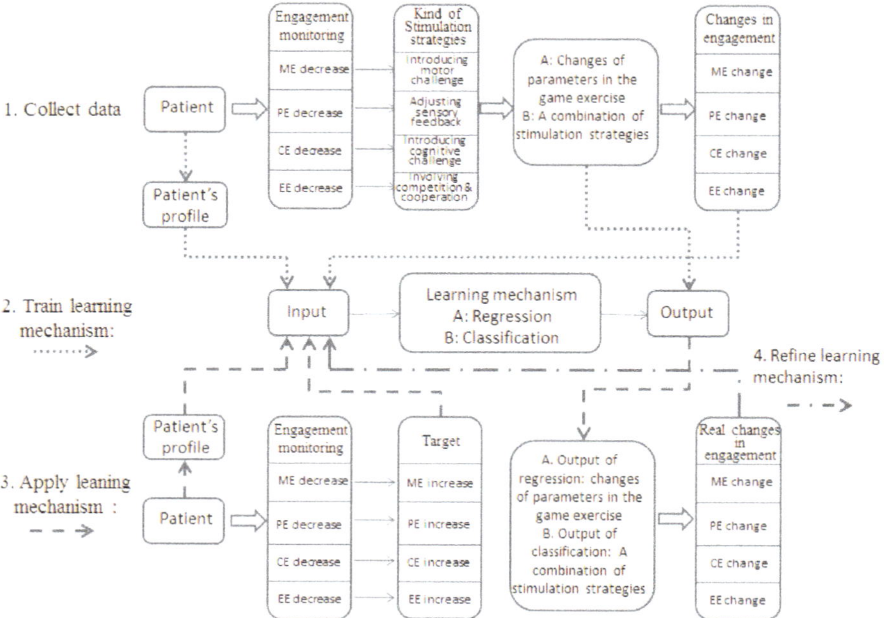

Figure 7. The overall information processing workflow including the smart learning mechanism.

The main findings of this research project can be summarized as follows: It has been found that the ratio of the root mean square of the measured electromyography (EMG) signal and the velocity of motion of the human limb was a reliable indicator of motor (function) engagement. However, the indicators introduced for measuring the motor, perceptive, cognitive and emotional engagements had to be taken into consideration simultaneously in order to achieve an optimal stimulation strategy. They had also to be interrelated in order to form a distinct measure. In the process of computing the applicable stimulation strategies, there was a need to consider the personal profiles of the patients in addition to their motor, perceptive, cognitive, and emotional engagement indicators. A neural network-based smart learning mechanism could be used to learn the effects of the different stimulations strategies on different persons and to propose personalized enhancement. Continuous monitoring of the state of the patient and learning the enhancement options lead to efficient, personalized and self-adaptive stroke rehabilitation training.

The identified engagement indicators were useful not only for enhancing engagement, but also to understand the limitations of the current engagement enhancing methods. Although the methodology developed for monitoring and enhancing engagement was dedicated to rehabilitation, this approach can be used in other fields as well, such as sports, driving, and education. The contribution of this project to revealing constituents necessary for procedural abstraction were: (i) building situation awareness based on input sensor data, (ii) devising a reasoning model run-time based on relative changes of state indicators, (iii) development of possible adaptation (stimulation) strategies by machine learning using the state indicator-based reasoning model, and (iv) operationalization of the adaptation plan based on changing the setting of the physical and computational effectors.

4. The Framework of Computational Implementation of Procedural Abduction

4.1. The General Workflow and the Underpinning Knowledge of Procedural Abduction

From a computational point of view, procedural abduction is seen as a recurrent sequence of eight processing stages: (i) run-time extraction of data/signals by sensing, (ii) recognition of change events, (iii) inferring about exiting operational situations, (iv) building awareness of the system's performance, (v) devising alternative performance enhancement strategies, (vi) designing adaptation of the system as a whole, (vii) planning the implied interventions, and (viii) actuating effectors and controls. The operational workflow (computational implementation) of procedural abduction (PA) is graphically shown in Figure 8. The activities (i)–(iv) enable the system to capture data about its momentary performance (outcome of system operation), to compare the data with those representing the assumable best objective of servicing, and to define the necessary and possible changes of the system. The activities (v)–(viii) enable the system to determine the necessary functional and/or architectural changes, to define the adaptations to be introduced, and to set the values of the actuator control variables accordingly. Each activity involves at least one, but typically multiple interacting computational algorithms, which are integrated into the overall reasoning mechanism of procedural abduction.

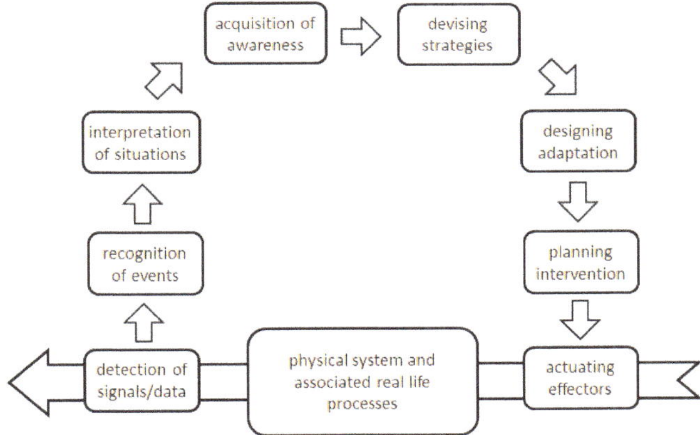

Figure 8. Graphical representation of the generic workflow of procedural abduction.

The basis of implementation of the computational workflow of PA is the predefined operation and servicing model (OSM) of the concerned CPS system, which specifies: (i) the default objectives, (ii) the values of the operational parameters, and (iii) the initial state of the system, as a reference. Symbolically, the OSM can be specified on a given level of resolutions as a septuplet (Equation (1)):

$$OSM = (IV, AC, PS, OD, AD, OV, TV) \tag{1}$$

where: IV is a finite non-empty set of input variables belonging to the total of interoperating system actors on a given level of resolution; AC is a finite non-empty set of architectural constituents realizing the system actors on a given level of aggregation; PS is a finite non-empty set of services generated by the system actors, OD is a finite non-empty set of operational dependences among system actors and the provided services; AD is a finite non-empty set of architectural dependences among system actors and the provided services; OV is a finite non-empty set of output variables describing the provided services, and TV is a finite non-empty set of the required/possible target (interval) values of operation and services of the system as a whole. An explicit incorporation of the operational and architectural dependences of the system actors is needed in view of the compositional synthesis of the system. (Compositional synthesis means that the operational and architectural manifestation of the system actors as constituents is simultaneously defined by the overall operational/servicing objectives of the system as a whole and by the manifestation of the closely interoperating other constituents.)

PA is not purely a logical propositions-based reasoning, but data-, information- and knowledge-based. The knowledge required for the implementation of the generic workflow of procedural abduction (shown in Figure 8) has two parts: (i) static knowledge and (ii) dynamic knowledge. The static knowledge is conveyed by the predefined OSM of the system (referred to as $OSM_{initial}$ in the rest of the article). The dynamic knowledge is derived based on the run-time acquired data and the modified OSM (OSM_{actual}), and thus by the change of the operational context and objectives. The concerned algorithms of the computational mechanism blend the parts of these two bodies of knowledge in each stages of processing. The overall scheme of using the knowledge sources is shown in Figure 9. From the point of view of realization logical/semantic reasoning by PA, there are four sources of knowledge considered: (i) the content of $OSM_{initial}$, (ii) the content of OSM_{actual}, (iii) the descriptive spatial, attributive and temporal data, semantic relations, and prescriptive constraints included in the permanent context $(P\hat{C})$ and dynamic context $(D\hat{C})$ representations, and (iv) specifications of conceptualizations in the associated system-used ontologies and data residing on the Web. The necessity of these sources of knowledge has been proven in our research, as well as why their sufficiency for a wide range of applications should still be tested experimentally.

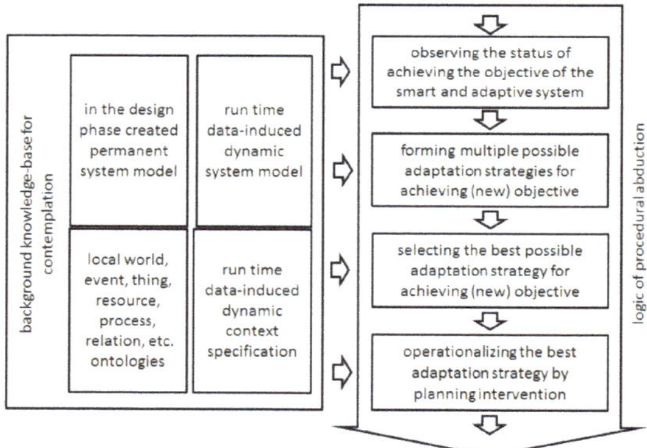

Figure 9. The knowledge assets used in procedural abduction.

4.2. From Enabling Operators to Transforming Algorithms

The computational mechanism of procedural abduction (MPA) has been interpreted as an arrangement of operators completing the computational activities in each stage of PA. MPA can symbolically be represented by the following formula (Equation (2)):

$$MPA = \xrightarrow{DD} \gg \xrightarrow{RE} \gg \frac{\xrightarrow{FI} \gg \xrightarrow{AP}}{\xrightarrow{IS} \gg \xrightarrow{AA} \gg \xrightarrow{DS} \gg \xrightarrow{DA}} \gg \xrightarrow{PI} \gg \xrightarrow{AE} \tag{2}$$

where:→ (rightward arrow) indicates an operator of the workflow; (horizontal line) is used to separate alternative sequences of operators (i.e., operation with or without adaptation). The symbol '$\gg$' indicates the orientation of the flow of information processing. The group of operators above the horizontal line allows self-adjustment of a system, while the group of operators below the horizontal line enables self-adaptation of a system. DD is the operator of detecting signals and elicitation of data, RE is the operator of recognizing and monitoring events, FI is the operator of providing feedback information for operational control; AP is the operator of adjusting the values of operational parameters; IS is the operator of interpretation of situations; AA is the operator of acquisition of operational awareness; DS is the operator of devising adjustment and/or adaptation strategy; DA is the operator of designing adaptation; PI is the operator of planning interventions; and AE is the operator of selective actuating of effectors.

Each element of the computational process, including those belonging to its decision making sub-process, can be implemented by two kinds of computational operators: (i) (physically-grounded) operators handling state changes in the physical realm, and (ii) operators for processing data, information and knowledge in the cyber realm. While the first kind of operators process changes in the continuous space and time, the second kind of operators work with digital representation and feature event-oriented execution. Ultimately, both capture state changes and are handled in similar manner as computational transformations, $T_{x,i}$, where: x is any operator of PA, and i is the identifier of a particular computational action belonging to x. The operators may include three sets of transforming algorithms, namely basic, auxiliary and interaction algorithms. These types are identified based on the purpose of the algorithms. Basic algorithms generate new information by reasoning/inferring, while auxiliary algorithms maintain information, for instance, by recording digital data in files. Interfacing algorithms avail information, e.g., receive manual data input, display data to enable human interaction, or convert data between cooperating system modules. Due to space limitation, only the basic transformations (sets of transforming algorithms) will be discussed.

4.3. Transforming Algorithms Needed for the Operators

The operator for detecting signals and elicitation of data (DD) includes the following transformations (Equation (3)):

$$DD = (T_{DD,1}, T_{DD,2}, T_{DD,3}, T_{DD,4}, T_{DD,5}, T_{DD,6}, T_{DD,7}, T_{DD,8}) \tag{3}$$

where: $T_{DD,1}$ is identification of active physical and action sensors in the environment; $T_{DD,2}$ is local sensing of the attributes of material flows; $T_{DD,3}$ is local sensing of the attributes of energy flows; $T_{DD,4}$ is local sensing of the attributes of information flows; $T_{DD,4}$ is collecting data from linked software sensors; $T_{DD,6}$ is transferring signals on the wired/wireless network; $T_{DD,6}$ is multiplexing analogue signals; $T_{DD,7}$ is converting analogue signals into digital data; and $T_{DD,8}$ is cleaning/filtering digital data.

Used for recognizing and monitoring events, the next operator (RE) includes transformation algorithms for the analysis of the sensed and input signals/data in order to detect something that happens or might happen at a given physical or logical place and time. The analysis considers the operational state of the system as a whole, of the constituents, and the set operation/servicing objectives. The recognizing and monitoring events operator is defined as (Equation (4)):

$$RE = (T_{RE,1}, T_{RE,2}, T_{RE,3}, T_{RE,4}, T_{RE,5}, T_{RE,6}, T_{RE,7}) \tag{4}$$

where: $T_{RE,1}$ is detection of change trends in digital data, $T_{RE,2}$ is obtaining information over operation modes of the system, $T_{RE,3}$ is detection of remarkable signal changes that may be associated with discrete change events, $T_{RE,4}$ is features-based investigation of the signal changes, $T_{RE,5}$ identification and classification of a recognized events according to their nature (space-related, attribute-related, or time-related), $T_{RE,6}$ is time stamping and recording of recognized events, and $T_{RE,7}$ is monitoring the life cycle of the recorded events. These transformations make it possible to recognize state changes in terms of deviation from the set objective and the preferred system states, respectively.

A situation has been defined as interactions of the recognized events, not matter if they concern the change in the objectives or in the system-internal states. The main goal of this activity is to identify the interacting events and to determine their relationships in space, time and logic. Thus, the operator for interpretation of situations (IS) includes the following transformations (Equation (5)):

$$IS = (T_{IS,1}, T_{IS,2}, T_{IS,3}, T_{IS,4}, T_{IS,5}, T_{IS,6}\ T_{IS,7}) \tag{5}$$

where: $T_{IS,1}$ is recalling all recognized events; $T_{IS,2}$ is investigation of the space of the individual events in a considered local world based on the location of the signal provider; $T_{IS,3}$ is investigation of the time stamps and durations of the individual events in the considered operation window; $T_{IS,4}$ is computation of spatial relationships of the events occurring in the considered local world; $T_{IS,5}$ is computation of temporal relationships of the events occurring in the considered time window; $T_{IS,6}$ is determining the set of correlated events and recording it as a situation; and $T_{IS,7}$ is monitoring the trend of change of an identified situation.

Having a grasp on the states of system operation and objective achievement, the reasoning mechanism intends to 'understand' the meaning and implication of the actual situation. This is based on learning, which allows the knowledge-enabled mechanism to acquire additional information and to build awareness. As discussed above, there are four sources of information: (i) the initial system model, (ii) the run-time acquired data, (iii) the dynamic context model, and (iv) additional data repositories. Building operational awareness allows making logical judgments and inferring conclusions. Thus, the operator for acquisition of operational awareness (AA) is composed of the following transformations (Equation (6)):

$$AA = (T_{AA,1}, T_{AA,2}, T_{AA,3}, T_{AA,4}, T_{AA,5}, T_{AA,6}, T_{AA,7}, T_{AA,8}, T_{AA,9}) \tag{6}$$

where: $T_{AA,1}$ is operationalizing the $OSM_{initial}$; $T_{AA,2}$ is determining the deviations from $OSM_{initial}$ in the given situation; $T_{AA,3}$ is computation of the operation/servicing indicators; $T_{AA,4}$ is generating implicit context information; $T_{AA,5}$ is generating spatial context information; $T_{AA,6}$ is generating temporal context information; $T_{AA,7}$ is generating attributive context information, $T_{AA,8}$ is computation of the dynamic context model; and $T_{AA,9}$ is unsupervised learning of the necessary control regime (that is, if maintaining $O_{initial}$ is needed or if there is a possibility for a more favoring $O_{possible}$). Together with the results of the transformations included in the operators AA and DA, the DS operator informs the reasoning mechanism about how and why a particular set of conclusions were made. That is, the sequence » AA » DS » DA » represents a local proposition-based abduction in the whole process of procedural abduction.

Depending on the control regime, a proper strategy is needed to achieve the set operational/servicing objective of the concerned CPS. This is the task of the devising adjustment and/or adaptation strategy (DS) operator, which includes the following computational transformations (Equation (7)):

$$DS = (T_{DS,1}, T_{DS,2}, T_{DS,3}, T_{DS,4}, T_{DS,5}, T_{DS,6}, T_{DS,7}) \tag{7}$$

where: $T_{DS,1}$ is initiation of computational actions according to the control regime ('observation'); $T_{DS,2}$ is investigation of operation/servicing indicators with regards to possible enhancement; $T_{DS,3}$ is devising. alternative operational strategies; $T_{DS,4}$ is devising feasible associated adaptation strategies ('hypotheses'); $T_{DS,6}$ is assessing the operational and adaptation strategies ('stratagems') considering the resources and the context of actions; and $T_{DS,7}$ is ranking the stratagems and selecting the best one.

While strategizing focusses on both the functional and logical aspects (i.e., what to change and why to change), architecture and operation adaptation concentrates on the technical and practical aspects of altering the system (i.e., on how to change and when to change). In this sense in produces a technical blueprint of the system alteration together with a course adaptation plan. This plan is the basis of intervention specification. The operator for designing adaptation (DA) is realized by the following computational transformations (Equation (8)):

$$DA = (T_{DA,1}, T_{DA,2}, T_{DA,3}, T_{DA,4}, T_{DA,5}, T_{DA,6}, T_{DA,7}, T_{DA,8}, T_{DA,9}) \tag{8}$$

where: $T_{DA,1}$ is investigation of the degrees of freedom in which the system can be adapted according to the best adaptation strategy; $T_{DA,2}$ is determining the necessary/possible operation/servicing adaptation; $T_{DA,3}$ is determining the necessary/possible architecture adaptations; $T_{DA,4}$ is computation of the OSM_{actual} based on the skeleton of $OSM_{initial}$; $T_{DA,5}$ is computational simulation (pre-playing) of the system's operation/servicing after introducing the adaptations; $T_{DA,6}$ is investigation of the impact of adaptation on the system's properties; $T_{DA,7}$ is adjustment of OSM_{actual} according to the findings and enhancement of the adaptation plan; $T_{DA,8}$ is identification of the outgoing and/or incoming system resources; and $T_{DA,9}$ is determining the sequence of the hardware, software and cyberware adaptation actions. As it can be seen, this operator is one of those that require the largest number of interoperating algorithms.

The DA operator can provide only a 'delayed' adaptation plan due to the necessary preliminary computational testing of the impacts of adaptation. The possibility of adaptation is influenced by then completion of certain operations of the CPS. This creates the need for the planning interventions operator. The objective of this operator is to operationalize a refined adaptation plan operation- and time-wise. Actually, it converts the adaptation blueprint into a transition blueprint, which considers the operation/servicing of the CPS and the conditions for this. The operator for planning interventions (PI) comprises the following transformations (Equation (9)):

$$PI = (T_{PI,1}, T_{PI,2}, T_{PI,3}, T_{PI,4}, T_{PI,5}, T_{PI,6}) \tag{9}$$

where: $T_{PI,1}$ is generation of a scenario for modification of the effectors; $T_{PI,2}$ is computation of control information for all motors; $T_{PI,3}$ is computation of the control information for all regulators; $T_{PI,4}$ is computation of the control information for all sensors; and $T_{PI,5}$ is computation of the control information for all information handling components, and $T_{PI,6}$ is computation of the control information for all computational effectors.

Finally, the operator for selective actuation of effectors (AE) is defined as (Equation (10)):

$$AE = (T_{AE,1}, T_{AE,2}, T_{AE,3}, T_{AE,4}, T_{AE,5}) \tag{10}$$

where: $T_{AE,1}$ is activating and setting rotary motors, stepper motors, servos and specialty motors; $T_{AE,2}$ is activating and setting linear actuators, effect transformers, regulators and transceivers; $T_{AE,3}$ is activating and setting environmental sensors, physical sensors and action sensors; $T_{AE,4}$ is activating and setting communicators, transceivers, modems, converters, cameras, and displays setting of system parameters; $T_{AE,5}$ is activating and setting computational effectors.

5. Some Conclusions and Future Research Opportunities

5.1. Reflection on the Approach

Humans typically apply the divide-and-conquer strategy combined with some informal reasoning to solve complex practical application problems. In cyber-physical systems, the complex application problem (e.g., providing multi-activity assistance in home care context) is to be decomposed to tasks that can be allocated to the active nodes of the system. Decomposition of complex problem typically needs informal (intuitive and semantics-based) reasoning. It also assumes certain level of autonomy and collaboration of the active nodes [85]. However, as Pease, A. and Aberdein, A. argued, a comprehensive theory of informal reasoning is not available and perhaps even not expectable [86]. Therefore, problem solving by cognitively enabled systems needs to be based on formal, computationally processable reasoning theories.

The objective of the presented research effort was to contribute to the progress in this field of interest. As a computational implementation of a formal theory that provides a flexible system-level reasoning capability for various smart CPSs, procedural abduction was proposed. Its idea emerged based on a conceptual synthesis of solutions for reasoning in various application contexts. The reasoning pattern of procedural abduction is straightforward:

> Some phenomenon concerning the operation of the system is observed;
> If a particular explanation would be true, then the phenomenon could be a matter of course;
> Hence, there is a reason to suspect that the particular explanation is proper (true).

Contrary to the above fact, implementation of the computational algorithms and data processing for PA is a complex and challenging undertaking. One of the challenges of implementation is that the computational reasoning should be knowledge-based, rather than just purely logics-based. On the other hand, PA offers affordances and benefits that cannot be expected from other approaches.

In the case of a smart CPS, the observed phenomenon is the relation of the state of the system to the set operational objective, or to a possible optimal operational objective. To generate proper explanations about this relation, the process of reasoning should include a part that collects information and builds awareness about the actual operational state of a system (contemplation), and another part that reasons about the necessary/possible adaptations towards a better operational objective (alteration). These are the functional backbones of the proposed procedural abduction mechanisms. PA exemplifies a form of reasoning that is ampliative in the sense that it aims at extending the domain of the actually existing system knowledge in every given state of its operation.

In principle, PA is application independent, but it should most probably be tailored according to the specificities of the considered application domains in order to achieve the best possible performance. By enabling deep penetration into real life processes and complementing model-based system control regimes with non-preprogrammed, run-time data acquisition-enabled, learning and reasoning, PA may be the reasoning mechanism for many physical, biological, medical, social, cognitive, etc. applications which includes processes of highly dynamically changing nature. Computational realization of PA necessitates a combination of a large number of conventional and specific artificial intelligence algorithms, which should be interconnected in a compositional manner [87]. Achieving compositional synergy in terms of the large number of interdependent algorithms, as well as in terms of the knowledge flow needed for system level problem solving, is found as a challenge for implementation. This problem is already a recognized one in the literature [88].

One of the aims of this article was to emphasize the significance of abduction as a computationally feasible problem solving process and to propose computational framework for procedural abduction. PA operationalizes the principle that systems and agents of cognitive problem solving should incorporate knowledge about the world (ontological commitment) and an abstract procedure (inferential commitment) for interpreting this knowledge towards constructing operation plans and taking informed actions. Clearly, implementation, application and validation of procedural abduction as an ampliative reasoning mechanism for varied cyber-physical systems are a work in progress. However, the development of its underlying theoretical framework and computational methodology has reached an advanced stage. At this time, it can be forecasted that its realization may come to fruition, though a fully-fledged implementation in a form of a platform, which is applicable in multiple CPSs in various contexts, still requires substantial work.

5.2. Future Research Opportunities

The results summarized in this article are related to the first phase of our research, which concentrated on exploring the elements of a feasible conceptual framework for procedural abduction. The on-going research efforts are made towards a fully-fledged computational implementation and integration. The development activities should extend to the refinement of all algorithms chosen or developed for system level reasoning. The ultimate objective is to use the abductive reasoning mechanism as pluggable module of smart CPSs, which can provide application independent reasoning and reduce the software and knowledge engineering work. Since high-fidelity computational replicas of complex mental representations are inherently compositional, conceptual frameworks and design methodologies fostering compositionality of smart CPSs are highly necessary. Towards this end, the issues associated with compositional design of reasoning mechanisms need to be addressed too. The importance of compositionality is well recognized in the literature, but further research is needed in the case of run time self-organizing systems. The issue of maintaining synergy between the initial system model and the (dynamically changing) actual system model should also be addressed. Even a partial modification of a well-tested system reasoning model by a run time developed reasoning model may create problems with a dependability and resilient operation of CPSs. The uncertainty created by each step of procedural abduction, such as inferring and interpreting the changing context of application, developing adaptation strategies, and performing adaptation of system models, should be treated with outmost care. Further explorative research is needed in this field too. Furthermore, specific methods that are able to verify adjusted system operation models at run-time are also needed. They should be able to investigate and forecast the effects of various operation strategies and system adaptations on the performance and behavior of CPSs in changing contexts. Real time formal verification of operational strategies is an essential feature of procedural abduction. While addressing formal verification at run time is in the focus of CPS research, methods have to be developed that would help address the concomitant challenges.

Funding: This research received no external funding.

Acknowledgments: The author gratefully recognizes the contribution of Shahab Pourtalebi, Santiago Ruiz-Arenas, Yongzhe Li, Chong Li, and Zoltán Rusák to the Cognitive Engineering of Cyber-Physical Systems project.

Conflicts of Interest: The author declares no conflict of interest.

References

1. MacKay, D.M. Mind-like behaviour in artefacts. *Bull. Br. Soc. Hist. Sci.* **1951**, *1*, 164–165. [CrossRef]
2. Poovendran, R.; Sampigethaya, K.; Gupta, S.K.; Lee, I.; Prasad, K.V.; Corman, D.; Paunicka, J.L. Special issue on cyber-physical systems. *Proc. IEEE* **2012**, *100*, 6–12. [CrossRef]
3. Rho, S.; Vasilakos, A.V.; Chen, W. Cyber physical systems technologies and applications. *Future Gener. Comput. Syst.* **2016**, *56*, 436–437. [CrossRef]
4. Knight, J.; Xiang, J.; Sullivan, K. A rigorous definition of cyber-physical systems. In *Trustworthy Cyber-Physical Systems Engineering*; CRC Press: Boca Raton, FL, USA, 2016; Chapter 3; pp. 47–74.
5. Sztipanovits, J.; Koutsoukos, X.; Karsai, G.; Kottenstette, N.; Antsaklis, P.; Gupta, V.; Goodwine, B.; Baras, J.; Wang, S. Toward a science of cyber–physical system integration. *Proc. IEEE* **2012**, *100*, 29–44. [CrossRef]
6. CPSOS Consortium. *Cyber-Physical Systems of Systems: Research and Innovation Priorities*; Initial Document for Public Consultation; European Union's Seventh Programme for Research, Technological Development and Demonstration; CPSOS Consortium: Düsseldorf, Germany, 2014; pp. 1–21.
7. Schätz, B.; Törngreen, M.; Bensalem, S.; Cengarle, M.V.; Pfeifer, H.; McDermid, J.A.; Passerone, R.; Sangiovanni-Vincentelli, A. Cyber-Physical European Roadmap and Strategy: Research Agenda and Recommendations for Action, CyPhERS Consortium Technical Report. February 2015. Available online: www.cyphers.eu (accessed on 11 October 2018).
8. Horváth, I.; Rusák, Z.; Li, Y. Order beyond chaos: Introducing the notion of generation to characterize the continuously evolving implementations of cyber-physical systems. In Proceedings of the ASME 2017 International Design Engineering Technical Conferences, Cleveland, OH, USA, 6–9 August 2017; pp. 1–14.
9. Sheth, A.; Anantharam, P.; Henson, C. Physical-cyber-social computing: An early 21st century approach. *IEEE Intell. Syst.* **2013**, *28*, 79–82. [CrossRef]
10. Engell, S.; Paulen, R.; Reniers, M.A.; Sonntag, C.; Thompson, H. Core research and innovation areas in cyber-physical systems of systems. In *Cyber Physical Systems. Design, Modeling, and Evaluation*; Springer International Publishing: Berlin, Germany, 2015; pp. 40–55.
11. Colombo, A.W.; Bangemann, T.; Karnouskos, S.; Delsing, J.; Stluka, P.; Harrison, R.; Jammes, F.; Lastra, J.L. *Industrial Cloud-Based Cyber-Physical Systems. The IMC-AESOP Approach*; Springer Science + Business Media: Berlin/Heidelberg, Germany, 2014; ISBN 3319056239 9783319056234.

12. Yue, X.; Cai, H.; Yan, H.; Zou, C.; Zhou, K. Cloud-assisted industrial cyber-physical systems: An insight. *Microprocess. Microsyst.* **2015**, *39*, 1262–1270. [CrossRef]

13. Gerritsen, B.H.; Horváth, I. Current drivers and obstacles of synergy in cyber-physical systems design. In Proceedings of the ASME 2012 International Design Engineering Technical Conferences, Chicago, IL, USA, 12–15 August 2012; pp. 1277–1286.

14. Ribeiro, L.; Barata, J. Self-organizing multiagent mechatronic systems in perspective. In Proceedings of the 11th International Conference on Industrial Informatics, IEEE, Bochum, Germany, 29–31 July 2013; pp. 392–397.

15. Palviainen, M.; Mäntyjärvi, J.; Ronkainen, J.; Tuomikoski, M. Towards user-driven cyber-physical systems—Strategies to support user intervention in provisioning of information and capabilities of cyber-physical systems. In *Industrial Internet of Things*; Springer International Publishing: New York, NY, USA, 2017; pp. 575–593.

16. Juuso, E.K. Integration of intelligent systems in development of smart adaptive systems. *Int. J. Approx. Reason.* **2004**, *35*, 307–337. [CrossRef]

17. Barile, S.; Polese, F. Smart service systems and viable service systems: Applying systems theory to service science. *Serv. Sci.* **2010**, *2*, 21–40. [CrossRef]

18. Zhuge, H. Cyber physical society. In Proceedings of the Sixth International Conference on Semantics Knowledge and Grid, IEEE, Beijing, China, 1–3 November 2010; pp. 1–8.

19. Hahanov, V.; Litvinova, E.; Chumachenko, S.; Hahanova, A. Cyber physical computing. In *Cyber Physical Computing for IoT-driven Services*; Springer: Cham, Switzerland, 2018; pp. 1–20.

20. Stankovic, J.A. Research directions for cyber physical systems in wireless and mobile healthcare. *ACM Trans. Cyber-Phys. Syst.* **2016**, *1*, 1–12. [CrossRef]

21. Leitão, P.; Colombo, A.W.; Karnouskos, S. Industrial automation based on cyber-physical systems technologies: Prototype implementations and challenges. *Comput. Ind.* **2016**, *81*, 11–25. [CrossRef]

22. Sobhrajan, P.; Nikam, S.Y.; Pimpri, D.; Pimpri, P.D. Comparative study of abstraction in cyber physical system. *Int. J. Comput. Sci. Inf. Technol.* **2014**, *5*, 466–469.

23. Russell, S.J.; Norvig, P. *Artificial Intelligence: A Modern Approach*; Prentice Hall: Englewood Cliffs, NJ, USA, 1995; ISBN 0-13-103805-2.

24. Tani, J. Model-based learning for mobile robot navigation from the dynamical systems perspective. *IEEE Trans. Syst. Man Cybern. Part B Cybern.* **1996**, *26*, 421–436. [CrossRef] [PubMed]

25. Domingos, P. *The Master Algorithm: How the Quest for the Ultimate Learning Machine Will Remake Our World*; Basic Books: New York, NY, USA, 2015; pp. 1–315.

26. Jamroga, W.; Ågotnes, T. Constructive knowledge: What agents can achieve under imperfect information. *J. Appl. Non-Class. Log.* **2007**, *17*, 423–475. [CrossRef]

27. Reed, S.K.; Pease, A. Reasoning from imperfect knowledge. *Cogn. Syst. Res.* **2017**, *41*, 56–72. [CrossRef]

28. McFarlane, D.; Giannikas, V.; Wong, A.C.; Harrison, M. Product intelligence in industrial control: Theory and practice. *Annu. Rev. Control* **2013**, *37*, 69–88. [CrossRef]

29. Meyer, G.G.; Främling, K.; Holmström, J. Intelligent products: A survey. *Comput. Ind.* **2009**, *60*, 137–148. [CrossRef]

30. Flasiński, M. Reasoning with imperfect knowledge. In *Introduction to Artificial Intelligence*; Springer: Cham, Switzerland, 2016; pp. 175–188.

31. Lees, B. Smart reasoning. In *Artificial Intelligence: Learning Models*; Pearson Education: London, UK, 2013; pp. 1–3.

32. Lumer, C.; Dove, I.J. Argument schemes—An epistemological approach. In Proceedings of the OSSA Conference Archive, Windsor, ON, Canada, 8–21 May 2011; Volume 17, pp. 1–5.

33. Resnik, M. New paradigms for computing, new paradigms for thinking. In *Computers and Explanatory Learning*; Springer-Verlag: Berlin/Heidelberg, Germany, 1995; pp. 31–43.

34. Ayodele, T.O. Types of machine learning algorithms. In *New Advances in Machine Learning*; IntechOpen: London, UK, 2010; pp. 19–48. ISBN 978-953-307-034-6.

35. LeCun, Y.; Bengio, Y.; Hinton, G. Deep learning. *Nature* **2015**, *521*, 436. [CrossRef] [PubMed]

36. Goodfellow, I.; Bengio, Y.; Courville, A. *Deep Learning*; The MIT Press: Cambridge, MA, USA, 2016; Volume 1, pp. 1–800, ISBN 978-0262035613.

37. Huang, G.B.; Wang, D.H.; Lan, Y. Extreme learning machines: A survey. *Int. J. Mach. Learn. Cybern.* **2011**, *2*, 107–122. [CrossRef]

38. Finocchiaro, M.A. Fallacies and the evaluation of reasoning. *Am. Philos. Q.* **1981**, *18*, 13–22.

39. Sigman, S.; Liu, X.F. A computational argumentation methodology for capturing and analyzing design rationale arising from multiple perspectives. *Inf. Softw. Technol.* **2003**, *45*, 113–122. [CrossRef]

40. Larvor, B. How to think about informal proofs. *Synthese* **2011**, *187*, 715–730. [CrossRef]

41. Marfori, M.A. Informal proofs and mathematical rigour. *Stud. Log.* **2010**, *96*, 261–272. [CrossRef]

42. Håkansson, A.; Hartung, R.; Moradian, E. Reasoning strategies in smart cyber-physical systems. *Procedia Comput. Sci.* **2015**, *60*, 1575–1584. [CrossRef]

43. Johnson-Laird, P.N. Deductive reasoning. *Annu. Rev. Psychol.* **1999**, *50*, 109–135. [CrossRef] [PubMed]

44. Psillos, S. Ampliative reasoning: Induction or abduction. In Proceedings of the ECAI96 Workshop on Abductive and Inductive Reasoning, Budapest, Hungary, 12 August 1996; pp. 1–6.

45. Hintikka, J. What is abduction? The fundamental problem of contemporary epistemology. *Trans. Charles S. Peirce Soc.* **1998**, *34*, 503.

46. Peirce, C.S. *Philosophical writings of Peirce*; Buchler, J., Ed.; Courier Corporation: Dover, NY, USA, 1955.

47. Aliseda, A. *Abductive Reasoning: Logical Investigations into Discovery and Explanation*; Springer Science & Business Media: Dordrecht, The Netherlands, 2006; Volume 330.

48. Meheus, J.; Verhoeven, L.; van Dyck, M.; Provijn, D. Ampliative adaptive logics and the foundation of logic-based approaches to abduction. In *Logical and Computational Aspects of Model-Based Reasoning*; Springer: Dordrecht, The Netherlands, 2002; pp. 39–71.

49. Satoh, K.; Inoue, K.; Iwanuma, K.; Sakama, C. Speculative computation by abduction under incomplete communication environments. In Proceedings of the Fourth International Conference on Multi Agent Systems, Boston, MA, USA, 10–12 July 2000; pp. 263–270.

50. Batens, D. A universal logic approach to adaptive logics. *Log. Univers.* **2007**, *1*, 221–242. [CrossRef]

51. Gauderis, T. Modelling abduction in science by means of a modal adaptive logic. *Found. Sci.* **2013**, *18*, 611–624. [CrossRef]

52. Menzies, T.J. An overview of abduction as a general framework for knowledge-based systems. In Proceedings of the Australian AI 1995 Conference, Canberra, Australia, 13–17 November 1995; pp. 1–8.

53. Hermann, M.; Pichler, R. Counting complexity of propositional abduction. *J. Comput. Syst. Sci.* **2010**, *76*, 634–649. [CrossRef]

54. Kean, A.C. A Formal Characterization of a Domain Independent Abductive Reasoning System. Ph.D. Dissertation, University of British Columbia, Vancouver, Canada, 1993; pp. 1–175.

55. Eiter, T.; Gottlob, G.; Leone, N. Abduction from logic programs: Semantics and complexity. *Theor. Comput. Sci.* **1997**, *189*, 129–177. [CrossRef]

56. Gottlob, G.; Pichler, R.; Wei, F. Bounded treewidth as a key to tractability of knowledge representation and reasoning. *Artif. Intell.* **2010**, *174*, 105–132. [CrossRef]

57. Poole, D.; Rowen, G.M. What is an optimal diagnosis? In Proceedings of the Sixth Conference on Uncertainty in AI, Cambridge, MA, USA, 27–29 July 1990; pp. 46–53.

58. De Campos, L.M.; Gámez, J.A.; Moral, S. Partial abductive inference in Bayesian belief networks using a genetic algorithm. *Pattern Recognit. Lett.* **1999**, *20*, 1211–1217. [CrossRef]

59. Psillos, S. Abduction: Between conceptual richness and computational complexity. In *Abduction and Induction*; Springer: Dordrecht, The Netherlands, 2000; pp. 59–74.

60. Pagnucco, M.; Foo, N. Inverting resolution with conceptual graphs. In Proceedings of the 13th International Conference on Conceptual Structures, Kassel, Germany, 18–22 July 1993; Springer: Berlin/Heidelberg, Germany, 1993; pp. 238–253.

61. Boutilier, C.; Becher, V. Abduction as belief revision. *Artif. Intell.* **1995**, *77*, 43–94. [CrossRef]

62. Kakas, A.; van Nuffelen, B.; Denecker, M. A-system: Problem solving through abduction. In Proceedings of the Seventeenth International Joint Conference on Artificial Intelligence, Seattle, WA, USA, 4–10 August 2001; Volume 1, pp. 591–596.

63. Verdoolaege, S.; Denecker, M.; van Eynde, F. Abductive reasoning with temporal information. *arXiv* **2000**, arXiv:cs/0011035.

64. Sangiovanni-Vincentelli, A. Quo vadis, SLD? Reasoning about the trends and challenges of system level design. *Proc. IEEE* **2007**, *95*, 467–506. [CrossRef]

65. Gabbay, D.M.; Woods, J. Formal approaches to practical reasoning: A survey. In *Handbook of the Logic of Argument and Inference: The Turn Towards the Practical*; Elsevier: Amsterdam, The Netherlands, 2002; Volume 1, pp. 445–478.

66. Oaksford, M.; Chater, N. Conditional probability and the cognitive science of conditional reasoning. *Mind Lang.* **2003**, *18*, 359–379. [CrossRef]

67. Van Nuffelen, B. A-system: Problem solving through abduction. In Proceedings of the 13th Dutch-Belgian Artificial Intelligence Conference, Amsterdam, The Netherlands, 25–26 October 2001; Volume 1, pp. 591–596.

68. Kakas, A.; Riguzzi, F. Abductive concept learning. *New Gener. Comput.* **2000**, *18*, 243–294. [CrossRef]

69. Bylander, T.; Allemang, D.; Tanner, M.C.; Josephson, J.R. The computational complexity of abduction. *Artif. Intell.* **1991**, *49*, 25–60. [CrossRef]

70. Pourtalebi, S. System-Level Feature-Based Modeling of Cyber-Physical Systems. Ph.D. Thesis, Delft University of Technology, Delft, The Netherlands, 2017.

71. Ruiz-Arenas, S. Exploring the Role of System Operation Modes in Failure Analysis in the Context of First Generation Cyber-Physical Systems. Ph.D. Thesis, Delft University of Technology, Delft, The Netherlands, 2018.

72. Li, Y. Modelling, Inferring and Reasoning Dynamic Context Information in Informing Cyber Physical Systems. Ph.D. Thesis, Delft University of Technology, Delft, The Netherlands, 2018.

73. Li, C. Cyber-Physical Solution for an Engagement Enhancing Rehabilitation System. Ph.D. Thesis, Delft University of Technology, Delft, The Netherlands, 2016.

74. Horváth, I. Procedural abduction as enabler of smart operation of cyber-physical systems: Theoretical foundations. In Proceedings of the 23rd ICE/IEEE International Conference on Engineering, Technology, and Innovation, Madeira Island, Portugal, 27–29 June 2017; pp. 124–132.

75. Horváth, I.; Tepjit, S.; Rusák, Z. Compositional engineering frameworks for development of smart cyber-physical systems: A critical survey of the current state of progression. In Proceedings of the ASME Computers and Information in Engineering Conference, Quebec City, QC, Canada, 26–29 August 2018; pp. 1–14.

76. Tavčar, J.; Horváth, I. A Review of the Principles of Designing Smart Cyber-Physical Systems for Run-Time Adaptation: Learned Lessons and Open Issues. *IEEE Trans. Syst. Man Cybern. Syst.* **2018**, *49*, 145–158. [CrossRef]

77. Pourtalebi, S.; Horváth, I. Information schema constructs for defining warehouse databases of genotypes and phenotypes of system manifestation features. *Front. Inf. Technol. Electron. Eng.* **2016**, *17*, 862–884. [CrossRef]

78. Pourtalebi, S.; Horváth, I. Information schema constructs for instantiation and composition of system manifestation features. *Front. Inf. Technol. Electron. Eng.* **2017**, *18*, 1396–1415. [CrossRef]

79. Ruiz-Arenas, S.; Rusák, Z.; Colina, S.R.; Mejia-Gutierrez, R.; Horváth, I. Testbed for validating failure diagnosis and preventive maintenance methods by a low-end cyber-physical system. In Proceedings of the 11th Tools and Methods of Competitive Engineering Symposium, Aix-en-Provence, France, 9–13 May 2016; Volume 1, pp. 1–11.

80. Ruiz-Arenas, S.; Rusá, Z.; Horváth, I.; Mejia-Gutierrez, R. Systematic exploration of signal-based indicators for failure diagnosis in the context of cyber-physical systems. *J. Front. Inf. Technol. Electron. Eng.* **2018**, 1–16, accepted.

81. Horváth, I.; Li, Y.; Rusák, Z.; van der Vegte, W.F.; Zhang, G. Dynamic computation of time-varying spatial contexts. *J. Comput. Inf. Sci. Eng.* **2017**, *17*, 011007-1-12. [CrossRef]

82. Li, Y.; Horváth, I.; Rusák, Z. Building awareness in dynamic context for the second-generation cyber-physical systems. *Future Gener. Comput. Syst.* **2018**, 1–34, submitted.

83. Li, C.; Rusák, Z.; Horváth, I.; Ji, L. Development of engagement evaluation method and learning mechanism in an engagement enhancing rehabilitation system. *Eng. Appl. Artif. Intell.* **2016**, *51*, 182–190. [CrossRef]

84. Li, C.; Rusák, Z.; Horváth, I.; Kooijman, A.; Ji, L. Implementation and validation of engagement monitoring in an engagement enhancing rehabilitation system. *IEEE Trans. Neural Syst. Rehabil. Eng.* **2017**, *25*, 726–738. [CrossRef] [PubMed]

85. Wolpert, D.; Wheeler, K.; Tumer, K. *Collective Intelligence for Control of Distributed Dynamical Systems*; Technical Report NASA-ARC-IC-99-44; NASA: Ames, IA, USA, 1999; pp. 1–8.

86. Pease, A.; Aberdein, A. Five theories of reasoning: Interconnections and applications to mathematics. *Logic Log. Philos.* **2011**, *20*, 7–57. [CrossRef]

87. Tripakis, S. Compositionality in the science of system design. *Proc. IEEE* **2016**, *104*, 960–972. [CrossRef]
88. Fouquet, F.; Morin, B.; Fleurey, F.; Barais, O.; Plouzeau, N.; Jezequel, J.M. A dynamic component model for cyber physical systems. In Proceedings of the 15th ACM SIGSOFT Symposium on Component Based Software Engineering, Bertinoro, Italy, 25–28 June 2012; pp. 135–144.

 *designs*

Article

Model Testing of Complex Embedded Systems Using EAST-ADL and Energy-Aware Mutations

Eduard Paul Enoiu * and Cristina Seceleanu *

Mälardalen University, Department of Networked Embedded Systems, 722 20 Västerås, Sweden
* Correspondence: eduard.paul.enoiu@mdh.se (E.P.E.); cristina.seceleanu@mdh.se (C.S.)

Received: 18 November 2019; Accepted: 13 February 2020; Published: 19 February 2020

Abstract: Nowadays, embedded systems are increasingly complex, meaning that traditional testing methods are costly to use and infeasible to directly apply due to the complex interactions between hardware and software. Modern embedded systems are also demanded to function based on low-energy computing. Hence, testing the energy usage is increasingly important. Artifacts produced during the development of embedded systems, such as architectural descriptions, are beneficial abstractions of the system's complex structure and behavior. Electronic Architecture and Software Tools Architecture Description Language (EAST-ADL) is one such example of a domain-specific architectural language targeting the automotive industry. In this paper, we propose a method for testing design models using EAST-ADL architecture mutations. We show how fault-based testing can be used to generate, execute and select tests using energy-aware mutants—syntactic changes in the architectural description, used to mimic naturally occurring energy faults. Our goal is to improve testing of complex embedded systems by moving the testing bulk from the actual systems to models of their behaviors and non-functional requirements. We combine statistical model-checking, increasingly used in quality assurance of embedded systems, with EAST-ADL architectural models and mutation testing to drive the search for faults. We show the results of applying this method on an industrial-sized system developed by Volvo GTT. The results indicate that model testing of EAST-ADL architectural models can reduce testing complexity by bringing early and cost-effective automation.

Keywords: model testing; mutation testing; energy consumption; EAST-ADL

1. Introduction

Embedded products are widely used in many industries. For example, embedded systems are used in automotive companies in the implementation of vehicle functions (e.g., ABS, electronic stability) [1]. Such functions contribute to the complexity of developing the entire vehicle system [2], making the verification and validation of new functions more problematic due to the interconnections between both functional and non-functional requirements posed on the whole system. For instance, a structural or behavioral update in the software or the replacement of a software or hardware part can influence the consumption of resources [1]. In this case, just showing the overall system's functional correctness is not enough. One would need to verify that the system meets its non-functional (also known as extra-functional) requirements, such as energy consumption, memory allocation and real-time perfomance. In addition, although testing is arguably the most used verification and validation technique, for these complex systems testing is highly expensive when performed on the actual system. Given the increasing demand in embedded systems for low-energy computing [3], early testing the energy consumption becomes an increasingly important issue. To ensure the quality of service of embedded systems and to estimate its performance early in the development process, testing the behavior of the system with respect to its supplied energy budget as well as testing for the worst-case energy consumption is very important.

In this paper we outline a method that targets these challenges by bringing extra-functional testing of these complex embedded systems by moving the creation, execution of test cases earlier in the development process at an abstract architectural modelling level. Architectural models are used to represent relevant aspects of system behavior, environment, structure, and properties and are used as a basis for test generation, performance, and analysis. Briand et al. [4,5] refers to such techniques as model testing, since this kind of approaches aim to identify faults by executing test cases on models and sampling the input space. This in contrast to other verification and validation techniques such as model checking that attempt to exhaustively explore the model state space and check the correctness of the models against some given properties. Our goal is to tackle the challenges of testing such complex systems by developing an approach that provides, early-on in the development process, confidence about the system resource consumption by identifying and executing a selected set of test cases, from the whole test execution space, where faults are more likely to lie. In the automotive domain, modelling the architectural aspects including the resource consumption of complex embedded systems at high levels of abstraction is necessary. Architectural description languages such as EAST-ADL (EAST-ADL stands for Electronic Architecture and Software Tools-Architecture Description Language.) [6] are used to represent both hardware and software functions and extra-functional information (e.g., timing properties and resource consumption). If we assume energy as the resource of interest, the annotations of energy consumption in EAST-ADL models can be used to create test cases for feasibility and worst-case energy consumption that can be useful in the detection of faults.

This article is an extended version of a conference paper [7] in which we have demonstrated how architectures described in the EAST-ADL language can integrated into a testing approach in order to evaluate energy properties. Based on this previous work, in this study we present a novel mutation-based approach for model testing and evaluate its efficiency and effectiveness on an industrial use case. In particular, by selecting test cases using mutation testing [8], we propose a method for automatically generating and selecting test cases based on the concept of energy-aware mutants–small architectural model syntactic modifications designed to mimic real energy faults. Test cases where a certain behavior can be distinguished from its mutations are sensitive to changes in the architectural model and are therefore considered good at detecting faults.

We apply this method on an embedded system modeled in EAST-ADL after transforming it into a network of priced timed automata [9]. In particular, we select test suites based on random model executions that show the energy cost using UPPAAL SMC [10], the statistical extension of the UPPAAL model checker. We show how to seed faults in the EAST-ADL model and evaluate each generated test suite's energy-related fault detection capability. To illustrate the efficiency and effectiveness of our test generation method, we carry out an evaluation, using an industrial system modeled in EAST-ADL architectural language. The results of this study suggest that model testing is efficient in terms of test generation time and number of generated and selected test cases.

To summarize, the main contributions of this paper are:

- The identification of energy-aware mutation operators for mutation testing of EAST-ADL models.
- An approach for mutation test generation of EAST-ADL models using a statistical model checker.
- An evaluation of the method on a Brake-by-Wire industrial system.

The rest of the paper is organized in the following sections. In Section 2 we overview the preliminaries needed to comprehend our contribution, including architecture-based testing and mutation testing, the EAST-ADL architectural language, UPPAAL SMC and priced timed automata. The main contribution of the paper is our method for automatically generating energy-aware test cases using EAST-ADL models described in Section 3, and its application on the Brake-by-Wire system as well as the experimental results presented in Section 4. We conclude the paper and present the future work in Section 7.

2. Background

Major aspects of architecture-based and mutation testing, the language of EAST-ADL and UPPAAL SMC will be discussed in this section. These aspects are put in the scope of the contributions of this study.

2.1. Architecture and System-Level Testing

Architectural models are created during system development using components and connectors representing the whole system and its high-level structure [11]. The aim of testing using architectural designs as input is to verify whether the system meets its design specifications. This type of testing is also known as system-level testing [12] and its main purpose is to discover early-on architectural design problems, but also the overall system behaviour. Testing, at this level, aims to address such test goals as overall functionality, real-time properties, robustness and performance [11]. Testing extra-functional properties (e.g., bandwidth, energy and memory) [13] are crucial aspects during development and need to be addressed continuously when testing. Specifically, we focus on testing for energy consumption based on architectural models. Due to the intertwining of software and hardware and the complex interactions with the external environment, it is challenging to apply conventional testing methods directly to the real embedded systems. When testing for extra-functional properties at the software architecture level, models are annotated with energy consumption properties.

The aim of testing at architectural level based on the energy consumption is to find faults in the performance of an actually developed system in terms of its subsystems and interactions before the actual code implementation. The use of such architectural models for testing enables the execution of a large number of test cases and increases the chances of uncovering faults.

2.2. Mutation Testing

Mutation testing is the technique used for creating a faulty implementations (usually in an automated way) to examine a test-suite's ability to detect faults [8]. A mutant is a new version of a program created by making in the original program a small change. For instance, a mutant is created in a program by replacing an operator with another, negating a variable, or changing a constant's value. The execution of a test suite on the resulting mutant will produce a different output than the original program, in which case the test suite kills the mutant. In order to measure the mutant detection capability of the written test suite, a mutation score is calculated using the automatically seeding all mutants and executing the test cases on each mutant. One can compute a mutation score based on an output-only oracle (i.e., expected outputs) against all the generated mutants by calculating the ratio of mutants killed to the total number of mutants. Just et al. [14] showed that if a test suite can detect or kill most mutants, it can also detect real software faults, thus providing evidence that the mutation score is a fairly good proxy for real fault detection ability. Mutation testing has been widely used at lower levels of testing and mostly on implementation models. Even if there are some studies that have applied this technique on specification models [15–18] for designing behavioral faults, there is a lack of methods that target architectural models and extra-functional aspects for model mutation testing. No attempt has been made to propose and evaluate mutation testing for EAST-ADL models. This motivated us to develop an automated approach to test generation and model testing using mutations aimed at this kind of architectural model.

2.3. EAST-ADL Architectural Language

EAST-ADL [6] is an AUTOSAR-compatible (AUTOSAR is a standard for AUTomotive Open System ARchitecture and was developed by several manufacturing companies.) architectural description language intended to be used in the development of automotive embedded systems. A system can be described at four levels of abstraction, as follows: (i) the Vehicle Level describes the external features at the highest level of abstraction, (ii) the Analysis Level describes the abstract functionality of the system, (iii) the Design Level describes more details in the functional representation of the architecture and the hardware allocation of these onto the platform, and (iv) the Implementation Level provides the AUTOSAR-compliant code.

At each level of abstraction, the model is composed using components (i.e., FunctionType) which describe the functionality of the system. Each FunctionType contains: (i) Ports that receive and provide data, (ii) a trigger (i.e., time-based or event-based), and (iii) an internal behavior. Each of these components is instantiated as a FunctionPrototype. The execution of each FunctionPrototype uses the "read-execute-write" semantics, and the internal behavior is defined using different languages (e.g., Simulink, UML, UPPAAL PORT timed automata [19,20]). In this study we use the models at the design level, containing the Functional Design Architecture (FDA) and Hardware Design Architecture (HDA) annotated with non-functional properties. The design model can be annotated with a GenericConstraint property representing the energy utilization.

2.4. UPPAAL SMC and Priced Timed Automata

UPPAAL SMC [10] is an extension of UPPAAL, that supports the analysis of non-functional properties for networks of priced timed automata with stochastic semantics. Statistical model-checking is used to generate stochastic simulations and estimate probabilities and probability distributions over time with a certain level of confidence, so the analysis scales better than symbolic model-checking in verification of realistic industrial models. Specifically, statistical model checking samples executions using statistical inference methods to decide whether the model executions satisfy a property given a certain confidence. In this paper we use statistical model checking and UPPAAL SMC in a black-box manner for execution of models as well as producing probabilistic estimates about the correctness of the generated models.

Priced timed automata (PTA) are used in UPPAAL SMC and are extensions of timed automata with cost variables that can evolve at integer rates (also $\neq 1$). These are used for representing the energy consumption. The energy usage is modeled using a function $P : (L \cup E) \rightarrow \mathbb{N}$, where L is a finite set of locations, and E is the set of edges, which assigns costs to both locations and edges. A network of PTA (NPTA) is described as a composition of n PTA over clocks and actions; the PTA use send–receive actions (i.e., send $b!$ is complementary to receive $b?$) and shared variables are used in guards.

Let X be a finite set of clocks and $B(X)$ the set of guards, which are finite conjunctions of atomic guards of the form $x \bowtie n$, where $x \in X$, $n \in \mathbb{N}$, and $\bowtie \in \{<, \leq, =, \geq, >\}$. A (Linear) PTA over clocks X and actions Act is a tuple $(L, l_0, X, V, I, Act, E, P)$ where L is a finite set of locations, l_0 is the initial location, X is set of clocks, V is a set of data variables, $I : L \rightarrow B(X)$ assigns invariants to locations, Act is a set of actions, $E \subseteq L \times B(X, V) \times Act \times R \times L$ is the set of edges (where R denotes the reset set, i.e., assignments to manipulate clock- and data variables), and $P : (L \cup E) \rightarrow \mathbb{N}$ assigns costs to both locations and edges. In the case of $(l, g, a, r, l') \in E$, we write $l \xrightarrow{g,a,r} l'$.

The semantics of PTA is defined as a transition system over states (l, u), with the initial state (l_0, u_0), where u_0 assigns all clocks in X to zero. There are two kinds of transitions:

(i) Delay transitions: $(l, u) \xrightarrow{d,p} (l, u \oplus d)$, where $u \oplus d$ is the result obtained by incrementing all clocks of the automata with the delay amount d, and $p = P(l) * d$ is the cost of performing the delay, and

(ii) discrete transitions: $(l, u) \xrightarrow{d,p} (l', u')$, corresponding to taking an edge $l \xrightarrow{g,a,r} l'$ for which the guard g is satisfied by u. The clock valuation u' of the target state is obtained by modifying u according to updated r. The cost $p = P(l, g, a, r, l')$ is the priced associated with the edge.

A network of PTA $A_1 \parallel ... \parallel A_n$ is expressed as a composition of n PTA over X and Act, using synchronization actions and shared variables that can be used in guards and transitions. UPPAAL SMC uses an extended Weighted Metric Temporal Logic (WMTL) [21] for performing hypothesis testing, which checks if the probability to reach a state ϕ within cost $x \leq C$ is greater or equal to a certain threshold $p : Pr(\Diamond_{x \leq C}\phi) \geq p$.

A trace σ of a PTA is a sequence of delays, actions, and transitions:

$$\sigma = (l_0, u_0) \xrightarrow{a_1, p_1} (l_1, u_1) \xrightarrow{a_2, p_2} \dots \xrightarrow{a_n, p_n} (l_n, u_n), \text{ where the cost of performing } \sigma \text{ is } \Sigma_{i=1}^{n} p_i.$$

3. A Model Testing Method for Energy-Aware Testing Using EAST-ADL

In this section we introduce our model testing method which uses energy consumption objectives to select test suites using a statistical model checker based on the created simulations. The framework is based on the transformation of the EAST-ADL model into a network of priced timed automata (PTA) [22]. It is composed of several steps, mirrored in Figure 1:

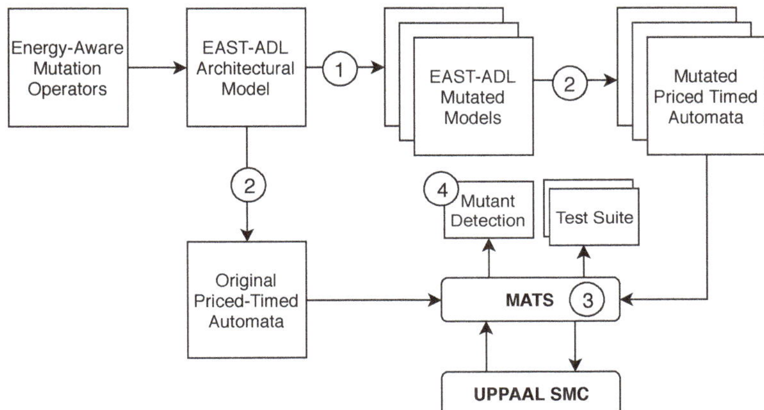

Figure 1. An overview of the test suite generation and evaluation method for energy consumption based on Electronic Architecture and Software Tools Architecture Description Language (EAST-ADL) architectural models and mutation testing.

1. Mutant Generation. This first step (described in detail in Section 3.1) is used for generating small syntactic changes (mutants) to the architectural description based on a set of mutation operators (e.g., mimicking architectural energy errors). The output of this step is a set of new versions of the original EAST-ADL model, each one containing an inserted change. When implementing this step, a set of operators needs to be available based on the desired types of mutants. Different set of operators were proposed in the scientific literature for both models and code [18,23] for mutation testing using the UPPAAL model checker. We show how mutant generation is implemented for the EAST-ADL language as an input for MATS and UPPAAL SMC.

2. EAST-ADL to PTA Transformation. The second step (described in detail in Section 3.2) is used for transforming the EAST-ADL model to PTA. The output of this step are PTA models containing the original structure and behavior of the EAST-ADL together with all inserted mutants and annotated with energy consumption information to be used by UPPAAL SMC for test-case generation and selection.

3. Test Suite Generation. The third step (described in Section 3.3) uses the MATS tool to generate a set of test cases by using the UPPAAL SMC ability to generate simulations. We show how a test simulation is obtained using a property expressed as a UPPAAL SMC simulation property.

4. Mutant Detection. The fourth step (described in Section 3.4) involves the instrumentation of the model with detection instructions for each mutant. This means that the monitor for mutation detection is used to record the execution and detection of each mutant.

We discuss these steps in further detail in the following sections by applying this approach to running examples.

Overall, we use an EAST-ADL system architectural model and mutation testing to automatically generate test suites for model testing the energy consumption. Our test generation method aims to use mutations in energy consumption in EAST-ADL to select test suites automatically using various random simulations.

3.1. Energy-Aware Mutant Generation

In this paper we assume a resource r for an EAST-ADL component represents the accumulated resource usage up to some point in time. By using this assumption, resources of this kind are categorized as discrete or continuous [24] . Energy in EAST-ADL is a continuous resource that can evolve linearly in time ($energy(t) = n \times t$, where $n \in \mathbb{N}$ and t is the time elapsed during system execution). In EAST-ADL components the resource usage is defined as the total energy consumption of the system as $energy_{total}(t) = \sum_{i=1}^{m} energy_i(t)$, where m is the number of components. Based on a predefined set of mutation operators, faults are injected. These mutants should represent naturally-occurring faults influencing the energy consumption. The general principle underlying mutation analysis is that the faults generated using the operators described in Table 1 represent the mistakes that architects often make while modelling in EAST-ADL that directly influence the energy consumption.

We propose a set of mutation operators and they are applied on all the EAST-ADL model elements that could influence the energy consumption in the following three categories:

- EAST-ADL resource annotation (i.e., Energy Replacement Operator (ero)). In this case, we insert a fault in the generic constraint of a function prototype by changing the original annotation. These types of mutations intend to model the errors in the energy consumed by each component.
- Timing Behavior of an EAST-ADL component (i.e., Period Replacement Operator (pro), Execution Time Replacement Operator (etro)) can be modified by changing the period and execution time constraints. The period and execution time value stand as integer values in the constraint.
- Functional Architecture Structure (i.e., Component Removal Operator (cro), Component Insertion Operator (cio), and Triggering Pattern Replacement Operator (tro)). We change the architectural elements in EAST-ADL by removing or inserting components that influence the energy consumption as well as modifying the triggering of each component.

In the case of the CRO mutation operator, a component is directly removed when we encounter an entry, computation and exit function prototype. An entry function prototype in an EAST-ADL model is a component that has at least one port receiving inputs externally. The computation function prototype has all input and output ports connected with other function prototypes in the same level of system abstraction. An exit function prototype has at least one output port sending data flows out of the actual system. When a component is removed, the connections must also be removed so the system remains well formed (compilable).

Table 1. Description of each mutation operator and the elements in EAST-ADL that are modified.

Mutation Operator	Description	EAST-ADL Element
Energy Replacement Operator (ero)	The operator is applied where an energy value occurs occurs, i.e., as an annotation of each component. The operator is applied by replacing a value of an energy constraint connected to a component (e.g., replacing a value ($value = 3$) with its boundary values (e.g., $value = 2$)).	GenericConstraint
Period Replacement Operator (pro)	The operator applies the mutations in each period constraint value. The operator is only applied in the components triggered periodically. The operator is applied by replacing a value of the period constraint connected to a component (e.g., replacing a value ($value = 20$) with its boundary values (e.g., $value = 19$)).	PeriodConstraint
Execution-Time Replacement Operator (etro)	The operator applies the mutations in each execution time constraint value. The operator is applied by replacing a value of the execution time constraint connected to a component (e.g., replacing a value ($value = 3$) with its boundary values (e.g., $value = 4$)).	ExecTimeConstraint
Component Removal Operator (cro)	This operator models errors related with missing components. This operator removes each component together with its constraints and connects the inputs of this component to the next component in the system.	FunctionPrototype
Component Insertion Operator (cio)	This operator models errors related with duplicated components. This operator adds a duplicated component together with its constraints and connects this component in the same configuration as the original one.	FunctionPrototype
Triggering Replacement Operator (tro)	The operator applies the mutations in each component triggering pattern. The operator is applied by replacing the periodic pattern with a event pattern connected to a component and vice versa.	FunctionPrototype, PeriodConstraint

In Figure 2 we show examples of mutations for each mutation operator in Table 1. For each category of mutations, a mutant operator was chosen to provide a small-scale example of the application to a EAST-ADL model. For example, in Figure 2d for the CRO operator changing the structure of the model within a system is a likely operation within an EAST-ADL project. Depending on how the mutation operator is used, the other inports and connections have to be updated. In this case, CRO removes FP2 and the connection between FP1.Port2 and FP3.Port1. To avoid compilation problems a component is removed together with its control and connection structures.

These mutation operators are systematically applied to the entire EAST-ADL model (i.e., components, ports, connections) each simulating one syntactic change resulting in a set of energy-aware mutants. During the execution of a test, energy consumption can be measured by the use of a statistical model checker, and represented as a consumption of a continuous resources where the rate of consumption over time is constant. A temporal sequence of energy values can have different shapes, depending on the sampling rate of the measurement and the energy consumption behavior.

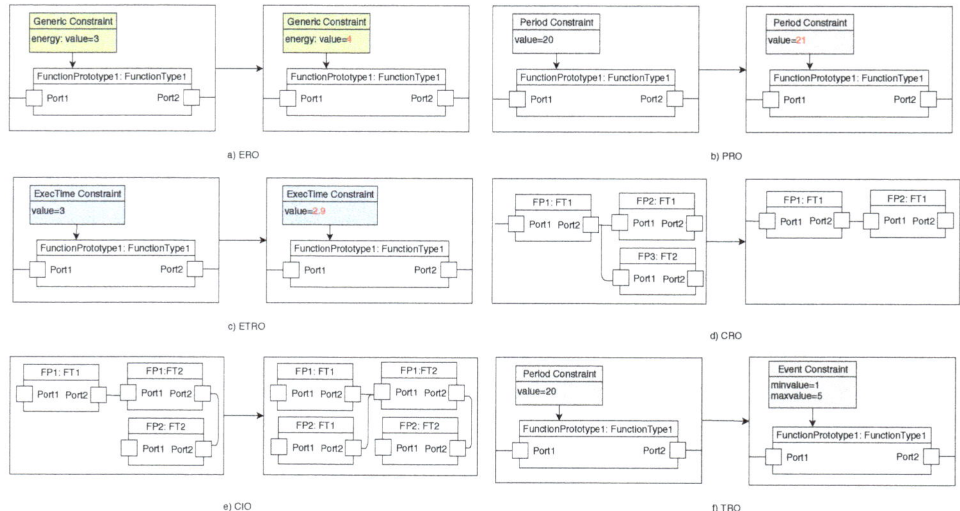

Figure 2. Examples of mutations for each mutation operator for EAST-ADL architectural models.

3.2. EAST-ADL to Priced-Timed Automata Transformation

We transform an EAST-ADL model (with annotations of energy consumption) into a PTA model. We use a small-case example in Figure 3 to show the transformation for a generic EAST-ADL FunctionPrototype. Every component in EAST-ADL is automatically converted into a network of two PTAs: An interface automaton representing the component interface, and a behavior automaton representing the internal behavior. The PTA interface contains the triggering of each component, timing information and energy annotations.

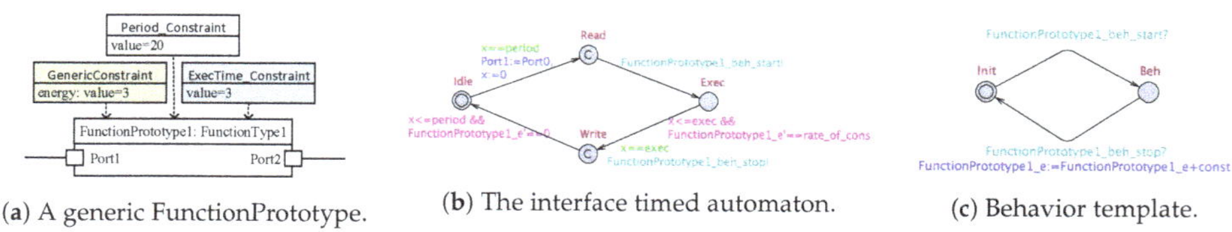

(**a**) A generic FunctionPrototype.   (**b**) The interface timed automaton.   (**c**) Behavior template.

Figure 3. An example of a generic interface timed automaton and a behavior template for an EAST-ADL component.

In practice, each FunctionPrototype is translated into a network of two syncronized automata (as shown in Figure 3): An interface automaton containing the ports of each EAST-ADL component and a behavior automaton representing the internal discrete and continuous behavior. Each *FunctionPrototype* is defined as an automaton with four locations: (i) *Idle*, (ii) a *Read* location used for updating the internal variables according to the values on the input ports, (iii) a *Exec* location used for triggering the internal *Behavior*, and (iv) a *Write* location allowing the update of output ports based on the internal variables values. Each interface is triggered based on the triggering annotation *Trigg* associated with each EAST-ADL *FunctionPrototype*. The energy starts to be consumed when information from the input ports is read until the component writes the information to the output ports. The energy consumed by each component increases with time during execution and is modeled as a cost "c" in the PTA ($c(t) = n_c \times t$, where $n_c \in \mathbb{N}$ is the rate of consumption over time t). When the component is idle ($c'(t) = 0$) energy is not consumed. In order to calculate the overall consumed energy we use an automaton to compute the energy used by the system based on the energy consumed by each component.

The GenericConstraint is used in the EAST-ADL model for annotating the energy consumption. By using *energy* in the GenericConstraint annotation for the behavior of the PTA model, we can measure the energy consumption inside each component. Specifically, we use a monitor automaton that includes all of the EAST-ADL model's energy annotations. This monitor is a PTA that supervises the system execution by using two synchronization channels: FunctionPrototype_beh_start and FunctionPrototype_beh_stop.

3.3. Test Suite Generation

To create executable test cases, we use traces obtained during simulation. At each predefined time unit, test values are obtained by extracting the input parameters and energy values from the these simulations. The generation of test suites is essentially the creation of signals using the generated simulations. Since the input signal search space is very large, we randomly select input signals that change over a certain predefined period of time using ordered signal sequences.

Extra-functional aspects in EAST-ADL, such as the energy consumption, are often more difficult to generate test than for functional properties. Embedded systems often require large amounts of energy to be consumed. Irregular and heavy use of energy could result in inadequate functionality of the system that keeps essential components running. The estimation of the allocated energy budget can be calculated using the energy consumption annotated in EAST-ADL. The system will complete its execution if the actual energy consumption of the system does not exceed its energy budget. Otherwise, the budget has been exhausted during the execution of the system. As a result, we concentrate on test queries that simulate the nominal but also the worst-case energy consumption for a EAST-ADL model.

While UPPAAL SMC's is used to provide statistical guarantees based on a series of system simulations, it is also suitable to use the input parameters and the consumed energy as values in test cases during individual system simulations. The simulation depends on the number of runs (n) and the upper time limit for the number of runs (*bound*). By extracting the input parameters and the energy values from these simulation traces, we create executable test cases using the MATS tool [25]. In practice, we use UPPAAL SMC's ability to generate simulation traces, which we transform into executable test cases using the MATS tool [25], by extracting the parameters and the energy values at predefined time points. Each test input is a vector of signals where the model's time-dependent behavior is executed using an ordered signal sequence. UPPAAL SMC is used by MATS to obtain traces of simulations over a predefined number of system model runs. A simulation can be formulated in UPPAAL SMC as the property:

$$simulate\ n[bound]\{E_1, .., E_k\},$$

where n is the number of simulations to be performed, *bound* is the time bound on the simulations, and $E_1, .., E_k$ are the monitored expressions.

We execute the generated test cases on each mutant and collect the simulation traces containing the energy values. On both the original model and its mutated versions, each test case is executed. We exclude test cases that do not contribute to the mutation score in order to minimize the final set of test cases [25]. The generated simulation traces are transformed into executable test cases sampling the simulation trace (as shown in the small-scale example in Figure 4a). Based on the generated data points we use intermediate values at predefined sample points and split the simulation trace in two: A set of sampled inputs used to trigger the system under test and a set of sampled expected energy consumption output.

In addition, during this phase we can generate test case with the worst-case energy consumption. Using UPPAAL SMC for statistical analysis we can obtain the peak energy value which eventually reaches a certain behavior in time. This problem is reduced to trying to maximize the energy cost function to satisfy the following property:

$$E[bound; n](max : energy)$$

where *bound* is the time bound of these simulation traces, n is the number of runs, and *energy* represents energy cost. The worst-case energy consumption analysis calculates the simulation cost for reaching a predefined system behavior. At this stage, feasibility analysis is used to verify whether the energy consumption is still within the maximum energy value provided by the worst-case analysis. The verification is accomplished by evaluating the energy distribution as a probability evaluation, as follows:

$$Pr[bound](\psi)$$

where *bound* is the time bound in all the simulation runs, and ψ is a property in the form "Eventually p", where p is a state predicate. We can select the required test suites based on this analysis. Since the tester may have limited time in practice to test all scenarios, one can choose from the randomly generated test cases the significant and potentially problematic simulations.

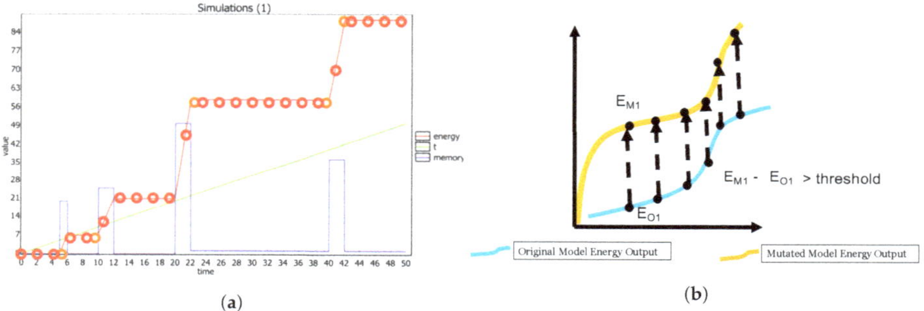

Figure 4. Test Generation and mutant detection for energy consumption. (**a**) An example of a sampled simulation trace obtained from UPPAAL SMC. (**b**) An example of applying mutation detection on two traces based on a predefined threshold.

3.4. Energy-Aware Mutant Detection

A fault is considered to be detected by a test suite if at certain time points the energy values differ drastically. In this way we increase the likelihood of evaluating a certain energy-related behavior (e.g., if the energy differs significantly from the expected result). We assume in this study that small deviations from the specified energy values can be acceptable, test engineers are likely to identify a fault if the energy deviations are substantial.

From the generated test cases, inputs and output values are extracted and used for mutation detection. The sequence of inputs in each test case is automatically inserted in a set of generated mutated models. The mutated models are simulated with the extracted inputs to obtain sets of outputs. The actual outputs extracted from these mutated models for each test case are compared to the expected outputs in order to determine the test case ability to kill (detect) any difference between the mutated models and the original one.

To exemplify this step, we show an example of a test suite that detects a mutant if the energy signal varies significantly from the expected energy values at certain time points (as shown in Figure 4b). In practice, we use a quantitative measure of mutant detection to measure the mutant-revealing capability of a test suite. Let a test case TC be created for a mutated system model M, and let $E_M = E_{M1}, ..., E_{MN}$ be the set of energy signals generated by simulating M for the inputs in TC and sampled at N time points. Let $E_O = E_{O1}, ..., E_{ON}$ be the corresponding expected energy signals. We use a threshold to verify if the distance is greater than this threshold between each value of *EO* and *EM* at each time point. If there is at least one energy value in *EM* for which the distance is greater than the expected threshold, we say that we can detect the mutant M. Otherwise, the injected fault is not detected by TC.

4. An Experimental Evaluation on the Brake-By-Wire System

We experimentally evaluate this model testing method by applying on an industrial system provided by Volvo Group Trucks Technology in Sweden. We perform experiments on a Brake-by-Wire (BBW) industrial system and evaluate the applicability of model testing in creating test cases based on energy-aware mutation testing. Additionally, by using automatically seeded faults, we examine the energy-related detection capability of the generated test suites. We start by injecting a set of faults into the original model to facilitate the assessment of fault detection. For the creation of faults, we rely on energy consumption mutation operators shown in Table 1.

Specifically, we use a PTA model, run the created test suites and collect the traces from simulations containing the energy values for each faulty model. Each test suite is executed on both the original model and its faulty counterpart to calculate the fault detection score. A fault is considered to be detected when energy results differ between executions. As an additional step, we can use the analysis result by removing the test cases that do not contribute to the mutation score of the entire test suite.

4.1. Case Description

The work proposed in this evaluation targets the Design Level in EAST-ADL (i.e., Functional Design Architecture (FDA) and Hardware Design Architecture (HDA) system aspects). The model can be extended with a GenericConstraint annotation, which allows the architect to model the energy consumption. Figure 5 presents a part of the BBW system at Design Level which is allocated to a pedal ECU. This model is extended with annotations for energy consumption as a GenericConstraint. The BBW system is a braking system that contains an anti-lock braking (ABS) feature and no mechanical connections between the brake pedal and the brake actuators. A brake pedal-mounted sensor reads its position, which is used to calculate the desired global brake torque. At each wheel, sensor values are used to calculate the wheel speed used by the ABS algorithm along with the brake torque and the approximate speed of the vehicle to determine the real brake torque to be sent to the actuators.

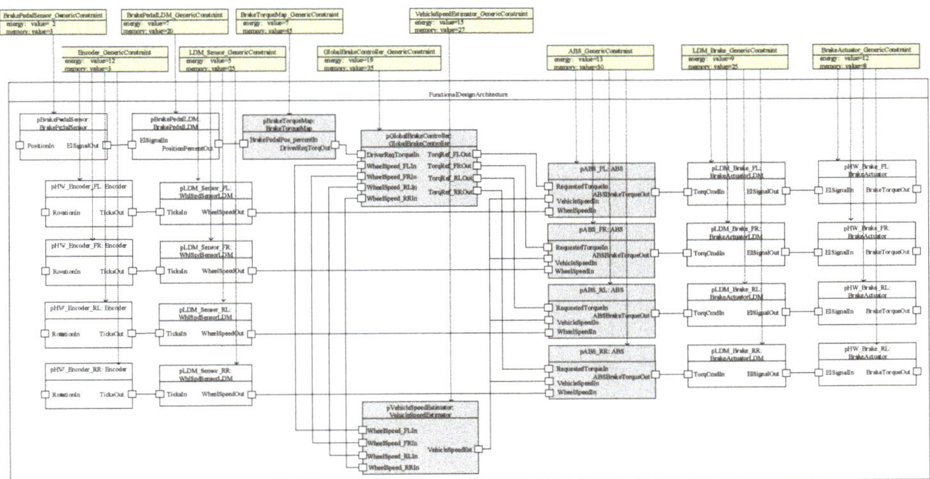

Figure 5. An overview of the Brake-by-Wire (BBW) system and its resource allocation.

The ABS algorithm calculates the slip rate *s* using the following equation:

$$S = (V - W \times R)/V,$$

where V is the vehicle speed, W is wheel speed, and R is the wheel radius. This coefficient has a nonlinear relationship with the slip rate: When s is increasing, the friction coefficient is also increasing, and its peak value is reached when s is 0.2. Further increase in s reduces the wheel friction coefficient. In this case, if s is greater than 0.2 the brake is released.

To exemplify the transformation of the BBW system modeled in EAST-ADL, we show a small-case example in Figure 6 that depicts the translation of the brake pedal sensor FunctionPrototype into a network of two synchronized PTA. For each component, the energy is calculated according to the rate of consumption monitored in the interface PTA ($pBrakePedalSensor\_e' == 2$), and a constant value is used in the behavior automaton ($pBrakePedalSensor\_e = pBrakePedalSensor\_e + 0.1$). The energy consumed by the entire system is computed using the sum of the energy consumed by each component.

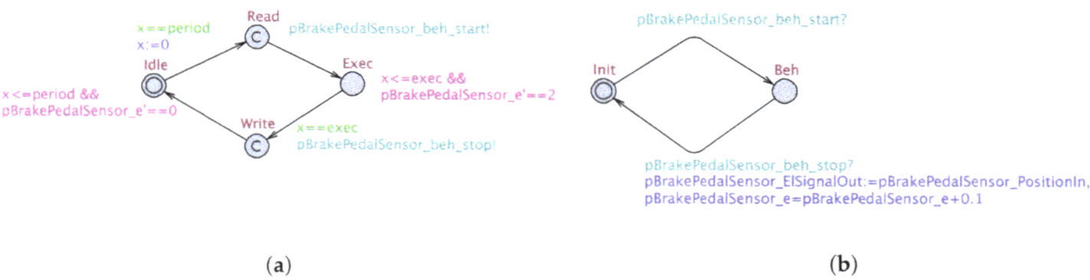

(**a**)   (**b**)

Figure 6. The network of Priced Timed Automata (PTA) for the brake pedal sensor component FunctionPrototype showing the transformed EAST-ADL architectural interface and behavior. (**a**) The interface timed automaton for the brake pedal sensor component. (**b**) The behavior timed automaton for the brake pedal sensor component.

4.2. Experimental Evaluation

We performed an experimental evaluation considering the creation of energy-aware mutations and model testing for the BBW system. We collected data for the following metrics: Generation time as a proxy for test efficiency, mutation score as a proxy for fault detection and the number of selected test cases as a measure of model testing cost reduction. In order to calculate the fault detection score, each test suite is executed on the faulty versions of the original EAST-ADL model to evaluate whether it detects the injected energy faults or not.

4.2.1. Test Suite Generation Results

As input for test generation, the method requires a standard EAST-ADL model. The generation stops searching for test inputs when it achieves the number of simulation runs. The MATS tool automatically runs the test cases on the BBW model and compares the expected outputs with the actual ones. We used different number of simulations (i.e., 25, 50, 100, 200, 400, 1000) to assess the test generation efficiency and effectiveness. As simulation time we used 64, value based on the BBW model and its full system execution and the calculated end-to-end deadline. In addition, we use 0.05 as the sample size for detecting the differences in the energy signal and 5 as the threshold delta. These values are selected based on our experience with verifying and analyzing the BBW system, as this is a realistic scenario and the values show significant differences in the energy consumption upon manual inspection of traces.

In Table 2, we present the results of applying model testing to the BBW case. As mentioned previously, in this experiment, we assume that the time is bounded to 64 time units. Table 2 lists—for each test suite and query to be checked—the time for test generation, as well as the mutation score achieved by each test suite. Regarding the generation of simulations for energy consumption, UPPAAL SMC is able to generate test cases between 17.2 s for 50 simulations and 563.4 s for 1000 simulations of the model.

In addition, to create a test suite demonstrating the worst-case energy consumption, we can use UPPAAL SMC's capacity to generate the maximum energy value. In this case, we generate simulations of the system over 200 runs, trying to maximize the energy during these simulations, with the query E[t<=64, 200](**max** : energy). The mean value generated by UPPAAL SMC is 447.2 energy units. Using this estimation, we use can use feasibility analysis for obtaining the probability for the energy value to remain within the threshold. For example, on the BBW system we are able to demonstrate, after 86 runs, that the energy consumption is lower than 447.2 with a 0.9 probability, and a confidence of 0.9. Given these results, an engineer can select test cases from the generated test suites showing the worst-case energy consumption and which can be executed on the actual system.

These results show the applicability of statistical model checking, for model testing the energy consumption using real architectural descriptions from an industrial system. We conclude that we have obtained experimental evidence that this is an efficient method for model testing applied on a real-world embedded system using its energy consumption information at the EAST-ADL architectural level. These results suggest that model testing is computationally inexpensive when used for mutation testing.

Table 2. Overall results showing the efficiency of the energy consumption test generation and mutant detection score for all generated test suites. In addition, we show the results of the test case (TC) selection based on mutation analysis.

Test Suite	SMC Query	Generation Time (s)	Mutation Score (%)	Selected TCs | Total TCs
TS1	simulate 25[<=64]{inputs[], energy}	17.2	30%	8|25
TS2	simulate 50[<=64]{inputs[], energy}	36.4	57%	15|50
TS3	simulate 100[<=64]{inputs[], energy}	72.1	75%	42|100
TS4	simulate 200[<=64]{inputs[], energy}	137.8	87%	59|200
TS5	simulate 400[<=64]{inputs[], energy}	277.5	100%	123|400
TS6	simulate 1000[<=64]{inputs[], energy}	563.4	100%	130|1000

4.2.2. Fault Seeding and Mutation Detection Results

The fault seeding procedure, results in 223 mutations (i.e., 50 ERO-based mutants, 48 PRO-based mutants, 50 ETRO-based mutants and 25 mutants for each CRO, CIO and TRO-based mutant operators). All mutants are versions of the original EAST-ADL model containing a single fault (i.e., each fault assumes one change in the system). All 223 faults are the result of applying all mutation operators. On each of the faulty versions and their original counterparts, the generated test suites are executed so that a fault detection score can be calculated. A mutant is deemed to be detected by a test suite if at some time points the energy values vary significantly. This is performed in order to increase the likelihood of detecting if the energy varies considerably from the expected result. Based on our experience in verifying and analyzing the BBW system, we set the threshold to 5 energy units.

The results in Table 2 show that for all test suites (i.e., TS1 to TS6) the achieved mutation score ranges from 30% for a test suite with 25 test cases to 100% for TS5 and TS6. Our results suggest that for test suites containing over 400 test cases all mutants are detected. Test suites TS5 and TS6 are assumed to be good at detecting faults in comparison to the other generated test suites. Even so, executing all 400 test cases is costly when performing testing on the actual BBW system and not all test cases contribute directly to the overall mutation score. This is extremely expensive from a computational point of view when considering using model testing on a complex embedded system. Therefore, we use the results of mutation analysis when executing test cases on the actual model to find the minimum number of test cases achieving the overall mutation score. In Table 2 we show the number of selected test cases out of the total number of test cases generated for each test suite (e.g., a subset of eight test cases can achieve a 30% mutation score). For all test suites, regardless of the number of generated test cases, one can reduce the number of test cases achieving the desired mutation score by over 40%. For TS5, 123 test cases are needed for achieving 100% mutation coverage with the rest of the test cases in TS5 not improving the overall score. This shows, that some test cases overlap in their exercised behavior of the BBW system and an improved generation strategy is needed to select the necessary test cases during model testing.

5. Validity Evaluation

Here we present a validity evaluation using the guidelines proposed by Runeson and Höst [26].

5.1. Construct Validity

Proper construct validity investigates the phenomenon that the researchers intended to study. The development of the method and the design of our experimental evaluation was based on certain assumptions. We have seeded energy-aware mutants automatically to calculate the ability of the selected test cases to detect energy errors. This process is carried out before test cases are generated in order to avoid a potential bias. It is possible that a larger number of naturally-occurring faults would yield different results. Adding real faults from previous projects should be employed in order to control the results more objectively early in the development process.

5.2. Internal Validity

For an experimental study as ours, internal validity relates to how credible the testing results and the mutation detection are. Detection of faults is based on a threshold of the energy budget and the time points selected to check the difference in the signal. This requirement is specific to the system and will not be enough to draw any strong conclusions. The effectiveness of this criterion depends on the energy difference interpretation and would clearly differ from one system to another. Because differences are characterized by the features of the signal shapes, we have checked at certain points in time whether the energy values differ substantially. This is a realistic situation with test engineers likely to find faults based on the measured energy consumption and the manual visual inspection.

5.3. External Validity

External validity relates to the study generalizability. Our method aims at designing and selecting a proper test suite based on a generic evaluation of the architectural model and the mutant detection score to reveal energy-related problems. However, unlike functional testing, which can use various metrics (e.g., code coverage, input space partitioning) for test generation, mutation testing for energy consumption is not as well studied. There is a need to develop and evaluate metrics capturing aspects of test effectiveness of energy consumption (such as those suggested in this paper) that can be used for test generation and selection.

6. Related Work

Recently, there has been a growing interest [27] in developing testing techniques focusing on architectural designs in software engineering. Testing based on software architectures has been explored in a considerable amount of work [28–32], leading to contributions in system and integration testing [28–30], criteria for architectural-based testing [31], and regression test selection [32]. Jiang et al. [33] compared several techniques that are used in performance and load testing of software intensive systems. For example, Zhang et al. [34,35] proposed the use of load testing of timing and resource requirements using system-level models. Compared to this work, we focus on energy consumption and we provide an efficient and effective model testing method; this aspect has received little attention in the literature.

Testing software based on extra-functional properties at the architectural level has received less attention [7,36–40] than the functional testing of such models. In our previous work [7], we explored the use of statistical model checking for analysis and performance testing of EAST-ADL models and used manually injected faults to measure test effectiveness. In this work, we build upon these results and consider how to automatically select and generate test suites for testing the energy performance of a system based on its architectural model and mutation testing.

One of the initial papers on non-functional testing [36] used a software architecture for selecting the parameters that directly influence the system performance. Among the approaches for non-functional test generation, only a few [41–44] consider robustness and performance aspects. Nebut et al. [41] and Shaukat et al. [42] proposed methods for automatic test generation that support robustness goals expressed in UML models. In contrast to these studies, our approach is tailored to an architectural language, which is an emerging notation in the automotive domain. We automatically select test cases that cover energy-aware injected faults on the model level early-on in the development process. In addition, in this paper we focus on selecting test cases for testing the performance of an existing Brake-By-Wire system by using the energy consumption information encoded in the architectural model.

7. Conclusions

In this paper we outline a method for testing EAST-ADL architectural models using energy-aware mutations. Furthermore this method uses UPPAAL SMC and MATS approaches to select test suites that contribute to the overall mutation score. The method makes use of energy requirements as expressed in EAST-ADL architectural models, transforms these requirements into priced timed automata together with the component interfaces, and uses statistical model checking to identify relevant test cases. We use simulations to create test suites containing input parameters and energy signals. We select test cases that maximize the mutation score. An experimental evaluation of this method, using a Brake-by-Wire system provided by Volvo Group Trucks Technology in Sweden, indicates that model testing of energy consumption is applicable for the automatic generation and execution of test suites at architectural level. The evaluation indicates that this method of creating test suites is efficient in terms of generation time. In this study, we proposed to evaluate the fault detection capability of these test suites by seeding modeling errors for energy consumption and altering the level of energy consumption over time. Our results suggest that an approach that selects test cases showing diverse energy consumption patterns can increase the fault detection ability.

Future work aims to extend our approach to generate tests for other types of resources and to apply it more thoroughly to real industrial cases to demonstrate its strengths and limitations by using naturally occurring faults.

Author Contributions: The first two authors contributed equally to the research, approach, study design, analysis and reporting of the research work. In addition, the first author contributed with tool development and data collection. All authors have read and agreed to the published version of the manuscript.

Funding: This research was supported by the Swedish Research Council (VR) through the "Adequacy—based testing of extra-functional properties of embedded systems" project. This work is partially funded from the Electronic Component Systems for European Leadership Joint Undertaking under grant agreement No. 737494, The Swedish Innovation Agency, Vinnova (MegaM@Rt2 and XIVT), as well as by the Swedish Knowledge Foundation (KKS), within the DPAC (Dependable Platforms for Autonomous Systems and Control) research profile.

Acknowledgments: We would like to thank Jonatan Larsson for his support in using the MATS tool and Raluca Marinescu for her valuable comments on this work.

Conflicts of Interest: The authors declare no conflict of interest.

References

1. Pretschner, A.; Broy, M.; Kruger, I.H.; Stauner, T. Software engineering for automotive systems: A roadmap. In Proceedings of the IEEE Computer Society on 2007 Future of Software Engineering, Minneapolis, MN, USA, 23–25 May 2007; pp. 55–71.
2. Hammond, J.; Rawlings, R.; Hall, A. Will it work?[requirements engineering]. In Proceedings of the IEEE International Symposium on Requirements Engineering, Toronto, ON, Canada, 27–31 August 2001, pp. 102–109.
3. Barroso, L.A.; Hölzle, U. The case for energy-proportional computing. *Computer* **2007**, *40*. [CrossRef]
4. Briand, L.; Nejati, S.; Sabetzadeh, M.; Bianculli, D. Testing the untestable: model testing of complex software-intensive systems. In Proceedings of the 38th International Conference on Software Engineering Companion, Austin, TX, USA, 14–22 May 2016; pp. 789–792.
5. González, C.A.; Varmazyar, M.; Nejati, S.; Briand, L.C.; Isasi, Y. Enabling model testing of cyber-physical systems. In Proceedings of the 21th ACM/IEEE International Conference on Model Driven Engineering Languages and Systems, Copenhagen, Denmark, 14–19 October 2018; pp. 176–186.
6. Blom, H.; Lönn, H.; Hagl, F.; Papadopoulos, Y.; Reiser, M.O.; Sjöstedt, C.J.; Chen, D.J.; Tagliabò, F.; Torchiaro, S.; Tucci, S. EAST-ADL: An Architecture Description Language for Automotive Software-Intensive Systems. *EAST-ADL White Paper* **2013**, *1*.
7. Marinescu, R.; Enoiu, E.; Seceleanu, C.; Sundmark, D. Automatic Test Generation for Energy Consumption of Embedded Systems Modeled in EAST-ADL. In Proceedings of the 2017 IEEE International Conference on Software Testing, Verification and Validation Workshops (ICSTW), Toyko, Japan, 13–17 March 2017; pp. 69–76.
8. DeMillo, R.A.; Lipton, R.J.; Sayward, F.G. Hints on test data selection: Help for the practicing programmer. *Computer* **1978**, *11*, 34–41. [CrossRef]
9. Behrmann, G.; Fehnker, A.; Hune, T.; Larsen, K.; Pettersson, P.; Romijn, J.; Vaandrager, F. Minimum-Cost Reachability for Priced Time Automata. In *Hybrid Systems: Computation and Control*; Lecture Notes in Computer Science; Springer: Berlin/Heidelberg, Germany, 2001.
10. Bulychev, P.; David, A.; Larsen, K.G.; Mikučionis, M.; Poulsen, D.B.; Legay, A.; Wang, Z. UPPAAL-SMC: Statistical Model Checking for Priced Timed Automata. In Proceedings of the Workshop on Quantitative Aspects of Programming Languages and Systems, Tallinn, Estonia, 31 March–1 April 2012.
11. Shaw, M.; Garlan, D. *Software Architecture: Perspectives on An Emerging Discipline*; Prentice Hall Englewood Cliffs: Englewood Cliffs, NJ, USA, 1996; Volume 1.
12. Ammann, P.; Offutt, J. *Introduction to Software Testing*; Cambridge University Press: Cambridge, UK, 2008.
13. Malavolta, I.; Lago, P.; Muccini, H.; Pelliccione, P.; Tang, A. What industry needs from architectural languages: A survey. *IEEE Trans. Softw. Eng.* **2013**, *39*, 869–891. [CrossRef]
14. Just, R.; Jalali, D.; Inozemtseva, L.; Ernst, M.D.; Holmes, R.; Fraser, G. Are Mutants a Valid Substitute for Real Faults in Software Testing? In *International Symposium on Foundations of Software Engineering*; ACM: New York, NY, USA, 2014.
15. Black, P.E.; Okun, V.; Yesha, Y. Mutation operators for specifications. In Proceedings of the ASE 2000. Fifteenth IEEE International Conference on Automated Software Engineering, Grenoble, France, 11–15 September 2000; pp. 81–88.
16. Aichernig, B.K.; Salas, P.A.P. Test case generation by OCL mutation and constraint solving. In Proceedings of the Fifth International Conference on Quality Software (QSIC'05), Melbourne, Australia, 19–20 September 2005; pp. 64–71.
17. Aichernig, B.K.; Jöbstl, E.; Tiran, S. Model-based mutation testing via symbolic refinement checking. *Sci. Comput. Program.* **2015**, *97*, 383–404. [CrossRef]
18. Lindström, B.; Offutt, J.; Sundmark, D.; Andler, S.F.; Pettersson, P. Using mutation to design tests for aspect-oriented models. *Inf. Softw. Technol.* **2017**, *81*, 112–130. [CrossRef]
19. Kang, E.Y.; Enoiu, E.P.; Marinescu, R.; Seceleanu, C.; Schobbens, P.Y.; Pettersson, P. A Methodology for Formal Analysis and Verification of EAST-ADL Models. *Reliab. Eng. Syst. Saf.* **2013**, *120*. [CrossRef]
20. Marinescu, R.; Kaijser, H.; Mikučionis, M.; Seceleanu, C.; Lönn, H.; David, A. Analyzing industrial architectural models by simulation and model-checking. In *International Workshop on Formal Techniques for Safety-Critical Systems*; Springer: London, UK, 2014; pp. 189–205.

21. Bulychev, P.; David, A.; Larsen, K.G.; Legay, A.; Li, G.; Poulsen, D.B. *Rewrite-Based Statistical Model Checking of WMTL*; Runtime Verification; Springer: London, UK, 2013.
22. Marinescu, R.; Enoiu, E.; Seceleanu, C. *Statistical Analysis of Resource Usage of Embedded Systems Modeled in EAST-ADL*; VLSI Symposium; IEEE: Piscataway, NJ, USA, 2015; pp. 380–385.
23. Aichernig, B.K.; Lorber, F.; Ničković, D. Time for mutants—model-based mutation testing with timed automata. In *International Conference on Tests and Proofs*; Springer: London, UK, 2013, pp. 20–38.
24. Seceleanu, C.; Vulgarakis, A.; Pettersson, P. REMES: A Resource Model for Embedded Systems. In Proceedings of the International Conference on Engineering of Complex Computer Systems, Potsdam, Germany, 2–4 June 2009.
25. Larsson, J. *Automatic Test Generation and Mutation Analysis using UPPAAL SMC*; Bachelor of Science Thesis Report; MDH Diva: Ascona, Switzerland, 2017.
26. Runeson, P.; Höst, M. Guidelines for conducting and reporting case study research in software engineering. *Empir. Softw. Eng.* **2009**, *14*, 131. [CrossRef]
27. Bertolino, A.; Inverardi, P.; Muccini, H. Software architecture-based analysis and testing: A look into achievements and future challenges. *Computing* **2013**, *95*, 633–648. [CrossRef]
28. Abdurazik, A.; Jin, Z.; White, L.; Offutt, J. Analyzing software architecture descriptions to generate system-level tests. In Proceedings of the Workshop on Evaluating Software Architectural Solutions, Irvine, CA, USA, 9 May 2000.
29. Bertolino, A.; Inverardi, P. Architecture-based software testing. In *International Software Architecture Workshop*; ACM: New York, NY, USA, 1996; pp. 62–64.
30. Richardson, D.J.; Wolf, A.L. Software testing at the architectural level. In *International Software Architecture Workshop and International Workshop on Multiple Perspectives in Software Development*; ACM: New York, NY, USA, 1996, pp. 68–71.
31. Jin, Z.; Offutt, J. Deriving tests from software architectures. In Proceedings of the International Symposium on Software Reliability Engineering, Hong Kong, China, 27–30 November 2001, pp. 308–313.
32. Harrold, M.J. Architecture-based regression testing of evolving systems. In Proceedings of the International Workshop on the Role of Software Architecture in Testing and Analysis, Marsala, Italy, 30 June–3 July 1998.
33. Jiang, Z.M.; Hassan, A.E. A Survey on Load Testing of Large-Scale Software Systems. *IEEE Trans. Softw. Eng.* **2015**, *41*, 1091–1118. [CrossRef]
34. Zhang, J.; Cheung, S.C. Automated test case generation for the stress testing of multimedia systems. *Softw. Pract. Exp.* **2002**, *32*, 1411–1435. [CrossRef]
35. Zhang, J.; Cheung, S.C.; Chanson, S.T. Stress testing of distributed multimedia software systems. In *Formal Methods for Protocol Engineering and Distributed Systems*; Springer: London, UK, 1999; pp. 119–133.
36. Weyuker, E.J.; Vokolos, F.I. Experience with performance testing of software systems: issues, an approach, and case study. *IEEE Trans. Softw. Eng.* **2000**, *26*, 1147. [CrossRef]
37. Denaro, G.; Polini, A.; Emmerich, W. Early performance testing of distributed software applications. In *ACM SIGSOFT Software Engineering Notes*; ACM: New York, NY, USA, 2004; Volume 29, pp. 94–103.
38. Jiang, Z.M.; Hassan, A.E.; Hamann, G.; Flora, P. Automated performance analysis of load tests. In Proceedings of the International Conference on Software Maintenance, Edmonton, AB, Canada, 20–26 September 2009; pp. 125–134.
39. Grechanik, M.; Fu, C.; Xie, Q. Automatically finding performance problems with feedback-directed learning software testing. In Proceedings of the International Conference on Software Engineering (ICSE), Zurich Switzerland, 2 June 2012; pp. 156–166.
40. Marinescu, R.; Saadatmand, M.; Bucaioni, A.; Seceleanu, C.; Pettersson, P. A model-based testing framework for automotive embedded systems. In Proceedings of the 2014 40th EUROMICRO Conference on Software Engineering and Advanced Applications, Verona, Italy, 27–29 August 2014; pp. 38–47.
41. Nebut, C.; Fleurey, F.; Le Traon, Y.; Jezequel, J.M. Automatic test generation: A use case driven approach. *IEEE Trans. Softw. Eng.* **2006**, *32*, 140–155. [CrossRef]
42. Ali, S.; Briand, L.C.; Hemmati, H. Modeling robustness behavior using aspect-oriented modeling to support robustness testing of industrial systems. *Softw. Syst. Model.* **2012**, *11*, 633–670. [CrossRef]

43. Yue, T.; Ali, S. Bridging the gap between requirements and aspect state machines to support non-functional testing: industrial case studies. In *European Conference on Modelling Foundations and Applications*; Springer: London, UK, 2012; pp. 133–145.
44. Garousi, V.; Briand, L.C.; Labiche, Y. A UML-based quantitative framework for early prediction of resource usage and load in distributed real-time systems. *Softw. Syst. Model.* **2009**, *8*, 275–302. [CrossRef]

Article

A full Model-Based Design Environment for the Development of Cyber Physical Systems

Roberto Manione

Media on Line, Caselette, 10040 Torino, Italy; roberto.manione@taskscript.com

Received: 30 September 2018; Accepted: 9 February 2019; Published: 13 February 2019

Abstract: This paper discusses a full model-based design approach in the applicative development of Cyber Physical Systems targeting the fast development of Logic controllers (i.e., the "Cyber" side of a CPS). The proposed modeling language provides a synthesis between various somehow conflicting constraints, such as being graphical, easily usable by designers, self-contained with no need for extra information, and to leads to efficient implementation, even in low-end embedded systems. Its main features include easiness to describe parallelism of actions, precise time handling, communication with other systems according to various interfaces and protocols. Taking advantage the modeling easiness deriving from the above features, the language encourages to model whole CPSs, that is their Logical and their Physical side, working together; such whole models are simulated in order to achieve insight about their interaction and spot possible flaws in the controller; once validated, the very same model, without the Physical side, is compiled and into the logic controller, ready to be flashed on the controller board and to interact with the physical side. The discussed language has been implemented into a real model-based development environment, TaskScript, in use since a few years in the development of production grade systems. Results about its effectiveness in terms of model expressivity and design effort are presented; such results show the effectiveness of the approach: real case production grade systems have been developed and tested in a few days.

Keywords: Cyber Physical Systems; Reactive Systems; Model-Based Design; Embedded Systems; Automatic Code Generation; IDE; Internet of Things

---

1. Introduction

CPS defines the class of systems where some Physical reality is observed and controlled by a software controller, with the purpose of obtaining the requested behavior(s) towards the accomplishment of the given task(s) [1].

This paper focuses on one of the major issues in the design of CPSs, that is, the development of the controller software, a job intrinsically complex because of its multidisciplinary nature, where deep competences ranging from Physics to Electronics to Computer Science are needed, and proposes a specific language and a complete toolchain to support the development of such software in an effective way.

To this goal, the model-based approach is proposed, a technique aimed at easing the design task by hiding most of the details; the envisioned result is to enable experts of the domain (i.e., the Physical side) to actively participate in all the phases of the controller design bringing in their domain competence, rather than leaving potentially impacting design decisions to the experts of the implementation.

"CPSs are integrations of computation with physical processes. Embedded computers and networks monitor and control the physical processes, usually with feedback loops where physical processes affect computations and vice versa". This definition, excerpted from [2], updates and generalizes the definition of Reactive Systems [3] introduced in the 1980s with a similar purpose, that is, providing a framework for the design of controllers for physical devices; while Reactive Systems focused exclusively on the Control Function leaving the Physical side outside their boundary, CPS consider both controller and controlled subsystems, and this allows for deeper understanding of the whole system.

Apart from this different way of drawing the boundary of the system under study, most assumptions within Reactive Systems are the same for CPS, the most important one being the "reactiveness" of the Control function. This function—the Cyber side in CPS jargon—must be able to receive and react to stimuli from the Physical side at any point in time; this constraint derives from the fact that the Physical side is generally made by several devices, each working at the same time, hence to be dealt with "simultaneously".

Notice that RS/CPS are quite different from Transformational Systems (i.e., a server, desktop computer, or a tablet), typically designed to process streams of data stored in files, or obtained from the network or from users, and there are no hard constraints upon the delay in the availability of their results; in a nutshell RS/CPS and TS have different requirements and capabilities: while a Transformation System is able to perform potentially highly sophisticated processing upon data available when the processor is ready to work on them, RS/CPS are able to perform the processing needed to maintain an ongoing interaction with their environment, exchanging asynchronous signals with it.

Model-based design is a methodology allowing designers to develop their projects through the use of primitives (i.e., elementary "components") chosen from a predefined set; the use of predefined components helps because it allows to overlook details and consequently to save effort and time.

Most MB design environments offer primitives sets chosen according to some analogy with the real world, so that designers are already familiar with such primitives. Designers can think they are using the real components that the primitives remind and expect similar results as if they built real systems with the real blocks which the primitives derive from.

The model-based technique presented in this paper is based upon a Domain Specific Language (DSL), with its execution semantics able to handle natively the relevant situations arising in CPS controller models, such as for example concurrency, timing, state transitioning, data handling.

The proposed language belongs to the class of Synchronous languages, pioneered by Esterel [4], Lustre [5], and Argos [6] (see [7] and [8] for a detailed discussion on synchronous languages) whose name derives from their analogy with the synchronous digital circuits. Models are executed cyclically, in a "read inputs–evaluate model–write outputs" loop that repeats forever; the execution of one cycle is considered instantaneous and time is advanced outside the cycle. Time is modeled according to the Timed Automata theory [9]: resettable clocks are available inside states and transitions to model time dependent behaviors.

One of the initial requirements of the presented project is to be implementable both in software/firmware, on off-the-shelf CPUs, and in microcode, on specialized hardware; to this end, the language has been defined following a bottom-up path. A dedicated virtual machine has been defined first, with its own instruction set natively supporting all the above features, with the additional requirement of being efficiently executable in all the possible implementations. For the time being the Machine has been implemented in its virtual version, i.e., in software and in firmware, on a number of off-the-shelf CPUs, with 32, 16, 8 bits; however, the very same instruction set is ready to be implemented as a real machine on special purpose microcoded hardware, to maximize the execution speed.

From the definition of the virtual machine, the DSL has been defined in such a way to maximally exploit the features of the virtual machine, and to be easily compiled into its primitives. As the Virtual Machine has primitives that explicitly handle the scheduling of concurrent activities, these aspects are exposed in the DSL so designers can take advantage of them within their models.

The resulting DSL has a clear execution semantics that allows to easily express the system to be implemented, without need of adding side information; the synthesized code is complete and optimized by construction, hence can be released in production without any manual intervention.

While the implementation of the microcoded version has not been tackled, an intermediate product has already been implemented along this path: the "controller-on-a-chip", which is actually an off-the-shelf CPU pre-flashed with the virtual machine. Once loaded with the code coming from the model compilation, such chip behaves as described in the model. This product can be used as a hardware building block to quickly build specialized CPS controller boards ready for Model-Based design; hardware designers can easily integrate the chip into their designs, using and interfacing only the needed I/Os.

The availability of a variety of virtual machines with different positioning in the (speed, consumed energy, arithmetics precision, type and number of I/Os, cost) hyperspace, combined with the fact that the system to be implemented is fully described by its model, so that a working implementation can be automatically generated, allows designers to explore trade-offs with little effort. A flavor of this is shown in the results discussion section, where a design has been implemented on two VMs for comparison purposes; for that simple case the two solutions trade off $4\times$ less consumed power with $7\times$ slower response time.

The paper is organized as follows: Section 2 briefly reviews the model-based bibliography and introduces the novelty of the proposed approach, Section 3 presents a reference architecture for CPS showing how the proposed approach can help in dealing with the complexity, Section 4 discusses the main aspects involved in MB design of CPSs analyzing the requirements for a DSL; Section 5 introduces and discusses the TaskScript Model-Based DSL; Section 6 addresses CPS modeling with the TaskScript toolchain; finally Section 7 reports and discusses results in terms of both modeling effort and performance of the generated code and Section 8 concludes the paper.

2. Previous Work and Novelties of the Presented Approach

In this section previous work is reviewed, mainly with respect to the ability to fulfill the requirements arising in implementing the reference architecture analyzed in Section 3.

Model-based strategies can conveniently be classified into lightweight and heavyweight (see [10]), according to the specificity of the adopted modeling language. Lightweight MBs extend or specialize existing general-purpose modeling languages (for example, the UML [11]) in order to inherit their theoretical foundations or execution semantics, and exploit the available tools.

On the other hand, heavyweight MBs define their ad hoc execution semantics and languages (DSL). At the cost of implementing a new toolchain, the expected advantages are: (a) specificity to the domain of interest; (b) modeling accuracy and easiness, with consequent reduced modeling effort; and (c) optimization of the produced results, that is the efficiency of the synthesized system.

The lightweight approach has been adopted by a number of design environments based upon the UML and its derivatives, like the SysML (System Modeling Language) [12] (see Rhapsody [13]), or MARTE [14]; the same approach has been adopted by Papyrus [15], based on the fUML [16], and others.

From the concurrency point of view, the execution semantics of UML is based upon concurrent capsules interacting via channels connected to capsule ports; capsules are scheduled and interfaced by the host O.S., where the application is deployed, which provides concurrence support. This was chosen by UML design team in order to preserve the genericity of the execution model across a variety of O.S.

However, due to the above genericity choice, the ability of such design environments to meet hard real-time constraints (i.e., the capability to deliver controllers able to react within a predictable, bounded amount of time, possibly in contexts of limited resources in terms of CPU power, memory, energy, etc ... .) can only be as good as the one of the host O.S.

The PSCS (Precise Semantics of UML Composite Structures) specification [17], which integrates fUML, allows finer control on the dispatching of events inside objects, allowing to introduce priorities and other mechanisms. However, this control is limited to intra-capsule activities; events flowing to and from different capsules are under the control of the O.S.

As a consequence, models must cope with the fact that there is no control upon the order of arrival of events coming from concurrent capsules; in applications where the result is sensitive to the order of arrival of data coming from concurrent capsules, this must be ensured at model level; as will be discussed later in this section, this technique could be named "constraints-abiding".

Another example of lightweight system is Yakindu [18], based upon StateCharts, that generates a C/C++ source code implementing the modeled state machine leaving the user with the task of filling in the code which performs the data handling. The choice about the execution semantics of concurrent activities is left to the users, who can allocate them (if any) to parallel threads/processes resorting to the OS for their scheduling, or implement an ad hoc scheduler, to have more control on it; in any case manual coding and debugging is needed.

Likewise, for the Matlab-Simulink-Stateflow [19] design suite, which produces code at source level; some intervention upon the generated code may generally be needed in order to achieve hard real-time capability, but this manual phase, in turn, needs further effort and can invalidate the correctness of the design as checked in the earlier stages.

On the other hand, heavyweight design systems take a different approach to be hard real-time capable and achieve the fine grain control over concurrent activities needed to control fast physical devices, as needed by the reference architecture. They implement their ad-hoc algorithms for the scheduling of concurrent activities in order to bypass the general purpose scheduling and Interprocess Communications mechanisms provided by the O.S.

In the following, a few examples are reported which follow this approach, to different extents. The SCADE design suite [20], based on a synchronous language, such as Esterel, compiles the input model into one large automaton to deliver the concurrent behavior of the model while executed on a sequential program; such sequential automaton is built with the Cartesian product of the state spaces of all the sequential threads that can occur in parallel.

AutoFOCUS-3 (Af3) [21] supports modeling of distributed systems through the definition of the timed message streams exchanged among the various components; the resulting network of interacting nodes is then allocated on the distributed platform according to a mapping expressed at the model level. This allows model designers to express constraints on allocation and to reach hard real-time capability.

LabView [22] is another example of a synchronous language; it is typically used to model networks of virtual instruments and have them operate upon real data coming from the physical world. However, dedicated hardware can be needed to execute the model with real-time data.

The proposed design suite follows the same approach and addresses the design of CPS control functions, mainly targeting embedded systems with limited resources in terms of CPU power, memory, etc. To this purpose, the target hardware is not equipped with a general purpose O.S., rather a special purpose lightweight multitasking kernel is provided. To maximize its effectiveness, the generated code is produced at executable level, optimized for the desired CPU/platform, and chosen among the supported ones.

The synthesized controllers provide deterministic durations for activities, bounded by estimates computed at compile time: hence they are hard real-time capable.

Unlike the Esterel approach, here the model is compiled into a set of small sequential automata, one for each sequential thread defined in the model: the concurrent behavior is delivered by the kernel executing concurrently all the sequential automata. This has the advantage of avoiding the generation of a large automaton, which would exceed the available memory of small and medium sized controllers.

Moreover, while Esterel is an imperative language, where states are implicit, in the proposed DSL states are defined explicitly; this makes the language declarative, and is thus more adaptable to a visual representation and easier to understand. This, however, was also the idea behind Argos.

One distinctive feature of the DSL is its fine-grain control upon the order of execution of the concurrent activities defined in the model; model designers can set the precedence in which actions belonging to the different concurrent threads are executed through DSL directives (see Section 5.6.1 for details).

The benefit of this feature is that a simpler execution semantics can adopted to handle data coming from concurrent threads. Where, in other MB languages, channels must be used in order to buffer data exchanged among concurrent threads in order to cope with the uncertainty in the order of their arrival, in the proposed DSL plain variables can safely be used, written and read by all the interested concurrent actors in the desired order. This opens a new, light-weight modeling approach for handling conflicts where the same variable is assigned by different concurrent activities. While the traditional way is "constraints-abiding", where conflicts are handled at model level, here an alternative is provided: "constraints-enforcing", where conflicts simply cannot occur; models exploiting this approach are generally simpler and more efficiently implemented (see Section 6.4 for a comparison among the two techniques on a simple example).

Finally, one unique feature of the DSL is the availability of "stateful primitives" (see Sections 5.4 and 5.5 for a detailed discussion). This mechanism, consistent with the synchronicity paradigm, allows the encapsulation of stateful resources through a simple "enable-done" interface; the enable input is used by the model to activate the resource and the done output is used by the resource to notify the model of the successful completion of its task.

Stateful primitives have their own internal state, which evolves concurrently and independently, and communicate with the model only through their enable-done interface. A number of resources have been integrated to the DSL as stateful primitives, to allow modeling controllers compliant with the reference architecture; this is, for example, the case of communications resources, such as ports and protocols.

3. Reference Architecture for CPS

The term CPS has been proposed to convey the concept of "smart" systems. Taking advantage of the communications infrastructure available in today's world, the implementation of systems exhibiting such a behavior is feasible leveraging the cooperation of multiple systems interacting remotely; within this picture, where a system is implemented as a network of geographically distributed subsystems, CPSs play the role of those remote subsystems interacting with some physical periphery with the purpose of observing it, or controlling it, or both, keeping periodic contact with the rest of the network.

A CPS must then be able to both observe and/or control its own physical periphery and at the same time communicate with one or more other systems using standard (i.e., http) protocols. Such two classes of tasks are very different with each another, as the former needs hard real-time capabilities, and the latter needs the ability to handle the many anomalous conditions that can arise in remotely distributed systems, such as communication errors, timeouts, temporary outages, and more.

Figure 1 proposes a reference architecture for the implementation of CPS, well suited to implement both stand-alone and connected devices; both internal (i.e., computation) and external (i.e., communication) aspects are addressed, with their respective peculiarities.

The atomic element of the architecture is the parallel "thread", interacting with the others through a system of private and shared variables. Each thread typically takes care of a single element of the Physical periphery, with the advantage that every thread deals with that element alone as if it was the only one in the system; this makes each thread easy to write and debug. Likewise, other threads handle the communication with the rest of the network, using the available communication resources.

The two main categories of threads can be summarized:

1. Local control: fast and independent from the network(s), robust against delays or temporary network outages, hard real-time capable;
2. Remote communication: here data are exchanged through the network(s) to update the remote nodes about the controlled process(es) and, possibly, to receive updated control constants.

The Concurrent Execution Environment has the responsibility of executing all the model-level threads concurrently, updating the I/O variables and the other internal resources, etc.

The aim of the DSL and associated IDE discussed in this paper is to effectively support the design of CPS according to the above reference architecture, through a model-based approach.

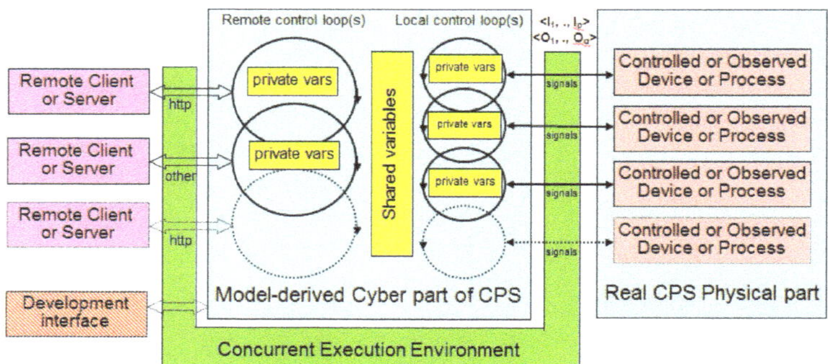

Figure 1. Reference architecture for Model-Based designed CPS.

4. Model-Based Design of CPS Controllers

CPSs integrate physical devices with processing and communicating devices; both Physical and Logical sides have their own internal state, $<P_1, \ldots, P_m>$ and $<L_1, \ldots, L_n>$ respectively; their internal state is continuously changed by two transformations operating the two sides, T_P (i.e., the transformation occurring in the Physical side) and T_L (i.e., the transformation occurring in the Logical side), the former involving exchange of energy and the latter involving exchange of information.

A CPS Controller is typically an embedded system hosting the Logical side of the CPS; in other words it is the place where $<L_1, \ldots, L_n>$ lays, along with the functions implementing T_L.

The Controller design is aimed at defining $<L_1, \ldots, L_n>$ and T_L in such a way that the CPS as a whole exhibits the requested behavior; in such an activity $<P_1, \ldots, P_m>$ and T_P are mainly used as part of the input requirements; the design of the Controller must ensure that the Physical side is correctly and timely observed and Controlled so that the whole system has the expected behavior.

However, in case controller designers realized that there is no way to implement the Logical side to keep up with the given Physical side, changes could be done to it; the following discussion implicitly assumes that such changes, if any, have already been made when tackling the controller design.

At the interface between the two sides, Physical and Logical (i.e., Cyber), two sets of signals are "continuously" exchanged, travelling in opposite directions: one set, $<I_1(t), \ldots, I_p(t)>$, contains signals produced within the Physical side, sensed and sampled into sequences for the Logical side; on the other hand, the other set, $<O_1(t), \ldots, O_q(t)>$, contains actuation signals for the Physical side, generated by converting sequences produced inside the Logical side.

The CPS Physical side operates, by its very nature, time-continuously, in full parallelism and total asynchronicity: all its components operate all the time, consuming the signals $<O_1, \ldots, O_q>$, transforming the internal state $<P_1, \ldots, P_m>$ according to the physical stimuli and $<O_1, \ldots, O_q>$ and T_P, i.e., the physical laws involved by the specific case, and producing the signals $<I_1, \ldots, I_p>$.

On the other hand, the Logical side consumes $<I_1, \ldots, I_p>$, changes its internal state $<L_1, \ldots, L_n>$ according to them and to T_L, i.e., the implemented control algorithm, and produces new $<O_1, \ldots, O_q>$.

The Controller implementing the Logical side must be designed to be able to read $<I_1, \ldots, I_p>$, and produce $<O_1, \ldots, O_q>$, in a correct and timely way; this statement is an alternative way to express the requirement that the sought solution is a hard real-time controller, able to cope with time-critical systems (a controller is defined hard real-time if it can guarantee to carry out given activities within a given amount of time); the following introduces how the above requirement is tackled in the proposed language. From the hardware point of view, the Controller operates cyclically: the various physical components are served by dedicated logics running on the processing node one after the other, according to the provided order, and this cycle is repeated forever: each iteration consumes one sample from all the input variables $<I_1(t), \ldots, I_p(t)>$ and produces one sample for all the output variables $<O_1(t), \ldots, O_q(t)>$.

This strategy is typical with Synchronous languages and ensures that the processing is deterministic, i.e., the same output sequence is always generated, out of the same input sequence.

The underlying hypothesis, necessary for the correctness of the results, is that the controller embedded CPU is fast enough to allow the controller perform all the computations contained in the model at the speed requested by the Physical side.

These conditions must be verified after the automatic program generation; a rule of thumb is that the cycle period is at least one order of magnitude shorter than the shorter significant time constant of the devices in the Physical side.

As will be shown in the results discussion, controllers generated with the presented approach have cycle times in the range of 20 to 50 microseconds for implementations with 16 bits CPUs and in the range of 100 to 250 microseconds for implementations with 8 bits CPUs.

According to the above figures, the feasibility of a controller can be decided before embarking on the modeling process: as an example, a project is feasible if the shorter significant time constant in the Physical side is slower than 0.5 milliseconds, if a 16 bit CPU is planned.

The proposed language is conceived in such a way that an accurate estimate of the execution time, that is the range (best case–worst case) can be automatically produced at model compile time.

The effectiveness of the MB approach in supporting designers heavily relies upon the fitness of the Modelling Language to the actual design domain; the chosen set of language primitives should consider, at least, the following (partly conflicting) constraints:

1. Effectively support system designers in modeling their actual systems;
2. Be easy to understand and remember by designers;
3. Lead to efficient implementations: while this requirement is of limited importance for the analysis and simulation use case, it becomes crucial in the code generation use case, where code efficiency, size, consumed energy and cost of the produced result play a significant role.

Here we concentrate upon modeling CPS Controllers; for this scenario a fundamental requirement has already been introduced, that is the ability to natively model execution parallelism: designers should be allowed to express their models in terms of individually designed blocks that operate "concurrently"; along with parallelism comes timing: typically, physical devices need to be interacted with in precise and accurate windows of time.

A further requirement is that models must be expressed visually, i.e., through schematic drawings, rather than in text, as the former is easier to understand and to remember and allows for more effective visualization techniques of simulated and run-time state.

4.1. Modeling the State Nature of Control Algorithms

States are a convenient abstraction helpful in modeling CPSs and, generally, parallel systems; modeling a system as a set of sequences of states, evaluated in parallel, is a way of applying the divide-and-conquer principle to this class of systems.

Key to CPS controller modeling is then the ability to capture the state nature of the needed control algorithm (if any), that is, the fact that a CPS exhibits different behaviors according to its actual state.

A number of formalisms have been defined over time to model the state nature of systems:

5. Finite State Machines (FSM): very basic formalism, where one and only one state is active at the same time; a transition from state I to state J triggers if state I is active the guard from I to J is true; no parallelism is allowed; easy to understand but too limited in expressivity.

6.   Petri Nets: very powerful formalism, where every place can have arbitrary marking, and this allows modeling parallelism and synchronization; however, it's difficult to map the real state of the system on the actual marking of the net.

7.   State Charts [3] and UML [11]: very general state machines formalisms, allowing hierarchy and parallel threads: they lay at the needed level of abstraction, but the fact that they allow hierarchical states and, specially, transitions among states belonging to different hierarchical boundaries, leads to heavy implementations, potentially unsuitable for low-end CPUs.

8.   GRAFCET: [23] a simplified Petri Net formalism, with boolean marking; with this restriction a GRAFCET is a kind of an augmented FSM, whose states map to places one-to-one; parallelism and synchronization are supported and it is efficiently implementable.

The most suitable choices are 3 and 4; in both cases the states defined at model level can be mapped onto lower level tasks, so that the multitasker executing on the target CPU will be able to process each and every active state. In the proposed language states are defined following the GRAFCET definition; states are named Steps, to remind that more that one of them can be active at the same time.

4.2. Modeling Data within Control Algorithms

Data Modeling, complementary to state modeling in the design of CPS controllers, defines the data used in controllers and their transformations along with its states.

Basic scalar types like Binary and Integer are obviously needed, possibly with options in their precision (i.e., number of bytes), to allow optimizing the implementation on CPUs with limited resources. Arrays of such data types are really needed, as they can store sequences of samples, in turn derived from sampling physical signals, and support iterations; strings are needed in order to support operations upon protocols and in particular the Internet ones; structured data are nice-to-have but not mandatory, as Controllers are not expected to perform very diversified processing.

Data variables should be differentiated according to their usage within the controller; the main needed classes are:

1.   Input and Output: data refreshed with signals sampled from the Physical side, that is $<I_1, \dots, I_p>$ and data updating signals sent to the Physical side, that is $<O_1, \dots, O_q>$; there are several options about when updating such variables, such as when-input-is-used, or when-output-changes, or synchronous; such choices have different impact upon the controller hardware and software complexity; the lighter weight choice is synchronous sampling and conversion, where all inputs are sampled at the same point in time and all outputs are converted at another same point in time. This is the actual choice.

2.   Internal: scratch variables, used for intermediate calculations

3.   Persistent: stored on long term memory in order to survive possible controller restarts, such as configurations or data related to the Control algorithm

4.   Communications: data exchanged with remote nodes

Input and Output variables data should have a global scope, to reflect the fact that they are available everywhere in the model; other types of variables could be differently scoped (i.e., global or Step-wide, etc . . .) to protect them from unwanted use. Variables contained in the same scope should be accessible by all model sections contained in the same scope.

4.3. Modeling Data Transformations within Control Algorithms

Modeling data transformations is about how designers express the assignment of variables within the Controller; before discussing how transformations are expressed, it is important to decide when they must occur within the control algorithm.

As the Control function is expressed in terms of states, data transformations can be associated to each and every controller state. A natural way to achieve this is to place data transformations inside each and every state in the Controller, so that at any point in time the Controller will execute the data transformations associated to all the states which are active at that point.

In case more than one state is active at the same time, this can lead to conflicts, in case the same variable is assigned to different values in the different states; the sequence in which active states are executed must be known in advance (the default choice could be the order in which states are defined), so this can be exploited when defining the control algorithm (see below).

In the choice of the modeling language for data transformations, two main alternatives are possible:

1. Imperative: this is the technique used in procedural programming languages (i.e., C, C++, Javascript, etc.) where assignment sequences like temp = a; a = b; b = temp; can be written. The adjective imperative denotes that the sequence of statements is strictly observed; the result of the above three statements would be different should them be executed in a different order. This choice would lead to sequences of statements, easy to write in text but not easily shown in a graphical way; however, the most important drawback of this choice in a state-based language is that the possibility to assign the same variable multiple times along with an assignment sequence allows to embed some "state related" behavior inside the data transformation section.

2. Declarative: data transformations are modeled through data-flow graphs; one graph can be defined for each assigned variable, defining in a functional way how the assigned result is obtained from its input variables; the graph for the expression f = (a + b)*(−e)*d is shown in Figure 2.

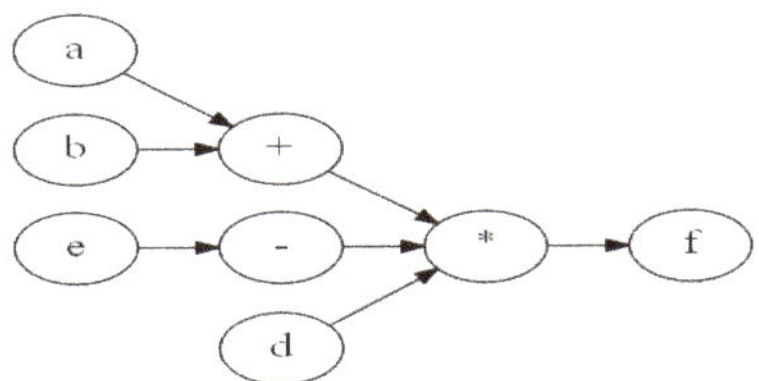

Figure 2. Data-flow graph representing the expression f = (a + b)*(−e)*d.

By definition, data flow graphs do not allow multiple assignments for the same variable and do not impose any order of assignments; for this reason they do not embed any state related information; they represent pure combinatorial behavior, hence this model is orthogonal to the state model; moreover they can be easily represented graphically; from a model-based point of view they can be interpreted as the (physical) network of arithmetic/boolean function blocks combined to produce the desired value for the assigned variable.

In the proposed language data transformation is modeled according to the data flow approach; besides the already mentioned advantage, data flows guarantee the time boundedness of their evaluation, as they are free of loops, and this property has been exploited in the implementation of the language (see below).

Another desirable property of data flows, although not yet exploited in the current version of the toolchain, is the easiness of application of Model Checking techniques to formally derive given properties for the model, such as liveness, reachability, etc, whose verification is mandatory for safety-critical systems.

5. The TaskScript Model-Based DSL

In the following, the proposed language is discussed; this is not, however, the complete reference of the language, which can be found on the web site (see [24]).

5.1. Data Modeling: The Context Diagram

The Context Diagram is the model "starting point"; it carries its general information, mainly made of:

1. Constants
2. Variables at the global scope; the TaskScript variable system is built upon the Boolean and Arithmetic types; variables of both types belong to a specific class, which designates how they are handled by the run-time executive; the classes are: (1) Input and Output, which be explicitly allocated, to map them on the physical connector on the controller board and are converted to and from digital or analog signals respectively; (2) Keep and Temp, the former keeping their values until they are changed by an assignment while the latter are reset at the beginning of every cycle (see below); finally, (3) Memory variables are persisted on long term memory (e.g., flash) (4) Remote allow to communicate with other nodes and (5) System allow to deal at the model level with some variables of the below run-time executive. Variables of any type and class can be scalar or arrays.
3. Model Tasks: Tasks are a structural abstraction level introduced for modularity purposes on top of state machines; each Task defines one sub-state machines, with one initial state and zero or more exit states (the term Task used here, at model level should not be confused with the atomic entity of the multitasking scheduler, which rather is the Step; to help against this confusion, model Tasks will be written with capital T).

Figure 3 shows a Context Diagram example; the shown model has four Boolean Input variables, goCW, goUp, NzeroR, NzeroH and one Analog Input Variable, energy; it has three Boolean Output variables, coil0R, coil1R, coil0H, coil1H and four Tasks; main, comm, StepMot, TM_env, each of which are defined by their Control Flow and set of Data Flows.

Task StepMot, for example, controls one stepper motor has six formal arguments, which will be bound to variables where it is instantiated, within a Control Flow (see below); task TM_env is tagged as "environment", hence used for simulation only. The other boxes of the diagram refer to internal variables (i.e., CWise, CCWise, Up, Down, . . .), and constants (i.e., NightEn, maxR, maxH) globally accessible to all the Tasks of the model.

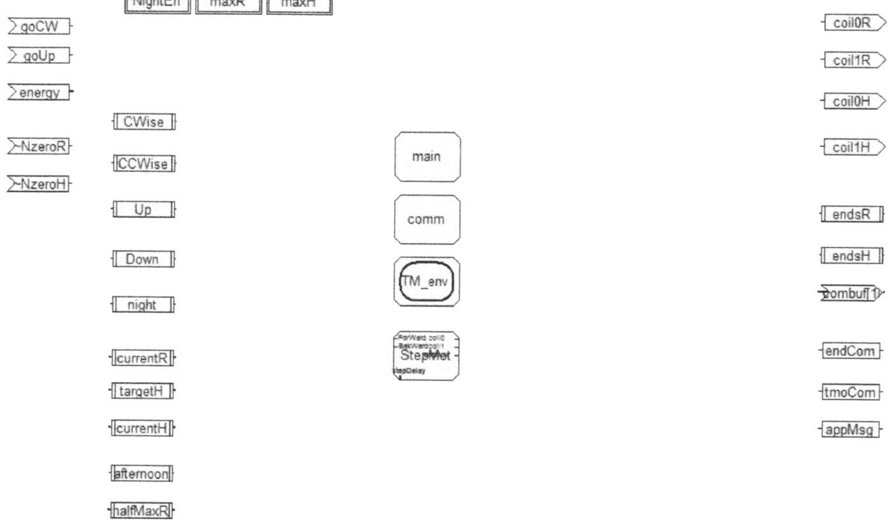

Figure 3. An example of a Context Diagram.

5.2. State Modeling: The Control Flow

The Control Flow language allows modeling the global behavior of the system, expressed as a generalized state flow diagram. CFs allow to visually define sophisticated flows of control with any degree of parallelism through a modular graphic formalism derived from the GRAFCET language. A CF is a graph made of Steps (i.e., places with Boolean marking) connected by guarded transitions: a transition triggers if and only if its upstream Step is active and its guard is true. After a transition has triggered its upstream Step is deactivated, and its downstream Step is activated; forks and joins are supported, which allow one transition to activate and deactivate more Steps at the same time, hence to open and close parallel flows.

Guards can contain timed conditions, of the type "clock C_i has expired": this allows to accurately model time-dependent behaviors inside the CF.

Tasks are activated by special Steps, depicted as rectangles with two horizontal lines inside; Task activation Steps contain nothing else than the Task activation and become notActive when the activated Task is running; When a Task terminates, its starting thread resumes from the Step after the Task Activation; notice that the guard that decides the termination of one Task lays within the Task itself.

Tasks can have private variables of Keep and Temp class; such variables are only available to the Steps belonging to the Task. Finally, Tasks can have arguments, allowing the activation of multiple instances of the same Task, each of them working upon a different set of variables.

Figure 4 shows the Control Flow whose Context Diagram is reported in Figure 3. Task StepMot, controlling a stepper motor, is instantiated 2 times (instance names: Rotation and Lift) with its arguments bound to different variables. In such a model Tasks never exit.

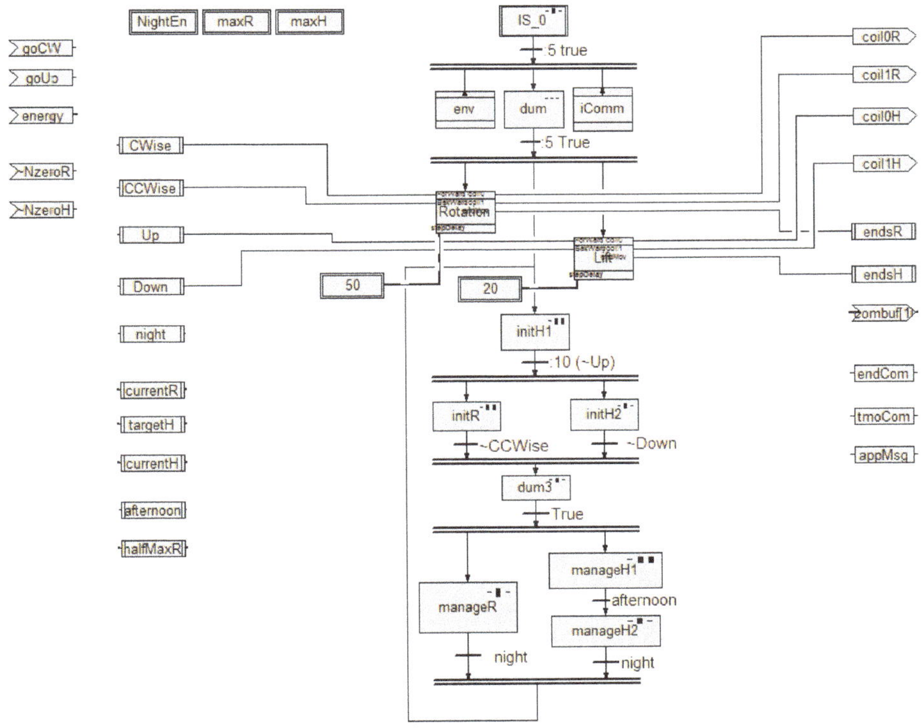

Figure 4. An example of Control Flow.

Double thick bars define forks, i.e., points where one execution thread splits into more simultaneous execution threads, and joins, i.e., points where multiple simultaneous execution threads synchronize into one single thread. The example in Figure 4 has three forks and two joins; for example, the first join, synchronizing initR and initH2 Steps through the guards ~CCWise and ~Down, waits for the two motors to reach their 0 position before going ahead.

Guards like the ones coming out from Step IS_0 (i.e., :5 True) or initH1 (i.e., :10 (~Up)) express time-dependent behavior; they mean "5 ms after the Step is Active" and "10 ms after variable Up has become false" respectively.

Figure 5 shows the Control Flow of the StepMot Task. Notice that ForWard, BakWard, StepDelay, coil0, coil1, endMov are arguments, bound to constants and global variables at instance time (i.e., the CF diagram of Figure 4). Readers with some knowledge about stepper motors would easily recognize a two-phase generator able to spin a motor in two directions, according to the parameters ForWard and BakWard.

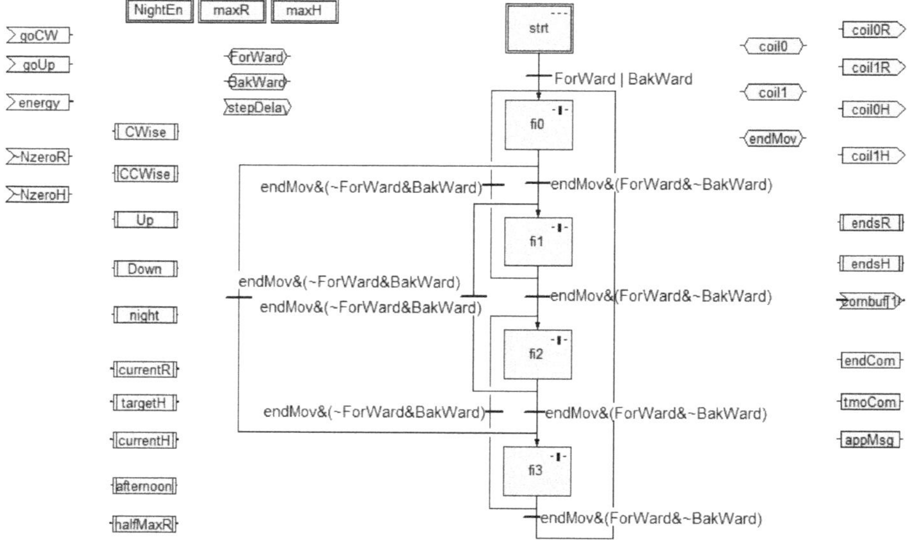

Figure 5. The Control Flow model of the StepMot Task.

Notice that global variables (such as the I/O variables) are shown for all diagrams in a model; conversely, Task-private variables are shown only in diagrams of the owning Task and its Steps.

In TaskScript CF Tasks private data belong to the specific Task instance as it occurs in Object Oriented languages such as ECMAScript/JavaScript, where function private data are stored in multiple copies, one per each function invocation; this is different from what occurs in traditional languages (i.e., C/C++), where function private data are placed on the stack and all the invocations of the function works on the same variables.

As already anticipated, Steps are the atomic entity for the multitasking scheduler; all the model Steps are placed in the execution sequence; the sequence is scanned from the first to the last Step, and the body of the each active Step is executed once.

The scheduler loops forever over the Step sequence; at the beginning and at the end of each scan, some housekeeping tasks are executed (see below).

To enhance the language expressivity, Step marking is "Colored Boolean", i.e., evolves through 4 phases: "notActive", "Active", "onEntry", "onLeave": at the system startup all states but the initial one of the "main" Task are in the "notActive" phase; when activated by an incoming firing transition

Steps go to the "onEntry" phase for exactly one scan and then they go to the "Active" phase; they stay in their "Active" phase until they are deactivated by an outgoing firing transition; in the scan immediately after the deactivation they go to the "onLeave" phase for one scan only, then they go to the "no:Active" phase until they are activated again.

5.3. Data Transformation: The Data Flow

The Data Flow language allows to express the local behavior of each and every Step in the system, expressed in terms of Data Flow Graphs, visualized by intuitive Function Block diagrams; the graphic formalism is straightforward and uses logical operators (i.e., &, |, ^) for Boolean variables and arithmetic operators (i.e., +, −, *, /, %) for Arithmetic ones.

Values held in variables "flow" through the network of primitives and combine with each other until they reach the assigned variable, which stores the current result; every data flow is evaluated once and only once for every cycle. For sake of clarity wires conveying Boolean and Arithmetic flows are drawn with thin and thick strokes, respectively; both Boolean and Arithmetic flows can be coexist in the same Data Flow sheet.

Each Data Flow is associated with one Step in the CF; it is executed if and only if the owning Step is active; more specifically, up to 3 different Data Flows can be provided for each and every Step, one to be evaluated when the Step is in its "onEntry" phase, one to be evaluated when the Step is in its "onLeave" phase, and one to be evaluated when the Step is in its Active phase. In the following, a specific DF associated with one Step will be referred to as the Step Transfer Function.

Table 1 reports the available operators listed with the variable type; the first part of the table contains stateless primitives (i.e., operators), while the second part contains stateful primitives (see below).

Table 1. Primitive Operator groups.

Group	Result Type	Source Type	Other Source Type	Notes
Logic	Bit/Word	Bit/Word		
Arithmetic	Word	Word		
Shift/Rotate	Word	Word		
Compare	Word	Bit		
Choice	Bit/Word	Bit/Word	Bit (selector)	
Composition	Word	Bit		
Extraction	Bit	Word	Word (selector)	
Array Subscripting	Bit/Word	Bit/Word	Word (subscript)	
Extrn. Function	Bit/Word	Bit/Word		
Edge	Bit	Bit		
Delay	Bit	Bit		Stateful
Counter	Bit	Bit	Bit (count)	Stateful
Communicators	Bit	Bit	Word (buf length)	Stateful

Figures 6–8 report a few examples of non-trivial usage of the above primitives within Data Flows (notice the thick wires, when Word variables are connected).

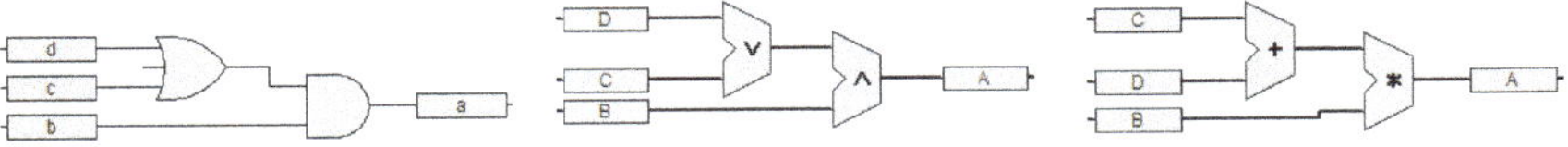

Figure 6. (bit) a = c&(c | d), (Word) A = B&(C | D), A = B*(C + D).

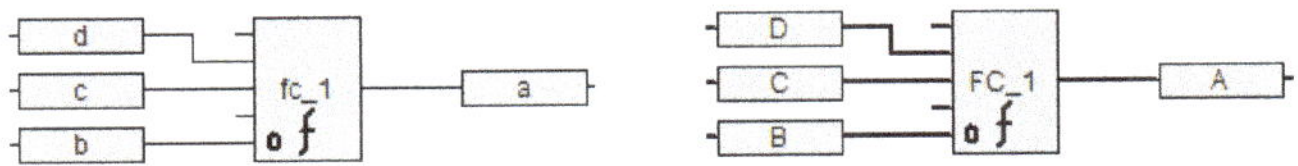

Figure 7. Function invocation: (bit) a = fc_1 (b,false,c,d) (Word) A = FC_1(B,0,C,D).

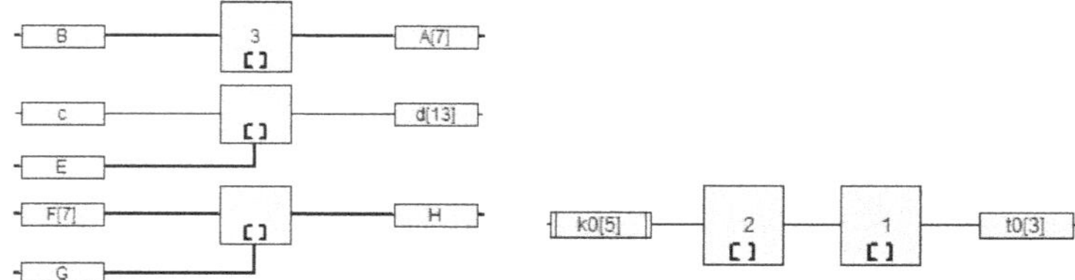

Figure 8. Examples of use of array subscripting: B = A [3]; c = d [E]; F [G] = H; K [2] = t0 [1].

Function invocation allows using, stateless, non-blocking functions, which map the set of input variables to its output.

As already pointed out, the DF graphic formalism is completely free from the order in which the primitives are placed on the sheet; the behavior is only influenced by the type of primitive and their interconnections flowing from the "source" to destination variables.

This aspect confirms the Model-Based nature of the TaskScript language: designers describe their systems as if they were designing physical pieces of hardware, following the very same rules, without need to add any further information; this is the reason why the declarative formalism was preferred over of the more "common" imperative one.

With this choice the DF only describes pure combinatorial behaviors; this is different in imperative languages where the possibility to (a) assign variables (b) use them, (3) change their value (d)use them again, or the presence of control statements such as conditionals and iterators allows modeling a stateful behavior.

The TaskScript Language makes a net separation: Control Flows abstract the sequential and parallel nature of the modeled system while each Data Flow abstracts a fragment of the combinatorial nature of the system.

This distinction adds clarity to models while leaving complete freedom to designers. While systems more sequential in nature could/should be modeled giving emphasis to the CF, the very same system could be modeled with less Steps in the CF, with Steps possibly having larger associated Data Flows; the tradeoff among the above cases is left to designers.

External Functions allow to encapsulate fragments of imperative behavior without breaking the DF rules for cases where a declarative implementation would be inefficient or difficult to read; for example: finding the maximum value in an array, calculate the current value for a PID regulation or co/decode a protocol. Functions can be either Boolean or Arithmetic, according to the type of the returned value. At present functions are implemented outside the language and provided as built-ins by the TaskScript run-time; future extensions of the TaskScript language could support user-defined in-model Functions; functions must have a time-bound execution duration.

5.4. Stateful Primitives

To enhance the language expressivity, some Data Flow primitives have been defined with a private internal state, controlled by a simple micro-state machine; they are Counters, Delays, Edges and Communicators. This is consistent with the MB hypothesis and designers can use them as the "regular" (i.e., stateless) primitives inside Data Flows.

Each Stateful primitive SP has (at least) one Boolean input, named 'enable', and one Boolean output, named 'done'; all Stateful primitives share the following basic behavior:

The primitive remains in its reset micro-state while its 'enable' input is false; when 'enable' goes to true, the primitive starts its operation (i.e., a "delay timer" starts to wait for its preset delay time); when the primitive reaches its final state (i.e., the preset delay has elapsed), it sets its 'done' output to true; should the 'enable' input become false before the primitive reached its final state, the operation aborts and the primitive goes back to its reset state. (in this case the 'done' output never goes to true).

The internal micro-states of these primitives evolve independently but are coordinated with the Step evolution: in particular they stay in their reset micro-state while the owning Step Transfer Function is not active, and start considering their inputs when the owning Step becomes active.

Figure 9 reports the micro state machine used by all the Stateful Primitives.

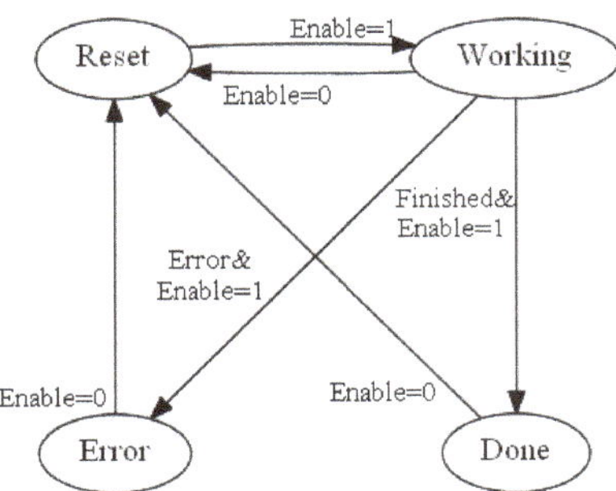

Figure 9. The micro state machine used by all Stateful primitives.

Figure 10 shows two variants of the delay primitive, the simplest case of Stateful Primitive: Ton and Toff; Tenab is the "enable" input of both the delays, while TOnDon and TOffDon are their respective "done" output. Ton output goes true after the preset delay has elapsed since its enable has been constantly true; Toff goes high as soon as its enable goes true and stay true until its preset delay has elapsed or its enable goes false, whichever comes first.

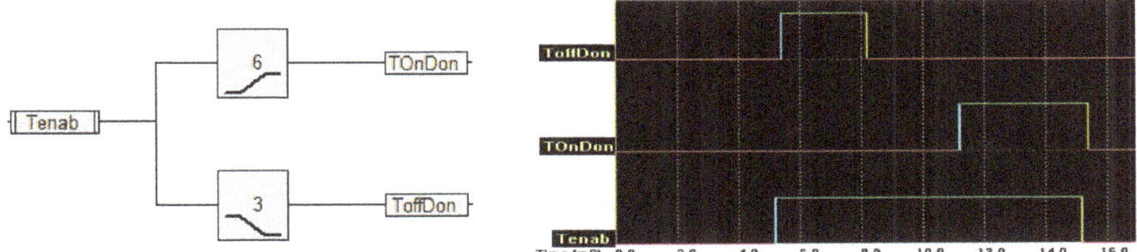

Figure 10. Example of Stateful primitives: delay timers, Ton, and Toff, with their behavior.

The Delay Ton primitive is the main way to model time-dependent behavior in TaskScript; for example, a Step S that has to be active for T units exactly is easily modeled with an output transition from S guarded by a Delay with T preset: as this delay belongs to the Transfer Function of S, it starts operating when S becomes active and makes the transition trigger as soon as the T delay has elapsed.

The Ton Delay behavior implements exactly the concept of clock as defined for Timed Automata [9].

5.5. Comunication Primitives in TaskScript

Communication primitives abstract to the model level communication resources available in Controller boards, so they can be used as black boxes inside models; they are implemented as Data Flow primitives, and build upon the above said concept of Stateful primitive, exposing an "enable"-"done" interface to synchronize their behavior with the rest of the model; in addition a Word array (of Remote type) is used to hold the exchanged payload. The first word of the payload has special meaning: it holds the number of bytes/words in the payload and other special information.

The communications related job is carried out by the kernel, which takes care of all the details that have been hidden at model level; this is particularly true for high level communication protocols, such as the http/tcp/ip, where the whole stack runs in a parallel thread with respect to the modeled system; the only points of contact with the model are the 'enable'-'done' interface and the payload buffer.

When the 'enable' input transits to true the communication port initializes according to the provided configuration parameters and gets ready to interact with its counterpart over the requested channel; interactions can evolve in the following two ways:

1. The communication operation terminates with success, with the buffer sent or received without errors; in this a case the 'done' output is set to true.
2. The communication reaches an error state (i.e., due to channel errors); in this case the 'done' output never goes to true; the rest of the model can detect this error condition testing the flags (i.e., System variables), timeouts and other conditions; in such a case it can reset the primitive, putting its 'enable' input to false and make a new attempt.

The above definition of Communication primitives makes their use simple and seamlessly integrated with the rest of the model; a few general patterns can be defined, usable in a wide range of communication situations; some of them are introduced below.

5.5.1. The Packet Communicator

The Packet Communicator is a model-level abstraction for communication over UART ports, such as RS-232 and RS-485, or SPI ports. Such ports allow one communication at a time; a channel parameter is used to designate which physical port is used, ranging from 0 to the N-1, where N is the number of physical ports contained in the specific board. This primitive operates at the OSI level is 2.

The use of Packet Communicator is exemplified in Figure 11, which shows a simple usage of the Packet communication over an RS-232 serial port, in Slave role it iteratively waits for a master to send packets, and store the received data locally.

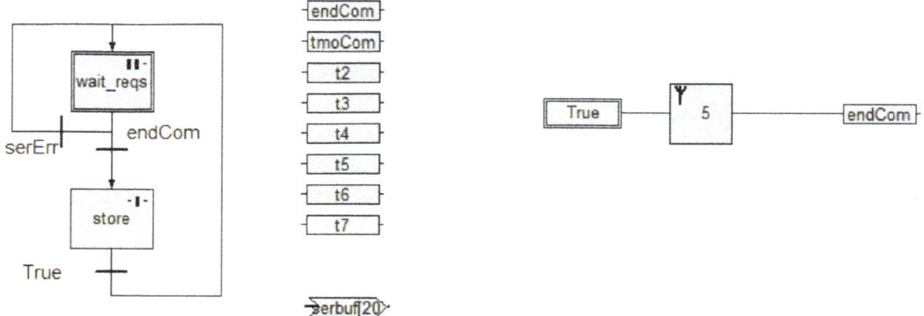

Figure 11. Example of use of the Packet Communication primitive.

The left hand side of the picture shows the CF (sequence of two Steps); the right hand side shows the DF associated with the wait_reqs Step (normal), where the Packet Communication-receive primitive is instantiated and started.

Notice the two Bit variables, serErr and tmoCom and the Word array, serbuf [20] used to store the received payload: this communication Task starts with the wait_rqs state, indefinitely waiting for the master to send its packet; the Step is left in two cases only: either one valid packet is received, in which case the endCom guard becomes true and the thread advances to store, where the received payload is saved, or an error condition is reached, in which casethe serErr guard becomes true, and the wait_reqs Step is re-entered, reinitializing the Communication primitive, and waiting for another packet from the master.

A similar pattern can be easily extended for two way communication, where a response is sent back to the Master, and also to implement the Master role.

5.5.2. The Network Communicator

The Network Communicator is a more sophisticated primitive, which abstracts communications at OSI level 7 occurring on Network ports (i.e., Ethernet), where more than one communication can take place at the same; in such a case there is one port only and the Channel parameter is used to correlate a message with its respective response.

The implemented Network Communicator abstracts the http GET verb; four functions are provided: WsGet, WsResp, WcGet, WcResp; they work in pairs, each one handling one direction of the interaction on the same Channel (i.e., the socket, which is kept open until completion of the interaction).

Server role

1. Web Server Get, WsGet (chan, url): waits until a socket is opened and an http GET query is received; if the path in the query string matches the provided 'url' argument, the primitive activates and fetches the expected number of arguments from the query string (if any). The primitive then parses the arguments and stores them in the payload buffer, then it raises its 'done' output. Notice that the socket is kept open for the reply (transparently from the model designer);
2. Web Server Response, WsResp (chan): can be invoked at a later time, passing the response data in the payload buffer. When activated (setting its 'enable' input to true) it formats the payload in a http response and sends it back to the client on the open socket; then it closes the socket and sets its 'done' output to true.

Client role

1. Web Client Get, WcGet (chan,url): opens a socket to the provided 'url' and sends its http GET query, formatting the data passed in the payload buffer (if any) to the query string; as soon as the packet is sent the primitive sets its 'done' output to true to notify of successful termination, keeping the socket open;
2. Web Client Response, WcResp (chan): waits on the socket left open by WcGet for the server response, then parses the received data into the payload buffer, closes the socket and sets is 'done' output to true.

The Network Communicator exchanges arrays of Arithmetic variables to/from remote Servers or Clients in the http query string. Likewise, variables sent back from the web server are copied back in the arrays, to be used by the rest of the model; variables are returned a fragment of XML, or JSON or JSON-P.

The above set of functions abstracts the definition of Web Service and Web Service Client functions at model level; more than one primitive can be used at the same time; each parallel thread using such primitives can support one interaction at a time, and the number of parallel threads is limited by the number of sockets simultaneously open set by the specific physical board (typically four or eight); possible conflicts on such resources (i.e., one client trying to access a Web Service when it is busy serving another client) are resolved by the tcp/ip own retry policies.

Figure 12 shows an example of a web client accessing the network to upload come data; this flow implements some tolerance over network outages accessing two different servers: Steps WC_Get and WC_Res try the "primary server"; in case no answer comes within the timeout, NetwErr becomes true and the "backup server" is attempted (WC_Gt1 and WC_Rs1); at the end of the communication (either successful or unsuccessful) the active Step is updStore; when variable netGo becomes true (actually after a delay of 0.1s), a new send is started; this variable is set to true in a parallel Step; this is a way to start network uploads at constant times (i.e., 30 seconds) regardless of the real time spent in the upload.

For sake of brevity, the discussion of the Communication primitives is only sketched here; a detailed discussion on the Communication primitives in the TaskScript language can be found in [25].

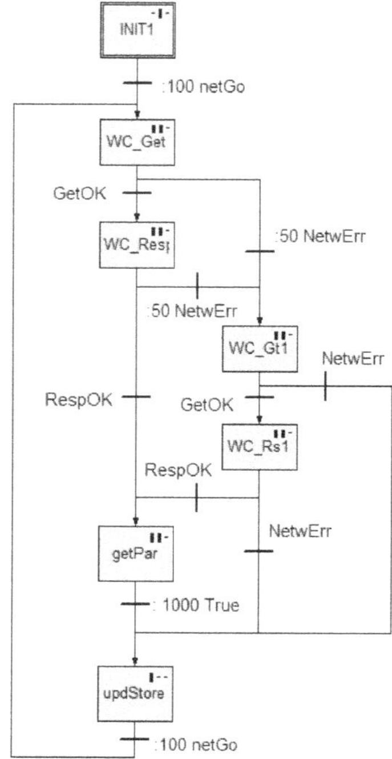

Figure 12. Example of use of the Network Communicator primitives.

5.6. The Model Execution Semantics

At its highest level, a TaskScript model is expressed as a generalized State Machine (modeled through CFs); data transformations take place within Step Transfer Functions, in turn active when the owning Step is active. Steps are evaluated cyclically and of each active Step its respective Transfer Function is executed atomically, once without interruptions.

The execution semantics of TaskScript models can be so summarized: Steps are evaluated cyclically; the state of each and every Step evolves according to the firing of its guards: when a Step becomes active, it stays active until one of its guards becomes true, in which case the current Step becomes inactive and the Step targeted by the guard becomes active; while a Step is active, variables are set according to the Transfer Function of that Step, as it would happen in a piece of hardware built according to that very same schematic.

Users can write their models relying upon the above definition.

From the execution point of view, TaskScript models are efficiently implementable into real controllers: Step Transfer Functions are evaluated according to a round-robin scheme; each Transfer Function is a list of Data Flows whose execution is time-bounded, is executed once per cycle; this can be implemented by a simple but efficient "cooperative multitasking" scheduler.

The variable system is updated by the TaskScript run-time through a number of "housekeeping" activities, coordinated with the model-derived activities.

One cycle, where each and every active Step is evaluated once, is shown in Figure 13; its most important operations are:

1. Evaluation of the current state of each Step
2. Copy of the actual Variables onto the "Last-scan" Variables
3. Refresh of Input Variables $<I_1, \dots, I_p>$ with new samples of the input signals
4. Reset of Variables belonging to the Temp and Output classes;

5. Evaluation of the Transfer Functions of the Steps not in NotActive State.
6. Refresh of the Stateful (i.e., Timer, Counter and Communicator) micro-state machines
7. Update output signals from the $<O_1, \ldots, O_q>$ Output Variables

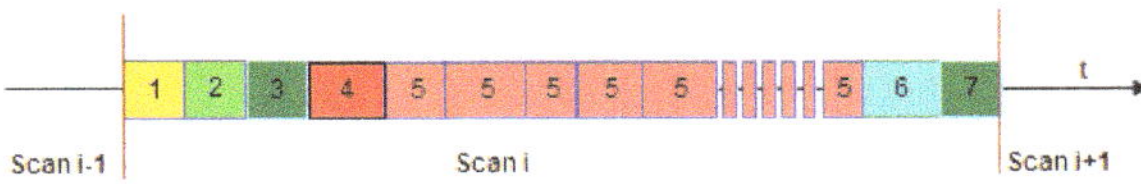

Figure 13. The sequence of scheduler operations within one cycle.

5.6.1. The Step Execution Order in Concurrent Threads

Having more than one Step active at same point in time, as it occurs when multiple threads are active at the same time, can lead to conflicts in the use of shared variables: a conflict arises when one variable is shared, that is it is assigned in one or more Steps and used in others at the same time: in such a case the order in which Steps are evaluated in the scan matters, as it changes the value that is found in that variable at the read time by the various Steps.

Shared variables, on the other hand, are used by threads to exchange data with each other. When one variable is written in one Step and read by another one, the impact is different for the various classes of variables:

1. Input: there is no impact, as they are assigned at the very beginning of each scan and cannot be changed by Transfer Functions within the scan
2. Keep: variables of this class keep their value until reassigned: if Step *Sa* assigns a variable and Step *Sr* reads it, the Step evaluation order is only marginally important, as the variable will always be read by *Sr* before the next assignment by *Sa*, which can occur either later in the same cycle if *Sa* is executed before *Sr*, or earlier in the subsequent cycle if *Sa* in executed after *Sr*.
3. Output, Temp: variables of these classes are reset at the beginning of each scan; if Step *Sa* assigns a variable and Step *Sr* reads it, the Step evaluation order is important: if *Sa* comes before in the cycle, then *Sr* will get the correct value but if *Sr* comes before *Sa* in the cycle, then it will always get 0.

Steps inside TaskScript Models have a constant and deterministic order of execution within the cycle; this order is known at model level and can be changed by designers, who can move Steps before or after in the execution sequence; this must be kept in mind when designing models, as it can simplify model designs.

This order is shown in the model tree view of the IDE, on the left side of the screen (see Figures 21 and 22 below); the Step evaluation order is the same order in which Steps appear in the tree; the model tree window in the IDE allows designers to move Steps up and down the tree.

Figure 14 shows the Steps Evaluation order; in the same chart, Steps are shown along with the indication whether they read, or write a given variable; in particular: an inner blue square indicates that variable is Read (or used within a transition guard), a thick blue border indicates that the variable is Written and a complete blue fill indicates that variable is both Read and Written.

This indication is important when using variables shared among concurrent threads, as it allows to manage conflicts. Section 6.4 below shows through an example how the Step ordering view can be used in managing a variable conflict among concurrent threads.

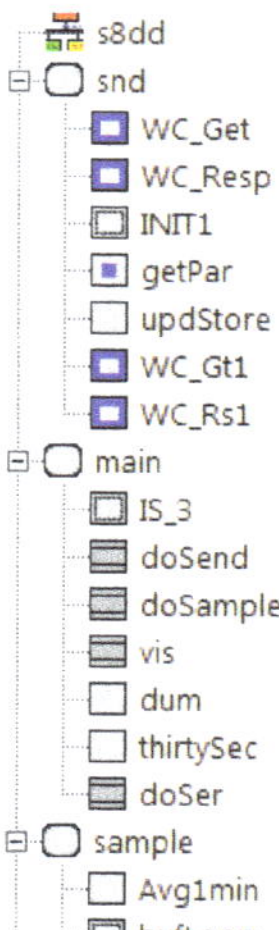

Figure 14. Model tree view, showing the order of evaluation of Steps and variable tracing.

6. CPS Modeling with the TaskScript Design Environment

A graphic IDE, TaskScript Studio, has been implemented, with a toolchain able to support the above presented language along the whole application development process, i.e., drawing, simulation, visualization, compilation into executable code and upload of the executable on the target system. One of its main benefits in the model drawing phase it the continuous consistence check between the Context Diagram, Control Flow(s) and the various Data Flows.

The IDE is in use in production contexts and is constantly evolved to support new CPUs, interface protocols and more. A variety of boards is available, pre-flashed with the TaskScript run-time, in turn implementing the Virtual Machine, ready to interact with the IDE via Ethernet or USB to upload the compiled models. At present TaskScript is available with a range of microcontrollers and PC boards; the most cost-effective and energy-efficient boards are equipped with 8 and 16 bit microcontrollers (i.e., Microchip PIC18 and PIC24 microcontroller families).

6.1. Getting Started

A TaskScript model of a CPS starts with the Context Diagram; such a diagram is built from the perspective of the CPS Logical side: the declared Input and Output variables are the ones exchanged among the Logical an Physical sides, that is $<I_1, \ldots, I_p> <O_1, \ldots, O_q>$ respectively; more variables will be used in case the controller needs to perform other tasks; for example, Remote variables will be declared if the Controller is to exchange data with other nodes.

The Context Diagram will contain at least one Task model, i.e., the "main" Task, the only one started by default at the beginning of the execution; should more Tasks be executed when in the main Task, they will be declared in the CD as well.

After the definition of the Context Diagram, the Data Flow Diagram of the "main" Task is drawn, starting from its Initial Step; Steps are then completed with the respective Transfer Functions (where needed).

TaskScript models can be simulated for validation. Once they are validated they are compiled into executable code ready to be flashed on the chosen target board; if the board has its Input/Output terminals connected to the Physical side, the whole CPS is expected to operate as envisaged in the model.

Although the TaskScript model is defined from the perspective of the CPS Logical side, the Physical side can be included in the model for simulation purposes; in such a case some model

Tasks tagged as "environment" in the model are dedicated to model the Physical side; such Tasks are considered during simulation but will be automatically excluded when the Controller code will be generated.

Adding the model of the Physical side and simulating the whole CPS operation provides more insight than just simulating the Logical side against a predefined set of stimuli and can help finding possible flaws where they are not expected; on the other hand, the level of detail in the modeling of the "environment" Tasks can be chosen according the needed simulation results; in order to save design time, such detail should be kept low, provided that it enables to stimulate the situations where Controller model needs to be verified.

6.2. A Simple Case Study

To provide a first example, putting most of the above discussed concepts together, a simple CPS will be modeled and simulated, ready to be flashed on a controller board: the "Lights" system; it has just one input, actually a pushbutton, and 2 outputs, lamp L1 and lamp L2, with power P1 and P2 = 2*P1.

> One push of the button turns L2 on; next push of the button turns L2 off, and so on; along with this basic behavior the controller has a more sophisticated one, when the button is pushed twice in a short period of time (say 0.3 s): the first double push turns both lights on, then switches only L2 on, then switches only L1 on, and so forth, providing 3 different levels of lighting. At any time when there is some lamp on, pushing the button once turns the lights off.

Figure 15 shows the Context diagram of the Lights system; notice the button and lamps I/O variables, plus a few internal variables (of temp class) and 2 Tasks; the Tuser Task (environment) models the relevant subset of the Physical side of the CPS, that is, the user pushing the button at certain points in time.

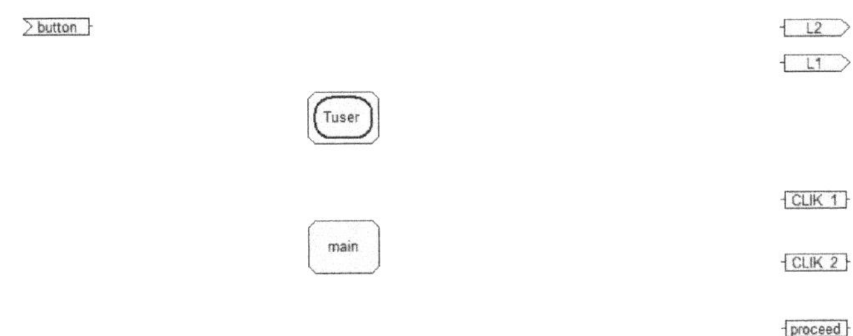

Figure 15. Context Diagram of the Lights example.

Figure 16 shows the Control Flow diagram of the main Task, made of three parallel threads: the leftmost one is dedicated to capture the button pulse sequence; the rightmost one implements the sub-state machine governing the Lamps; finally, the middle one starts the Tuser Task.

The Lamps sub-state machine has four states: all_OFF, all_ON, high_ON, low_ON and two events: CLIK_1 and CLIK_2; notice that from any of its states, CLIK_1 sends to all_OFF, hence switching all lamps off, and CLIK_2 sends to the next state, hence changing the lighting level, in a circular fashion.

The Transfer Functions of the four states of this machine simply turn on the respective light, as shown in Figure 17 for Step all_ON, where both lights are turned ON; the Step high_ON turns on only L2, the Step low_ON turns on only L1 and the Step all_OFF does not turn on any light. This Button-pulse sub-state machine captures the fact that the button has been pressed once or twice within 350 ms; this machine delivers its response setting one of the two variables CLIK_1 or CLIK_2; such variables are of temp class, hence they "last" only one scan.

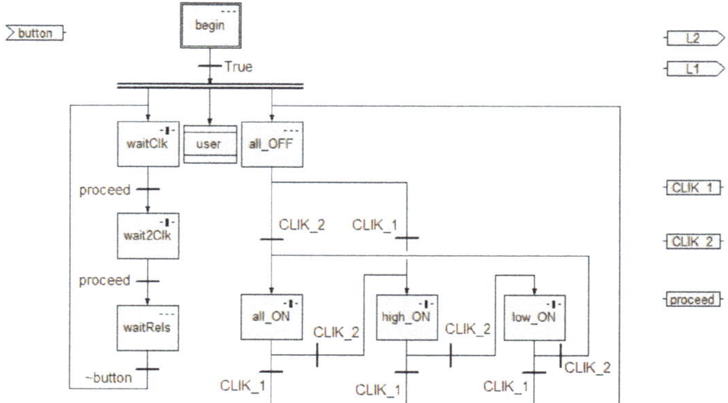

Figure 16. Control Flow of the Main Task.

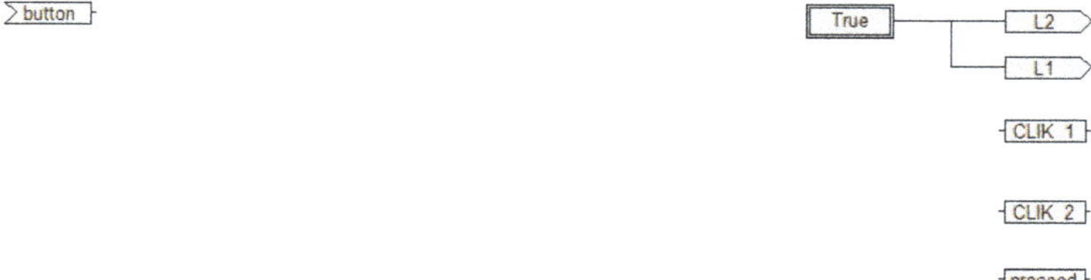

Figure 17. Data Flow of Step all_ON (normal case).

The Button-pulse sub-state machine condition the evolution of the Lights sub-state machine, setting the two variables CLIK_1 and CLIK_2,: when one of them is set, the Lights sub-state machine makes one transition to the respective next state, then, before the next scan, the variable is reset and no further transition occurs until the next pulse comes.

This behavior is simply obtained by the two Transfer Functions shown if Figure 18; when, in Step waitClk, the button is pressed, after a 30 ms debounce delay, the proceed temp variable is set, producing a transition to the wait2Clk Step; here the decision whether the button is pushed one or times is taken: if nothing happens after 350 ms, CLIK_1 is set and proceed is set to leave the Step; in case the button is pressed again within the 350 ms, CLIK_2 is set instead, and the Step is left as well; the Edge operator, placed next to the button input returns true if its input made a low-to-high transition: this is used to wait for the end of the first push before checking for the second push.

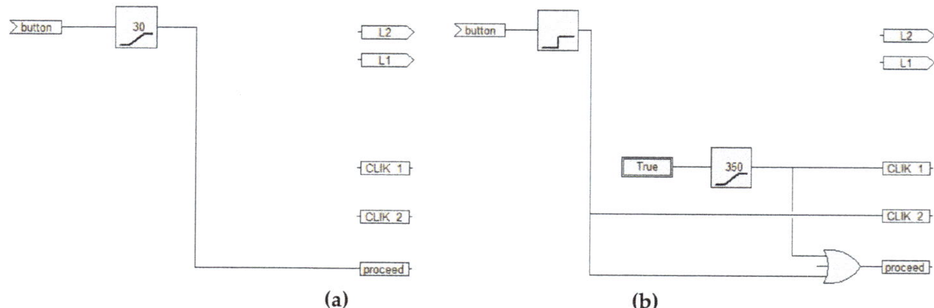

(a) **(b)**

Figure 18. Data Flow of Steps waitClk (**a**) and wait2Clk (**b**).

Finally, the Tuser Task is shown in Figure 19; it is tagged as "environment", hence used for simulation purposes and automatically excluded from the real controller; it contains a single Step (i.e., its initial one) which is never quit, after its start.

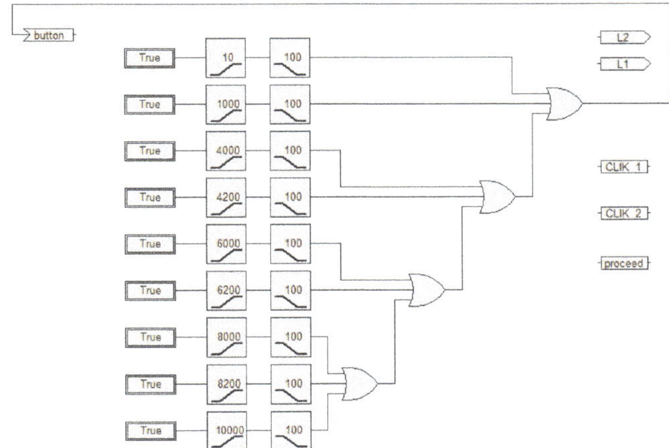

Figure 19. Transfer Function of the user Step, the only Step of the TUser Task.

The Step contains an array of delays, with the purpose of building pulses of 100 ms width at different delays, to stimulate the Controller, as if a user really pushed the button; notice that the "button" input is driven here; this is forbidden in regular Tasks, since the Input variables can only be set by the run-time as per the actual input signals; likewise, "environment" Tasks can read Output variables, which is forbidden as well for regular tasks; this last feature is not used here, as no feedback is needed from the Controller to the environment.

Figure 20 reports the temporal visualization of a simulation trace lasting about 10.5 seconds, including all the user simulated inputs: the bottom trace shows the simulated "button" input, and its pulses can easily be correlated with the pulses generated by the Tuser Task; the two traces above show the lamps; L2 is first switched on and off by the two first single pulses, at t = 0 and t = 1000ms; afterwards, from t = 4s on, all the pulses are provided in pairs, spaced of 200ms each other, hence advancing the lamps sub-state machine.

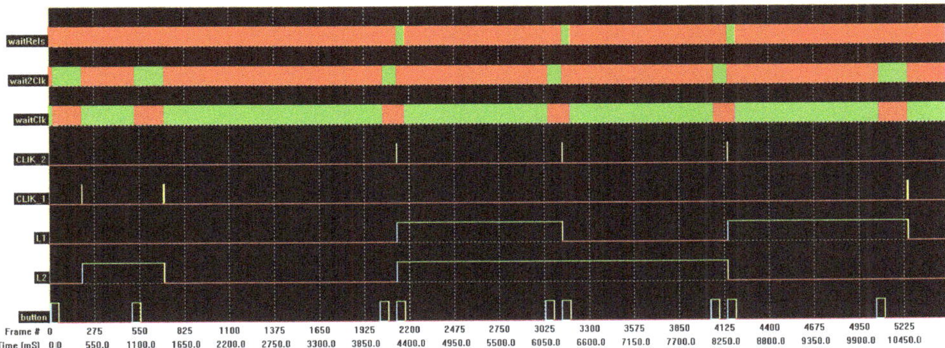

Figure 20. Temporal visualization of simulation results.

The two traces above report CLIK_1 and CLIK_2 internal variables, which "last" 1 only scan. Finally, the upper 3 traces report the state of the Steps waitClk, wait2Clk and waitRels, showing when they are active.

6.3. Structural Visualization

Figures 21 and 22 show another perspective of the same Lights system, derived from the same simulation trace as before: one specific point in time is selected and the state of variables and Steps is shown "in place", i.e., using the very model diagrams, for that point.

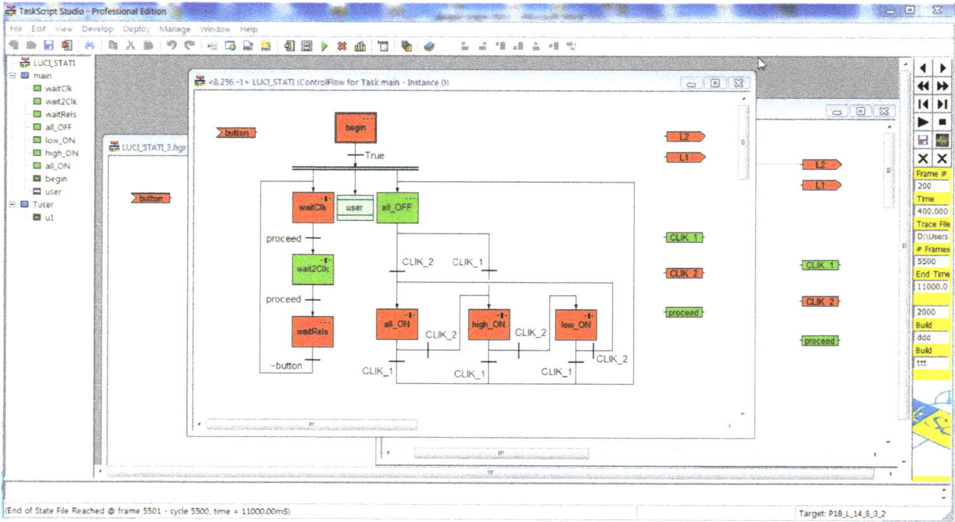

Figure 21. Structural visualization of simulation results.

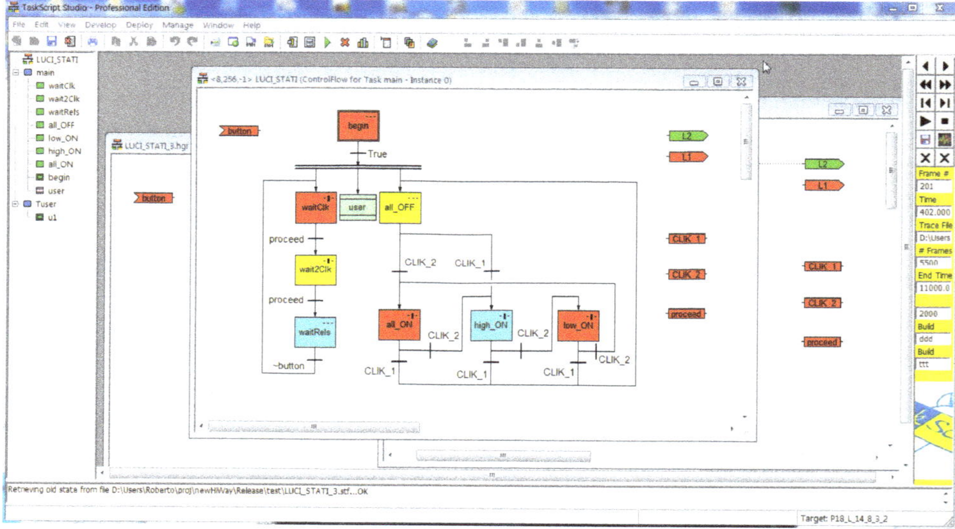

Figure 22. Structural visualization of simulation results–next frame.

Steps colored in Green are in their Active State, Steps colored in Red are in "notActive" state; steps colored in Cyan and Yellow are in their "onEntry" and "onLeave" state, respectively. The figures report two adjacent scans, around t = 4402 ms, just after the second pulse in a sequence has been recognized (see the CLIK_2 and proceed variables, colored in Green); the second frame shows the transitions occurring due to those variables: Step wait2Clk goes from "Active" to "onLeave", Step waitRels goes from "notActive" to" onEntry", Step all_Off goes "onLeave" and Step all_ON goes "onEntry", hence turning L2 on.

Structural visualization is of great help in deeply understanding how the model works; this kind of visualization is meaningful only for full Model-Based languages, where the executable code is generated automatically from the model and no changes are done.

Structural visualization can also be performed on real Controllers, executing on physical boards; in such a case, specific "snapshots" can be taken upon specific model-level conditions, i.e., when specific events occurs (i.e., one Step or variable changes its state to a given value); this feature can be useful to diagnose faults on the Physical side when the Controller is deployed in the field, in the maintenance phase, without any physical intervention: the Controller plays the role of a possibly remote first diagnosis tool.

6.4. The Added Value of Knowing and Changing the Step Evaluation Order

The previously discussed model has been built robustly against any possible execution order of Steps, in particular of the ones belonging to the Button-pulse sub-state machine with respect to the Lights sub-state machine (see Figure 15 and the following explanation); in other words is follows the constraints-abiding approach.

Variables CLICK_1 and CLICK_2 are shared among the two concurrent threads, so a conflict arises; in particular, the Button-pulse sub-state machine writes the two variables according to the detected button behavior and the Lights sub-state machine uses them as transition guards and immediately resets them, to prevent other transitions.

The two sides of Figure 23 show two possible orderings among Steps for the very same project; in the two charts, Steps are shown along with the indication whether they read, or write a given variable (in this case, CLICK_1 is considered, but the chart would be identical for CLICK_2).

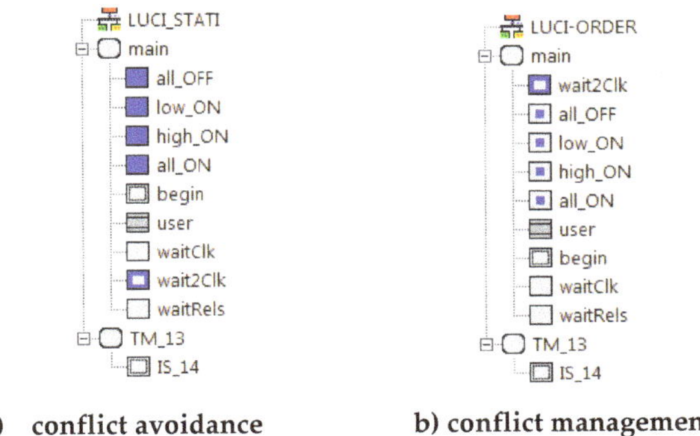

a) conflict avoidance   **b) conflict management**

Figure 23. Different orderings for Step evaluation (with variable tracing shown).

The two versions follow the two different approaches in the management of the variable conflict: the version on the left follows the constraints-abiding approach, and works with any Step ordering: CLICK_1 and CLICK_2 variables are of the Keep type (their values are retained until reassigned) and each and every Step of the Light state machine resets both variables when entered, to prevent further transitions.

On the other hand, the version on the right follows the constraints-enforcing approach: Step wait2Clk is executed first, so the variable is set to the correct value before it is considered by the other Steps; CLICK_1 and CLICK_2 variables are of the Temp type, so their values are automatically reset at the end of every cycle and there is no need for the Steps of the Light machine to reset them.

This shows that the precise knowledge of the Step evaluation order and the possibility to change it where needed allows to adopt the constraints-enforcing approach instead of the constraints-abiding one, which is the only available choice when there is no notion of Step evaluation ordering.

Choosing the constraints-enforcing approach, in the general case, leads to simpler models.

7. Results and Discussion

Three real life, production grade CPSs will be used in the evaluation of the TaskScript Language, the associated IDE and the set of physical Controller boards; they are summarized below, just to provide enough context to understand the reported results; they are: "TransfertMon", "EnergyGateway" and "LithiumBMS".

TransfertMon

This is an observer which monitors a number of boolean signals gathered from a transfer plant and sends the gathered values over the internet to a webservice every 15 seconds, using an Ethernet port available on the Controller, attempting send to two servers for network unavailability tolerance; the difficulty here is that some signals are pulses; for two of them their frequency must be evaluated with an accuracy of 1 ms (the controller CPU returns the time the signal was low or high in the last period); for others it must be evaluated whether they have been constantly 0, 1, or they have changed their value in the observation period.

EnergyGateway

This is a gateway able to sample up to eight boolean variables generated by S0 pulse energy meters, which send one pulse per every measured wh (watt*hour); for each channel the gateway must: store the actual input value, count the number of received pulses and store it permanently, evaluate the moving averages of pulse frequency (which provides the instantaneous measure of the power consumption on that channel) and of the pulse duty cycle; periodically, i.e., every 30 seconds, it must send the current values over the internet to a webservice, using an Ethernet port available on the Controller, attempting send to 2 servers, for network unavailability tolerance.

LithyumBMS

This device controls its physical periphery besides than observing it; its periphery is a series of 16 lithium batteries powered by solar cells, and an inverter; its job is to keep the voltage of each and every cell in the array within safety limits, activating individual resistive shunts to discharge the elements whose voltage exceeds a safety value; aside tasks are switch the inverter on and off according to the state of charge of the battery array, measure the battery in/out current and update a remote web service about the current readings. In order to interface the battery array, an analog front end IC is interfaced via an SPI port available on the Controller board and abstracted to the model level through a Packet Communicator; the inverter is interfaced through a RS-485 serial port available on the Controller abstracted through another Packet Communicator; the serial payloads are co-decoded as MODBUS RTU by means of a protocol library (functions) available within the TaskScript IDE.

The four CPSs: Lights, described in detail along the paper, TransfertMon, EnergyGateway and LithiumBMS are compared in the below Tables 2–4; the last three systems have been implemented all the way from model design to the physical, production grade level using the most appropriate among the available TaskScript physical boards and integrating the real Physical side to the implemented controller. The level of knowledge of the developers on the TaskScript Modeling Language and Studio IDE was fair to good; the level of domain knowledge of the developers was good.

Table 2. Model-level characteristics of the different CPSs.

Model features	Lights	TransfertMon	EnergyGateway	LithyumBMS
# Tasks	2	4	5	6
# Steps	8	23	44	42
# parallel threads	2	5	11	5
# discrete I/O	3	23	8	12 dig + 2 anlg
# serial ports	0	0	0	1 SPI + 1 UART
# of network ports	0	1	1	1
# of protocols	0	1	1	2
# internal variables	3	22	40	25

Table 3. Controller Performance for the different CPSs.

Controller Performance	Lights-8bit	Lights-16bit	TransfertMon	EnergyGateway	LithyumBMS
Controller board	P18L_14_8_3_2	P24_S_8_4_3_3	P18_E_SP_FTC_485 16_5_2_2	P18_E_SP_FTC_485 12_13_9_2	P18_E_SP_FT_485 8_8_10_3
Board size	~15cm^2	~15cm^2	~76cm^2	~55cm^2	~76cm^2
Controller CPU	PIC18F25K22	TS-P24EP-GP202	PIC18F87J60	PIC18F87J60	PIC18F87J60
CPU Clock	64MHz	140MHz	~42MHz	~42MHz	~42MHz
CPU bits	8	16	8	8	8
Controller Power	<50mW	<0.2W	<2W	<1W	<2W
Avg. scan time (ms)	<0.1ms	<15us	<0.2ms	<0.2ms	<0.25ms

Table 4. Development effort in person-hours for the different CPSs.

Development Effort (pers-hrs)	Lights	TransfertMon	EnergyGateway	LithyumBMS
modeling of logic	1	4	8	16
modeling of communication	-	4	4	8
domain knowledge	-	4	4	16
simulation-debug	1	4	8	16
physical setup	1	8	4	16

Discussion of Results

Apart from the "Lights" tutorial, the three reported cases are real world examples, two of them deployed in the field, one in testing.

They have been compiled through the TaskScript Studio IDE v1.10 and uploaded to TaskScript controller boards equipped with an 8 bit microcontroller, the Microchip®PIC18F87J60, delivering 10.66 MIPS, pre-flashed with the TaskScript v1.10 run-time (integrated with the Microchip tcp/ip protocol stack), with Ethernet and RS-485 interfaces. Figure 24 shows one of such boards.

Once deployed on the target board via TFTP (integrated in the Studio IDE) the model starts its execution interacting with both the Physical side and the remote web Server. The sampling rate of the signals from the field was in the order of 10 to 50 milliseconds; the web server refresh time ranged from 3 to 60 seconds, independently from the actual network delay time.

The "Lights" example was tested on two physical boards, one equipped with 8 bits CPU and another one equipped with a 16 bits CPU to show that the cycle time scales down by almost a decade; this is due to the higher clock frequency but also to the most powerful the instruction set which allowed to perform multibyte arithmetics in one instruction, that led to a simplified implementation of the Virtual Machine on the more powerful CPU.

The TaskScript run-time provides very stable timings; while the evaluation of the Steps has variable duration, according to the size of the respective Transfer Function, the Delay primitives allow to accurately model the time related behaviors, used within the model to provide signal sampling rates, refresh delays on the Network, and so on.

Figure 24. The P18_E_SP_FTC_485 12_13_9_2 board.

The development effort turned out to be low and consequently the turnaround time was short; the main reasons seem to be:

1- The ability to use parallelism in a very natural way: every parallel task was modeled as if it was the only task in the system, regardless of the other ones; however, some Steps needed to have a global view of the system;

2- the ability to simulate the whole CPS system, that is Logical and Physical sides, interacting together as a whole, to gain insight upon critical parts in the model during the design phase, hence saving debugging with the real Physical part, which would cost more effort; notice that the Physical side is not modeled/simulated accurately: this is out of scope of the TaskScript language; rather, an abstraction of the input-output causality relationships the Phisical side can be modeled inside Environment tasks at the purpose of "closing the loop" with the Logical side, in a way that they can interact together and provide evidence that the Logical side is working properly. This is the purpose of the TUser task in the example.

3- the ability to visualize the simulation results "in place" on the designed model, through the Structural visualization, and to navigate it over time (i.e., the frames) and space (i.e., the various sheets defined in the model);

4- the ability to browse the state of the real Controller while it is connected to its real Physical side (this action interrupts the real operation of the Controller for a few milliseconds, spent packing the state and sending it to the IDE via TFTP); this feature can also be used to test the Physical side of systems already deployed in the field.

5- The ability to change the model and have the new version compiled and loaded again on the controller in a few seconds; this allows designers to experiment with model variants in order to find the most suitable.

Overall, the choice of a heavyweight Model-Based approach turned out very effective to reach the initial goals: the defined DSL proved to be simple enough to be understood by domain experts, not only by computer scientists, and this was key to have them involved in the model design, not only in the requirements phase, and this in turn led to better quality designs.

The toolchain implemented in the IDE produces highly efficient implementations; as shown in the evaluation examples, relatively short cycle times are achieved for medium sized models running on low-end 8 bit CPUs, and almost a tenfold gain is achieved for 16 bits CPUs.

8. Conclusions

A model-based environment and its DSL language, specifically conceived to easily model CPS control functions has been presented. It belongs to the class of synchronous languages and is hard real-time capable.

Its typical use case is the implementation of network-interconnected CPS, acting as "smart" terminals of cloud-based applications, able to observe and control some physical periphery, with complexities ranging from simple smart sensors to controllers of large physical subsystems with remote interactions.

Such applications are typically constrained in terms of space, CPU power, memory, energy and more. The discussed technology is available in a number of "sizes" to suit design needs: a number of virtual machines are available starting with 8 bits CPUs; this variety allows to instantiate the very same model on different virtual machines without effort, in order to find the best point in the {speed, consumed energy, (arithmetic) precision, type and number of I/Os, cost} hyperspace among the available ones.

Besides boards, the virtual machine can be provided to custom designs thanks to the "controller-on-a-chip" approach, which allows the technology to be used in the context of ad-hoc hardware, designed according to specific design needs.

Regarding the DSL, the choice to bring the Step evaluation sequence at model level allows designers to accurately model the causality relationships among the various concurrent activities avoiding the need of higher level, less efficient synchronization mechanisms, such as channels, queues or semaphores.

The concept of stateful primitive, introduced at the Data Flow level, allows a variety of communications interfaces and protocols to be made available at the model level and this help developing network-aware models with little effort.

The easiness of the DSL also easily allows the modeling of an abstraction of the CPS's physical side within which the controller will interact, and simulates the whole controller and periphery in order to get a deeper insight of the model. This environmental simulation is more effective than the usual simulation, where external stimuli are provided, in finding out potential design flaws.

The reported results show that the proposed IDE allows designers to implement production-grade controllers within a few man hours of effort.

Funding: This research received no external funding.

Conflicts of Interest: The author declares no conflict of interest.

References

1. Griffor, E.R.; Greer, C.; Wollman, D.A.; Burns, M.J. Framework for Cyber-Physical Systems: Volume 1, Overview. *NIST SP* **2017**, *1*. [CrossRef]
2. Sangiovanni-Vincentelli, A.; Damm, W.; Passerone, R. Taming Dr. Frankenstein: Contract-Based Design for Cyber-Physical Systems. *Eur. J. Control* **2012**, *18*, 217–238. [CrossRef]
3. Harel, D. StateCharts: A Visual Formalism for Complex Systems. *Sci. Comput. Progr.* **1987**, *8*, 231–274. [CrossRef]
4. Berry, G.; Gonthier, G. The Esterel Synchronous Programming Language: Design, Semantics, Implementation. *Sci. Comput. Program.* **1992**, *19*, 87–152. [CrossRef]
5. Halbwachs, N.; Caspi, P.; Raymond, P.; Pilaud, D. The synchronous data flow programming language LUSTRE. *Proc. IEEE* **1991**, *79*, 1305–1320. [CrossRef]
6. Maraninchi, F.; Rémond, Y. Argos: an automaton-based synchronous language. *Comput. Lang.* **2001**, *27*, 61–92. [CrossRef]
7. Halbwachs, N. *Synchronous Programming of Reactive Systems*; Springer Science & Business Media: Berlin/Heidelberg, Germany, 1993; ISBN 978-0-7923-9311-5.

8. Ptolemaeus, C. System Design, Modeling, and Simulation Using Ptolemy II, Ptolemy.org. Available online: http://ptolemy.org/books/Systems (accessed on 12 February 2019).
9. Alur, R.; Dill, D.L. A theory of timed automata. *Theor. Comput. Sci.* **1994**, *126*, 183–235. [CrossRef]
10. Passerone, R.; Ben Hafaiedh, R.; Benveniste, A.; Cancila, D.; Cuccuru, A.; Damm, W.; Ferrari, A.; Gerard, S.; Graf, S.; Josko, B.; et al. Meta-models in Europe: Languages, Tools and Applications. *IEEE Des. Test Compt.* **2009**, *26*, 38–53. [CrossRef]
11. Object Management Group (OMG). Unified Modeling Language (UML) Specification (version 2.5.1). Available online: http://www.omg.org/spec/UML/ (accessed on 12 February 2019).
12. Object Management Group (OMG). System Modeling Language Specification v1.5. Technical Report. Available online: http://www.sysml.org (accessed on 12 February 2019).
13. A UML Profile for MARTE (version 1.1)," in OMG Document Number: formal/2011-06-02, Jun 2011. Available online: https://www.ibm.com/us-en/marketplace/rational-rhapsody (accessed on 12 February 2019).
14. Object Management Group. Available online: https://www.omg.org/omgmarte/ (accessed on 12 February 2019).
15. Gérard, S.; Dumoulin, C.; Tessier, P.; Selic, B. 19 Papyrus: A UML2 Tool for Domain-Specific Language Modeling. In *Model-Based Engineering of Embedded Real-Time Systems. MBEERTS 2007. Lecture Notes in Computer Science*; Giese, H., Karsai, G., Lee, E., Rumpe, B., Schätz, B., Eds.; Springer: Berlin/Heidelberg, Germany, 2007; pp. 361–368. ISBN 978-3-642-16277-0.
16. Object Management Group. About the Semantics of a Foundational Subset for Executable UML Models Specification. Version 1.3. Available online: https://www.omg.org/spec/FUML/1.3 (accessed on 12 February 2019).
17. Object Management Group. About the Precise Semantics of UML Composite Structures Specification. Version 1.1. Available online: https://www.omg.org/spec/PSCS/About-PSCS/ (accessed on 12 February 2019).
18. Yakindu Web Site. Available online: https://www.itemis.com/en/yakindu/state-machine/ (accessed on 12 February 2019).
19. Simulink Web Site. Available online: https://mathworks.com/products/stateflow.html (accessed on 12 February 2019).
20. SCADE Web Site. Available online: https://www.ansys.com/products/embedded-software/ansys-scade-suite (accessed on 12 February 2019).
21. AutoFOCUS-3 Web Site. Available online: https://af3.fortiss.org/ (accessed on 12 February 2019).
22. LabView Web Site. Available online: http://www.ni.com/academic/students/learn-labview/ (accessed on 12 February 2019).
23. David, R.; Alla, H. Du Grafcet aux réseaux de Petri. In *Traité des Nouvelles Technologies/Automatique*; Hermès Science Publications: Paris, France, 1992; ISBN 978-2-8660-1325-7.
24. TaskScript Web Site. Available online: http://www.taskscript.com (accessed on 12 February 2019).
25. Manione, R. Abstraction of communication resources in Model-Based design frameworks for CPS. In Proceedings of the SPIE Microelectronics Conference, Barcelona, Spain, 21 May 2015. [CrossRef]

Article

A Lazy Bailout Approach for Dual-Criticality Systems on Uniprocessor Platforms

Saverio Iacovelli [†] and Raimund Kirner [*,†]

The School of Engineering and Computer Science, University of Hertfordshire, College Lane,
Hatfield AL10 9AB, UK; savio.iacovelli@yahoo.com
*   Correspondence: r.kirner@herts.ac.uk; Tel.: +44-1707-28-4125
†   These authors contributed equally to this work.

Received: 20 October 2018; Accepted: 28 January 2019; Published: 1 February 2019

Abstract: A challenge in the design of cyber-physical systems is to integrate the scheduling of tasks of different criticality, while still providing service guarantees for the higher critical tasks in the case of resource-shortages caused by faults. While standard real-time scheduling is agnostic to the criticality of tasks, the scheduling of tasks with different criticalities is called mixed-criticality scheduling. In this paper, we present the *Lazy Bailout Protocol* (LBP), a mixed-criticality scheduling method where low-criticality jobs overrunning their time budget cannot threaten the timeliness of high-criticality jobs while at the same time the method tries to complete as many low-criticality jobs as possible. The key principle of LBP is instead of immediately abandoning low-criticality jobs when a high-criticality job overruns its optimistic WCET estimate, to put them in a low-priority queue for later execution. To compare mixed-criticality scheduling methods, we introduce a formal quality criterion for mixed-criticality scheduling, which, above all else, compares schedulability of high-criticality jobs and only afterwards the schedulability of low-criticality jobs. Based on this criterion, we prove that LBP behaves better than the original *Bailout Protocol* (BP). We show that LBP can be further improved by slack time exploitation and by gain time collection at runtime, resulting in LBPSG. We also show that these improvements of LBP perform better than the analogous improvements based on BP.

Keywords: real-time systems; Fixed-Priority Preemptive Scheduling (FPPS); mixed-criticality systems; cyber-physical systems

1. Introduction

Cyber-physical systems (CPS) typically require the integration of services of different criticality. At the same time, it is important that tasks of lower criticality have limited leverage to influence the schedulability of tasks with higher criticality in the case of resource shortages. Traditional real-time scheduling protocols, such as *rate-monotonic scheduling* (RMS) or *earliest deadline first* (EDF) [1], give priority to jobs with the most strict timing requirements. This approach works well as long as it can be assured that enough resources are available to schedule all tasks. However, in cases where availability of enough resources cannot be guaranteed, traditional real-time scheduling methods miss the flexibility to prioritise the resources to tasks of higher criticality.

Research on mixed-criticality scheduling protocols [2,3] has been started to overcome this limitation. The basic idea of mixed-criticality scheduling protocols is that, as long as enough resources are available, the scheduling priorities are defined by a real-time scheduling protocol. In the case of a resource shortage, e.g., a job overrunning its estimated worst-case execution time (WCET) [4], the tasks' criticalities are used as the primary criterion to allocate resources. A task's criticality can be derived from different aspects. One possibility is to express the relative importance or relative utility of different services in a system as their criticality [5]. Another possibility is to express the relative level of assurance, for example, dictated by different development standards for safety critical or relevant systems, such as DO-178C [6] in the avionics domain, ISO26262 [7] in the automotive domain, or IEC 61508 [8] in the automation domain as different levels of criticality. However, the meaning of criticality is still sometimes subject of discussion, with Esper et al. assuming different execution modes [9] not originally described by Vestal [2]. In this paper, we do not mandate a specific procedure for defining criticality levels, as this is an orthogonal issue to the mixed-criticality scheduling discussed in this paper.

To apply mixed-criticality scheduling, at least two levels of criticality have to be defined, typically labelled as LO (low-criticality) and HI (high-criticality). A common approach is to assume for LO tasks only the knowledge of easy to derive optimistic WCET estimates while for HI jobs also a higher level of assurance based on safe upper WCET bounds is assumed. The active research challenge is to find ways to effectively combine the resource prioritisation based on criticalities with the scheduling priorities based on real-time constraints.

Recent mixed-criticality scheduling approaches are the *Bailout Protocol* (BP) by Bate et al. [10] and its extension that exploits the system slack time, named *Bailout Protocol-Slack* (BPS). The authors afterwards presented further extended versions of the BP, aiming at a higher utilisation of LO jobs. Such extensions use a dynamic approach to deploy gain times in order to reduce the duration and number of times the system switches to high-criticality execution mode and are denoted as *Bailout Protocol with Gain Time* (BPG) and *Bailout Protocol-Slack and Gain Time* (BPSG) [11].

This article contains the following contributions:

1. *Lazy Bailout Protocol* (LBP), which is a mixed-criticality scheduling protocol inspired by the *Bailout Protocol* (BP) from Bate et al. [10,11], is introduced. Compared with BP and its derivatives, LBP does not abandon jobs immediately but rather keeps them for potential later execution during idle periods of the processor.
2. A formal criterion to compare different mixed-criticality scheduling protocols with priority given to high-criticality jobs is defined.
3. LBP is combined with the complimentary techniques used in BPG, BPS and BPSG, resulting in LBPG, LBPS and LBPSG, respectively, proving that LBP and its derivatives perform better than their corresponding BP-based protocols according to such a formal criterion.
4. The comparison and evaluation of BP, LBP and their derivatives protocols in a hard real-time setting is presented.

Section 2 presents an overview of the state of the art in mixed-criticality scheduling. A precise presentation of the scheduling problem is presented in Section 3. We present a new mixed-critcality approach named LBP in Section 4 that does not suddenly abandon LO task instances during resource shortages. In Section 5, we derive formal properties of LBP and its derivatives. Section 6 provides an experimental evaluation of the performance of the LBP-based approaches compared with other methods. Section 7 concludes the article.

2. Related Work

Most of the works about mixed-criticality systems that have been published by the real-time scheduling research community is based on a model proposed by Vestal [2]. The system model consists of a set of periodic tasks that perform functions having different criticalities and requiring different levels of assurance. Each task may have a set of alternative worst-case execution times, with each assured to a different level of confidence. The more confidence one needs in a task execution time bound, the larger and more conservative that bound tends to become in practice. The final aim was to guarantee that safety-critical task instances do not miss their deadlines.

Crespo et al. reviewed the challenges of applying mixed-criticality in control systems and studied the possibility of using virtualisation as basis for building mixed-criticality partitioned software architectures [12]. Their work reviews the challenges connected to systems with virtual partitions having different criticality that are executed in an independent way. Such systems are based on a hypervisor that provides temporal, spatial and fault isolation among partitions that contain components that have to be guaranteed at different assurance levels and on hierarchical scheduling as strategy to process jobs.

Ernst and Natale provided an explanation about the meaning of criticality and a review about the mixed-criticality model in current real-time research [13]. They highlighted how functional safety standards usually provide the basis to design industrial mixed-criticality systems. In fact, all industrial safety standards classify different levels of concern, called *Safety Integrity Levels* (SIL) in IEC 61508, *Automotive Safety Integrity Levels* (ASIL) in ISO 26262 or *Design Assurance Level* (DAL) in DO-178C. Each level involves a certain likelihood to perform successfully the required functions under certain conditions and within a stated period. In such standards, the definition of criticality levels is usually obtained as a result of a *Failure Modes, Effect and Criticality Analysis* (FMECA) process. However, these standards focus on the safety targets while engineers normally focus on metrics such as cost, performance and power consumption that are often in conflict with safety requirements. Such contrast grows with the autonomous driving and with the integration challenges derived from cyber-physical systems and Internet of Things.

Burns and Davis published a survey of research on mixed-criticality systems [14].The review contains an historical introduction of the topic and the challenges faced in developing better mixed-criticality on both single- and multi-processor systems. The key question emerging from their work is how to reconcile the conflicting requirements of partitioning for safety and sharing for efficient resource usage. Lastly, the review contains criticisms and limits of the current mixed-criticality approaches.

In 2011, Baruah et al. extended Vestal's model by proposing a refinement named *Adaptive Mixed-Criticality* (AMC) protocol together with related mixed-criticality response-time analysis techniques [15]. Such mixed-criticality schedulability tests have been recently extended for task sets containing tasks with arbitrary deadlines [16]. In 2013, Fleming et al. extended the response time bound techniques and the AMC protocol to work with multiple criticality levels [17]. The AMC protocol assumes two execution modes, a low-criticality mode (indicated as LO) and a high-criticality mode (indicated as HI). Once the system goes into the high-criticality mode, all LO task instances are abandoned and the system remains in that mode. However, to move mixed-criticality research into industrial practice, it is important to implement protocols whose runtime behaviour is acceptable for system engineers. Abandoning all LO tasks in high-criticality mode is not an acceptable behaviour and the system should return to the low-criticality mode, where all functionalities are provided, as soon as conditions are appropriate. Therefore, a simple but necessary extension to AMC is to allow a switch back to the starting mode when the system experiences an idle instant.

However, going back to the low-criticality starting mode only in case of idle instants leads to a high amount of LO tasks interrupted or abandoned and this is still not satisfactory. Different complementary ways of guaranteeing a higher level of service for LO tasks have been proposed, e.g., extending their periods and/or deadlines such as in the elastic task model [18] or reducing their execution times by switching to a simpler version of the software [19].

The *Priority May Change* (PMC) strategy has been proposed to better manage the overload situations in which higher priority LO tasks could preempt lower priority HI tasks [20]. The AMC algorithm assigns a single priority to each task by considering together both low- and high-criticality modes, whereas PMC computes priorities in two steps. Firstly, priorities are assigned to tasks according to some predefined policy such as deadline monotonic [21] and such priorities are used while the system is in low-criticality execution mode. Once the system switches to high-criticality mode, HI task priorities are re-assigned according to a priority ordering policy that is optimal for tasks with release jitter [22]. However, PMC does not dominate the standard AMC but it has performances similar to it.

In 2014, Fleming and Baruah proposed a scheme in which system designers can assign to lower critical functionalities a utility that is used to decide in which order their instances have to be suspended during an overload occurrence [23]. Such method allows the system designer to control how non-critical functionalities degrade after the most critical ones overrun their optimistic time threshold. The utility value is assigned as an ordinal scale [24] to provide a predefined order in which LO task instances are abandoned, with least important task instances being abandoned first. The authors adapted the Audsley priority assignment technique [25] to assign lower priority to lower utility LO tasks. Such protocol allows increasing performances for LO tasks and processing them for a significantly increased amount of time.

Somehow, the former methods considered thus far allow for LO task invocations to execute after a criticality mode change but they are mainly best effort and do not have a predefined minimum threshold guaranteed for lower critical tasks. Since most hard real-time systems could miss some deadlines provided that it happens in a known and predictable way, the *Adaptive Mixed Criticality with Weakly-Hard constraints* (AMC-WH) was introduced in 2015 [26] and represents an extension of AMC [15] that integrates the notion of weakly-hard constraints. The definition of weakly-hard real-time system was given in 2001 [27] to indicate systems in which hard real-time tasks are permitted to miss some deadlines as long as the number of missed deadlines is strictly bounded. The AMC-WH is a scheduling policy that allows a number of consecutive instances per LO task to be skipped during the high-criticality execution mode. This reduces the overall system load, frees more resources for highly critical tasks and provides a degraded but guaranteed minimum quality of service for LO tasks upon a criticality mode change. The number of skips permitted and the number of subsequent deadlines that must be met could be a requirement deduced either from the design of a control algorithm [28] or from physical properties of the system. Even if AMC-WH allows scheduling more LO task instances if compared with previous policies, it does not provide a fast recovery from the high-criticality execution mode since it is still necessary to wait for idle instants to go back to the starting mode. This leads to unnecessary abandonments of LO instances.

Such problem was considered with the *Bailout Protocol* (BP) [10]. The BP still represents an AMC refinement and hence exhibits both low- and high-criticality execution modes. The low-criticality mode is named *Normal* mode while the high-criticality execution mode is represented by both the *Bailout* and *Recovery* modes. Similar to AMC, the system starts its execution in the LO criticality mode, *Normal* mode, and whenever a HI job exceeds its optimistic WCET, then it switches to the Bailout mode. The protocol aims to restore the normal execution mode as soon as possible to minimize the number of LO instances that miss their deadlines or are not executed at all. LO tasks are still abandoned during the high-criticality execution but they contribute to make the switch back to the starting execution mode faster by means of a *Bailout Fund* (BF). In fact, if BF becomes not strictly positive during the Bailout mode, the system enters the Recovery mode to allow the lowest priority pending HI job to complete its execution before going back to the *Normal* mode without waiting for an idle instant. Once the system is back to the starting mode, all lower critical functionalities start again to be processed with their full timely behaviour. The strength of this protocol is that to provide an effective control mechanism to go back to the low-criticality mode, where all jobs can start and being processed. However, the main weakness of BP is that to immediately drop low-critical instances during the high-criticality modes. Because of this, the percentage of LO jobs that miss their deadlines is still high.

An orthogonal approach to improve the overall service for LO tasks is based on a method introduced by Santy et al. [29]. This approach was subsequently refined by Burns and Baruah [19]. They scaled up the optimistic WCETs of HI tasks using sensitivity analysis until the system is schedulable. If used together with the BP the resulting protocol is named *Bailout Protocol-Slack* (BPS). More recently, Bate et al. further refined BP with a second complementary technique [11]. Such approach consists of an update of the optimistic time budget made at runtime by collecting the so-called gain time, i.e., the spare CPU time not required at runtime by task instances. These techniques allow reducing both the number of times and the duration the system executes in high-criticality modes. By combining the online gain time collection with BP, the authors introduced two new scheduling protocols that are named *Bailout Protocol-Gain Time* (BPG) and *Bailout Protocol-Slack and Gain Time* (BPSG).

3. System Model

In the following, the system model used for task sets is described. A dual-criticality system, which consists of multiple tasks, where each task has a criticality $l \in \{LO, HI\}$ with HI being of higher criticality than LO, is assumed. As discussed in Section 1, the criticality of a task can be derived by different means but no specific interpretation of criticality is assumed, as this is orthogonal to the scheduling method presented in this paper. Furthermore, it is assumed that the processor is the only resource that is shared among tasks, and that the overheads due to the scheduling operations and context switches can be bounded by a constant included within each task worst-case execution times.

We consider a set of independent and sporadic tasks τ that has to be processed on uniprocessor systems and that consists of two sub sets:

$$\tau = \tau_{LO} \cup \tau_{HI} \tag{1}$$

with

$$\tau_{LO} = \{\tau_i \in \tau \mid l_i = LO\} \tag{2}$$

$$\tau_{HI} = \{\tau_i \in \tau \mid l_i = HI\} \tag{3}$$

where τ_{HI} is the subset of tasks that are highly critical and τ_{LO} is the subset of tasks that are not highly critical within the system.

The tasks represent scheduling units that the system has to perform. An individual task $\tau_i \in \tau$ is represented by the following tuple:

$$\tau = \langle T, D, C_{LO}, C_{HI}, L \rangle$$

where T is the period, D is the relative deadline, C_{LO} and C_{HI} are, respectively, the optimistic and the pessimistic worst case execution times and $L \in \{LO, HI\}$ refers to the criticality. In this paper, for simplicity, we assume implicit deadlines, i.e., tasks with deadlines equal to their periods: $D = T$.

A job is an instance of a task at runtime, i.e., a job represents the actual object processed by the scheduler and inherits almost all properties from the task that generates it plus the arrival time A as below:

$$j_i = \langle A, D, C_{LO}, C_{HI}, L \rangle$$

The LO tasks and, as a consequence, their relative jobs do not have a known safe WCET bounds C_{HI}, since safe worst-case execution times are rather costly to obtain and thus provided only for HI tasks. Once it finishes its execution, each job j_i has got a computation time $et(j_i)$ that can vary for each specific job of the same task. The job set produced by an individual task τ_i is indicated by $J(\tau_i)$ while $J(\tau)$ is the job set produced by all tasks belonging to the task set τ. Therefore, τ represents the set of activities that have to be performed by the system while J represents the set of concrete process instances that have to be considered by the scheduler.

The jobs produced via the task set are scheduled according to the standard fixed-priority fully pre-emptive real-time scheduling. However, the traditional fixed-priority scheduling is unaware of criticality of task instances and scheduling decisions are only made according to priority that indicates the job timing requirements. Therefore, it is also used a protocol that considers the task's criticality to meet the mixed-criticality requirements. The following assumptions are made about the task set and the underlying real-time scheduler, i.e., fixed priority fully pre-emptive scheduling:

Assumption 1. *All HI and LO jobs together are schedulable with the underlying real-time scheduling method with respect to their C_{LO}.*

Assumption 2. *All HI jobs alone are schedulable with the underlying real-time scheduling method with respect to their C_{HI}. Since C_{HI} is a safe WCET bound, i.e., et $\leq C_{HI}$, this assumption also implies that the HI jobs alone are schedulable with respect to their actual execution time.*

Assumption 3. *All HI jobs are schedulable with respect to their C_{HI}, while also assuming the execution of all LO jobs arrived in Normal mode with respect to their C_{LO}.*

Note that Assumption 3 is required so that, while LO tasks are allowed to run within their C_{LO}, it is still ensured that all HI tasks are still schedulable within their C_{HI}. Assumption 3 is based on jobs rather than tasks as it covers the moment in time when a HI task overruns its C_{LO}. In addition, note that Assumption 2 is just a weaker case of Assumption 3, without the LO tasks considered.

4. The Lazy Bailout Protocol

The standard BP is an adaptive protocol to schedule mixed-criticality job sets. The strength of BP is providing an effective and fast control mechanism to go back to the low-criticality mode, where all jobs can start and being processed. However, the main weakness of BP is immediately abandoning LO jobs in case of resource shortage, which leads to a high percentage of jobs that miss their deadline.

The *Lazy Bailout Protocol* (LBP) is built upon BP and inherits from it the following three execution modes that work as specified below:

1. *Normal*: It is the starting system execution mode. It corresponds to a low-criticality mode where all jobs within the system are supposed to be processed correctly according to the C_{LO} threshold.
2. *Bailout*: It is the emergency mode that is entered whenever a HI job overruns its C_{LO}.
3. *Recovery*: It is the emergency mode that is entered to allow the last pending lowest priority HI job to complete before going back to *Normal* mode.

Figure 1 shows the components of LBP. The LBP filter is responsible for changing the execution modes. The system has two ready queues for jobs: the *high-priority queue* represents the BP ready jobs queue while the *low-priority queue* keeps the LO jobs that have been released during Bailout or Recovery modes or that have exceeded their C_{LO}. Note that LO jobs inserted into the low-priority queue run until their deadline and *only* when the high-priority queue is idle. Thus, such jobs cannot lead to any deadlines being missed. There are two job monitors to check, respectively, LO and HI jobs that overrun their C_{LO}. *ET-MonLO* signals to the real-time scheduler the LO jobs that have to be inserted within the low-priority queue while *ET-MonHI* communicates to the LBP filter when a HI job exceeds its optimistic WCET to switch the execution mode to Bailout.

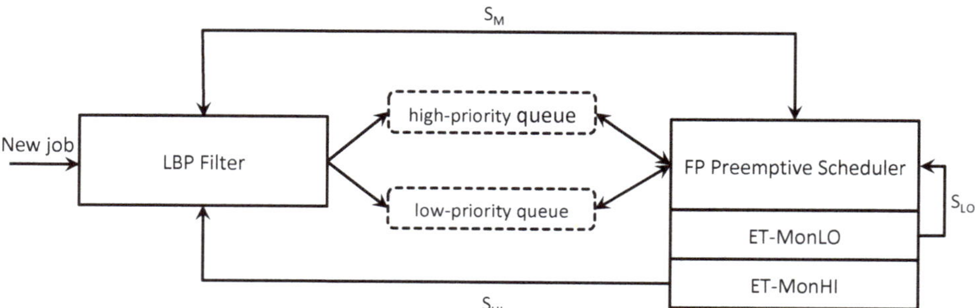

Figure 1. LBP architecture.

Similar to BP, LBP inherits from AMC the system execution behaviour, i.e., the system starts in a low-criticality execution mode and whenever a HI job exceeds its optimistic WCET, the system switches to a high-criticality execution mode where any LO job execution is prevented. Finally, the system goes back the starting execution mode in case of idle instant. Furthermore, LBP inherits from BP the control mechanism that is in charge of the execution mode changes that permits a fast recovery from the Bailout/Recovery modes back to the Normal mode. Such mechanism is based not only on the detection of a idle instant but also on the value of a decision variable named *Bailout Fund* (BF). It is worth noting that LBP, as well as AMC and BP, implement dispatching policies that are independent and separated from the priority assignment used. Moreover, since a fixed-priority scheduler is used, no priority change is allowed. Figure 2 shows how the execution mode changes in the scheduling protocol. It contains the events that trigger the switch to a different execution mode together with the related update of the BF value. The system starts in Normal mode and then, if any HI job overruns its C_{LO}, the BF variable is initialised and there is a change to Bailout mode. Once the system is in this mode, the BF variable is updated with the earlier completion of jobs, the release of new LO jobs or the HI jobs overrunning their C_{LO}. If an idle instant occurs, then Normal mode is entered, whereas, if the BF becomes zero, then Recovery mode is entered. After the pending lowest priority HI job completes its execution in Recovery mode, the system goes back to Normal mode.

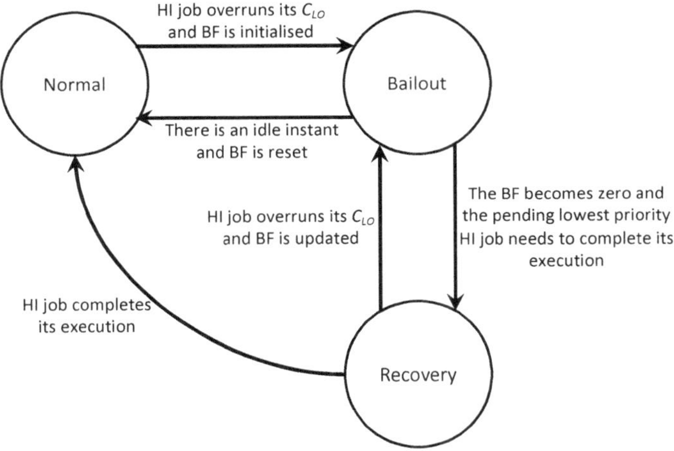

Figure 2. Execution mode changes in LBP.

The difference between LBP and BP is that LO jobs released in Bailout and Recovery modes or exceeding their C_{LO} are inserted into the low-priority queue instead of being abandoned. This allows increasing the amount of LO jobs scheduled without interfering with the execution of jobs in the high-priority queue. In fact, LO jobs in the low-priority queue run until their deadline when the high-priority queue is idle. The essential difference in scheduling behaviour is that, in those cases where BP would be idle, LBP might have some tasks preserved in the low-priority queue that can now successfully be executed.

Whenever a job in the low-priority queue misses its deadline, it is removed. LO jobs released in Normal mode can continue to execute in both Bailout and Recovery modes and they could even overrun their deadlines as long as they do not exceed their C_{LO}. Below, is a detailed description of LBP in each of its execution modes:

Normal mode:

- While all HI jobs execute for no more than their C_{LO} values, the system remains in this mode.
- If any HI job overruns its C_{LO} without signalling completion, then the system switches into the *Bailout* mode and the BF is initialised to $BF = C_{HI} - C_{LO}$.
- LO jobs that overrun their C_{LO} are interrupted and inserted into the low-priority queue.

- LO jobs that have been inserted into the low-priority queue are executed during idle instants. If they do not complete within their deadlines, then they are removed from the low-priority queue.

Bailout mode:

- If any HI job executes for its C_{LO} without signalling completion, then the bailout fund is updated by its maximum extra time budget: $BF = BF + (C_{HI} - C_{LO})$.
- If any HI job completes with an execution time e, with $e \leq C_{LO}$, then its time left is donated to the bailout fund: $BF = BF - (C_{LO} - e)$.
- LO jobs released in *Normal* mode that complete with an execution time of e, with $e \leq C_{LO}$, donate their time left to the bailout fund: $BF = BF - (C_{LO} - e)$.
- If any HI job that already exceeded its C_{LO} completes with an execution time of e, with $C_{LO} < e \leq C_{HI}$, then it donates its extra time left, reducing the bailout fund: $BF = BF - (C_{HI} - e)$.
- LO jobs released in *Bailout* mode are not started but inserted in the low-priority queue to be executed during idle instants in *Normal* mode. Furthermore, when the scheduler would otherwise have dispatched such a job, the job's budget of C_{LO} is donated to the bailout fund: $BF = BF - C_{LO}$.
- If the BF becomes zero, then the lowest priority HI job that did not complete its execution (let this job be j_k) is recorded and the *Recovery* mode is entered.
- If an idle instant occurs, then a transition is made to *Normal* mode, and BF is reset to zero.

Recovery mode:

- LO jobs released in this mode are not started but inserted within the low-priority queue to be executed during idle instants in *Normal* mode.
- If any HI job executes for its C_{LO} value without signalling completion, then the system switches back to *Bailout* mode and BF is initialised: $BF = C_{HI} - C_{LO}$.
- When the job j_k noted at the point when *Recovery* mode was last entered completes, then the system switches to *Normal* mode.

Figure 3 shows how the same task set is scheduled according to the On the one hand, the standard BP abandons all the LO jobs released during the HI criticality execution modes while the lazy approach allows to recover and schedule more LO jobs. In particular, in Figure 3b, jobs B_1 and B_4 are released, respectively, at times $t = 6$ and $t = 24$ and they have the highest priority. Such jobs are inserted in the low-priority queue to be removed, respectively, at times $t = 12$ and $t = 30$ when they miss their deadlines and the next instance of the same task arrives. Furthermore, the LO jobs B_2 and B_5 released respectively at times $t = 12$ and $t = 30$ are executed afterwards in Normal mode since there are idle instants to exploit before their deadlines. Such example highlights how LO jobs that are delayed, instead of being abandoned, are executed during idle instants in Normal mode to not influence the real-time behaviour of jobs in the high-priority queue. Overall, compared with LBP, the standard BP results in a decrease of the system utilisation because, whenever there is interference among HI and LO jobs released in Bailout or Recovery modes, LO jobs are simply abandoned. On the other hand, LBP increases the processor utilisation by exploiting the system idle time and, by doing this, it improves the overall service provided to LO tasks, which is achieved by increasing the number of LO jobs that are processed.

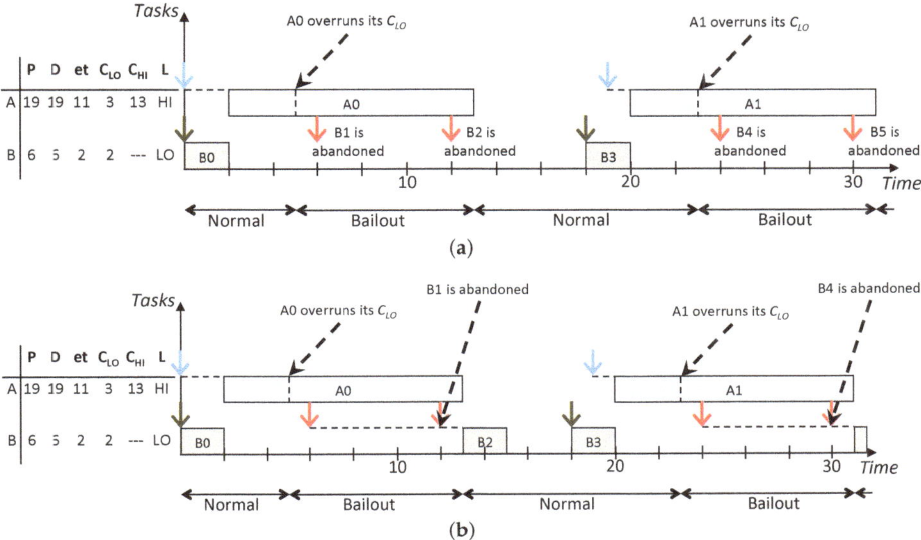

Figure 3. Comparison between BP and LBP: LBP schedules more LO jobs than BP: (**a**) BP abandons all LO jobs released in Bailout mode; and (**b**) LBP rescues the LO jobs *B2* and *B5*, while *B1* and *B4* are abandoned after they miss their deadlines.

5. Proofs

In this section, we formalise a criterion to compare different mixed-criticality systems. Below are definitions and predicates used to prove the theorems afterwards.

STS, τ, JS:

STS is a set of task sets τ. τ is an individual scheduling problem consisting of tasks. JS is a set of jobs created at runtime by scheduling a task set.

METHOD:

This is the scheduling method applied, which can be BP, LBP or some of their derivatives resulting from the integration with the offline sensitivity analysis or with the online gain time collection.

HI(τ), LO(τ): $\tau \rightarrow \tau$:

HI(τ) is a subset of τ containing only tasks of high-criticality. LO(τ) is a subset of τ containing only tasks of low-criticality.

Scheduled(mtd, τ): METHOD $\times \tau \rightarrow$ JS:

The list of jobs generated from a task set τ, which are *successfully* scheduled by method *mtd*, i.e., jobs which completed within their deadline.

ScheduledHI(mtd, τ): METHOD $\times \tau \rightarrow$ JS:

This includes only those jobs from Scheduled(mtd, τ) which are derived from tasks with high-criticality.

ScheduledLO(mtd, τ): METHOD $\times \tau \rightarrow$ JS:

This includes only those jobs from Scheduled(mtd, τ) which are derived from tasks with low-criticality.

Failed(mtd, τ): METHOD $\times \tau \rightarrow$ JS:

The list of jobs generated from a task set τ, which are *not successfully* scheduled by method *mtd*, i e., jobs which were not completed within their deadline.

Abandoned(mtd, τ): METHOD $\times \tau \rightarrow$ JS:

This predicate returns the list of jobs generated from a task set τ that were never forwarded by the mixed-criticality scheduling method *mtd* to its underlying real-time scheduler. This is a special case of failed jobs:

$$Abandoned(mtd, \tau) \subseteq Failed(mtd, \tau)$$

Abandoned jobs are also different from *dropped jobs*, which are jobs that failed after having started execution with the underlying real-time scheduler.

`LORated`(mtd, τ): METHOD $\times \tau \to$ JS:

This predicate returns the list of LO jobs, which were re-queued from the default high-priority queue to the low-priority queue (LBP-based methods only).

`IsBetterMCS`(mtd_1, mtd_2, τ): METHOD$^2 \times \tau \to$ BOOL:

This predicate tests whether a scheduling method mtd_1 is better than method mtd_2 for a task set τ with respect to mixed-criticality scheduling, which is formally defined as:

$$IsBetterMCS(mtd_1, mtd_2, \tau) \Rightarrow$$
$$= \begin{cases} \text{TRUE} & \text{if } (ScheduledHI(mtd_1, \tau) \supset ScheduledHI(mtd_2, \tau)) \vee \\ & ((ScheduledHI(mtd_1, \tau) == ScheduledHI(mtd_2, \tau)) \wedge \\ & (ScheduledLO(mtd_1, \tau) \supset ScheduledLO(mtd_2, \tau))) \\ \text{FALSE} & \text{otherwise} \end{cases}$$

This tests whether mtd_1 has a better performance than mtd_2 for HI jobs, or equal performance for HI jobs but better performance for LO jobs.

`IsBetterMCS`(mtd_1, mtd_2): METHOD$^2 \to$ BOOL:

This predicate tests whether a scheduling method mtd_1 is better than method mtd_2 for all task sets with respect to mixed-criticality scheduling, which is formally defined as:

$$IsBetterMCS(mtd_1, mtd_2) \Rightarrow$$
$$\exists ts \in \text{STS.} \quad IsBetterMCS(mtd_1, mtd_2, \tau) \wedge$$
$$\nexists \tau \in \text{STS.} \quad IsBetterMCS(mtd_2, mtd_1, \tau)$$

It is worth noting that $IsBetterMCS(mtd_1, mtd_2, \tau)$ and $IsBetterMCS(mtd_1, mtd_2)$ are transitive:

$$IsBetterMCS(m_A, m_B) \wedge IsBetterMCS(m_B, m_C) \Rightarrow IsBetterMCS(m_A, m_C)$$

5.1. Comparison between BP and LBP

Theorem 1. *LBP has the same success rate of HI tasks than BP, which can be formally written as:*

$$\forall \tau \in \text{STS. } ScheduledHI(BP, \tau) == ScheduledHI(LBP, \tau)$$

Proof. (Theorem 1) BP and LBP behave the same way regarding the handling of HI jobs:

1. If a HI job is overrunning its C_{LO}, it is granted an execution budget until C_{HI}.
2. If a HI job does not finish within C_{HI} or within its deadline, then it is dropped.

The only difference between BP and LBP lies in the handling of LO jobs, where LBP puts them in a lower priority scheduling queue instead of abandoning them immediately when released in Bailout/Recovery modes or dropping them after the overrun of their C_{LO} as BP does. The content of the low-priority scheduling queue of LBP cannot influence the scheduling of the default scheduling queue. Thus, for any task set τ, it follows that $ScheduledHI(BP, \tau) == ScheduledHI(LBP, \tau)$. $\square$

Theorem 2. *LBP can have a better success rate of LO tasks than BP, but never worse, which can be formally written as:*

$$\forall\, \tau \in \text{STS. } ScheduledLO(BP, \tau) \subseteq ScheduledLO(LBP, \tau)$$
$$\exists\, \tau \in \text{STS. } ScheduledLO(BP, \tau) \subset ScheduledLO(LBP, \tau)$$

Proof. (Theorem 2) The only difference between BP and LBP lies in the handling of LO jobs, where LBP puts them in a low-priority scheduling queue instead of abandoning them immediately or dropping them after the overrun of their C_{LO}. Hence, we have:

$$\forall\, \tau \in \text{STS. } Abandoned(BP, \tau) \subseteq LORated(LBP, \tau)$$

Thus, to prove Theorem 2, we only have to show that, among the tasks that BP abandons, there is at least one task that with LBP instead gets put into the low-priority queue and finally successfully scheduled:

$$\exists\, \tau \in \text{STS. } Abandoned(BP, \tau) \cap LORated(LBP, \tau) \cap Scheduled(LBP, \tau) \neq \varnothing$$

which means it is sufficient for the proof to show by example that it is possible to have task sets where LO-rated jobs can be scheduled within an idle time of the default scheduling queue. To do so, we use the following task set consisting of one HI task A and one LO task B:

Task	P	D	et	C_{LO}	C_{HI}	L
A	15	15	5	3	10	HI
B	4	4	2	2	-	LO

Task A is assumed to have an execution time $et = 5$, which always causes an overrun of the optimistic WCET estimate. The first time, job A_0 exceeds its C_{LO} at $t = 7$ and the system switches into Bailout mode. The LO job B_2 is released at time $t = 8.0$ during Bailout mode. Hence, the bailout fund BF is decreased by a quantity equal to the C_{LO} of B_2. However, BF remains positive. After A_0 is completed, the system experiences an idle instant and this causes a switch back to Normal mode.

As shown in Figure 4a, BP immediately abandons job B_2 at its arrival time during the high-criticality execution mode. In contrast, as shown in Figure 4b, LBP moves such job into the low-priority queue at its arrival and executes it when the default queue becomes idle. Thus, this example demonstrates the existence of a task set τ such that

$$\exists\, \tau \in \text{STS. } Abandoned(BP, \tau) \cap LORated(LBP, \tau) \cap Scheduled(LBP, \tau) \neq \varnothing$$

which demonstrates that there are cases where LBP can successfully schedule more jobs than BP. Here, we have to remind the fact that those jobs which are successfully scheduled by BP are processed exactly the same way by BP and LBP, meaning that, whenever BP successfully schedules a job, so does LBP. This property together with the existence of above example completes the proof of:

$$\forall\, \tau \in \text{STS. } ScheduledLO(BP, \tau) \subseteq ScheduledLO(LBP, \tau)$$
$$\exists\, \tau \in \text{STS. } ScheduledLO(BP, \tau) \subset ScheduledLO(LBP, \tau)$$

$\square$

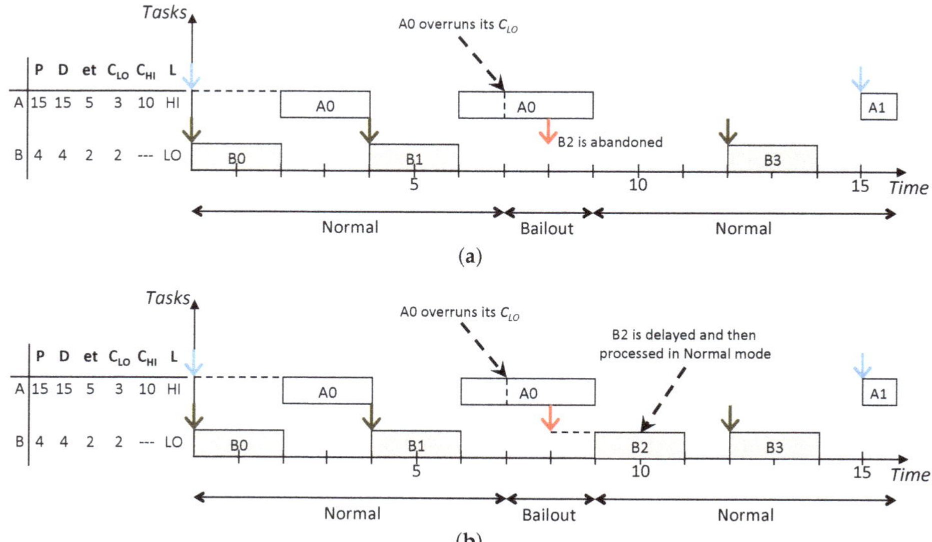

Figure 4. (Proof of Theorem 2) Example in which LBP successfully executes LO jobs that are abandoned by BP: (**a**) BP abandons LO jobs that are not released in Normal mode; and (**b**) LBP provides a delayed execution for job B_2.

From Theorems 1 and 2, it follows that:

Corollary 1. *LBP has a better mixed-criticality performance than BP, which can be formally written as:*

$$IsBetterMCS(LBP, BP)$$

5.2. Comparison between BPG and LBPG

Theorem 3. *LBPG has the same success rate of HI tasks than BPG, which can be formally written as:*

$$\forall \tau \in \text{STS. } ScheduledHI(BPG, \tau) == ScheduledHI(LBPG, \tau)$$

Proof. (Theorem 3) BPG and LBPG behave the same way regarding the handling of HI jobs:

1. If a HI job is overrunning its C_{LO}, it is granted an execution budget until C_{HI}.
2. If a HI job does not finish within C_{HI} or within its deadline, then it is dropped.

Moreover, any job that completes before its optimistic WCET during Normal mode gives its gain time to the next highest priority job in the ready queue. The only difference between BPG and LBPG lies in the handling of LO jobs that exceed their optimistic WCETs or that are released during Bailout and Recovery modes. BPG abandons such jobs while LBPG inserts them in the low-priority queue for later execution. Furthermore, the gain time collection only happens among jobs in the high-priority queue and no gain time is passed or happens among jobs in the low-priority queue. This guarantees that BPG and LBPG process and schedule jobs in the high-priority queue the same way. Thus, for any task set τ, it follows that

$$ScheduledHI(BPG, \tau) == ScheduledHI(LBPG, \tau).$$

□

Theorem 4. *LBPG can have a better success rate of LO tasks than BPG, but never worse, which can be formally written as:*

$$\forall \tau \in \text{STS. } ScheduledLO(BPG, \tau) \subseteq ScheduledLO(LBPG, \tau)$$
$$\exists \tau \in \text{STS. } ScheduledLO(BPG, \tau) \subset ScheduledLO(LBPG, \tau)$$

Proof. (Theorem 4) The only difference between BPG and LBPG lies in the handling of LO jobs, where LBPG puts them in a low-priority scheduling queue instead of abandoning them immediately or dropping them after the overrun of their C_{LO}. Hence, we have:

$$\forall \tau \in \text{STS. } Abandoned(BPG, \tau) \subseteq LORated(LBPG, \tau)$$

to prove Theorem 4 we only have to show that among the tasks that BPG abandons, there is at least one task that with LBPG instead gets put into the low-priority queue and finally successfully scheduled:

$$\exists \tau \in \text{STS. } Abandoned(BPG, \tau) \cap LORated(LBPG, \tau) \cap Scheduled(LBPG, \tau) \neq \emptyset$$

which means it is sufficient for the proof to show by example that it is possible to have task sets where LO-rated jobs can be scheduled within an idle time of the default scheduling queue.

We use the task set in Figure 5 to show that LBPG outperforms the standard BPG. The HI task A is assumed to have an execution time $et = 9$, which always causes an overrun of the optimistic WCET estimate. On the other hand, the LO task B always has an execution time of $et = 3$ apart from its first instance that runs for only two time units which allows to have a gain time of 1. Job B_0 completes earlier at time $t = 2$ and gives its gain time to job A_0 for which the optimistic time budget is now updated to A5. A_0 enters the Bailout mode at time $t = 7$ and then runs until its completion. No other gain time is collected during the schedule showed in the figure. Figure 5a,b show, respectively, that BPG abandons job B_1 and B_3, while LBPG runs them in Normal mode during idle time. Thus, this example demonstrates the existence of a task set τ such that

$$\exists \tau \in \text{STS. } Abandoned(BPG, \tau) \cap LORated(LBPG, \tau) \cap Scheduled(LBPG, \tau) \neq \emptyset$$

which completes the proof. $\square$

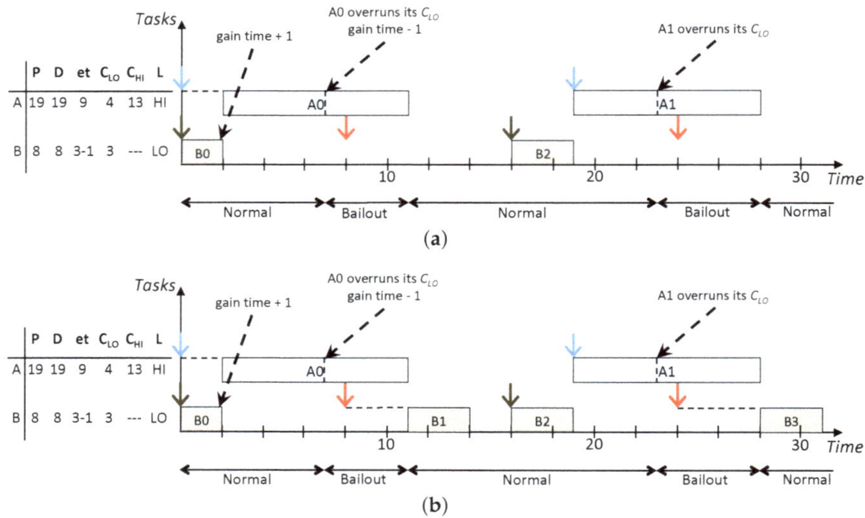

Figure 5. (Proof of Theorem 4) Example in which LBPG successfully executes LO jobs that are abandoned by BPG: (**a**) BPG abandons LO jobs in Bailout mode; and (**b**) LBPG provides a delayed execution for job B_1 and B_3

From Theorems 3 and 4 it follows that:

Corollary 2. *LBPG has a better mixed-criticality performance than BPG, which can be formally written as:*

$$IsBetterMCS(LBPG, BPG)$$

5.3. Comparison between BPS and LBPS

Theorem 5. *LBPS has better mixed-criticality performance than BPS, which can be formally written as:*

$$IsBetterMCS(LBPS, BPS)$$

Proof. (Theorem 5) The proof has two parts:

1. Proving that there is no task set τ such that

$$IsBetterMCS(BPS, LBPS, \tau)$$

2. Showing by concrete example that there exists a task set τ such that

$$IsBetterMCS(LBPS, BPS, \tau)$$

Part 1:

The strength of BPS over BP consists of the scaling up of the optimistic WCETs of HI tasks to increase the duration of Normal mode and to decrease the amount of time the system runs in Bailout mode. On the one hand, this leads to abandon a smaller amount of LO jobs due to the decrease in high-criticality mode duration. On the other hand, the increase of the Normal mode duration allows releasing and processing more LO jobs. The only difference between BPS and LBPS is in the handling of LO jobs released exceeding their C_{LO} or released in high-criticality modes, i.e., BPS suddenly abandons them while LBPS inserts them in a low-priority queue for later execution during system idle instants.

Therefore, LBPS keeps the BPS advantage, but it adds also the Lazy Bailout approach, which allows recovering LO jobs during high-criticality modes execution. This increases the amount of LO jobs processed and eventually scheduled during the Normal mode execution. This means that LBPS can never have worse performances than BPS.

Part 2:

To prove the second part, it is necessary to show that there exists one task set in which LBPS schedules more LO jobs than BPS. As an example, we use the task set given in Figure 6, for which the optimistic WCET C_{LO} of the HI task A has been already scaled up by sensitivity analysis. BPS increases the duration in which the system runs in Normal mode. However, it still abandons LO jobs released during high-criticality execution modes. Conversely, LBPS runs them afterwards during idle instants. Figure 6 displays how the LO job B_1 released at time $t = 9$ is abandoned with BPS while LBPS manages to execute it later at time $t = 10$.

This concludes the proof of Theorem 5. □

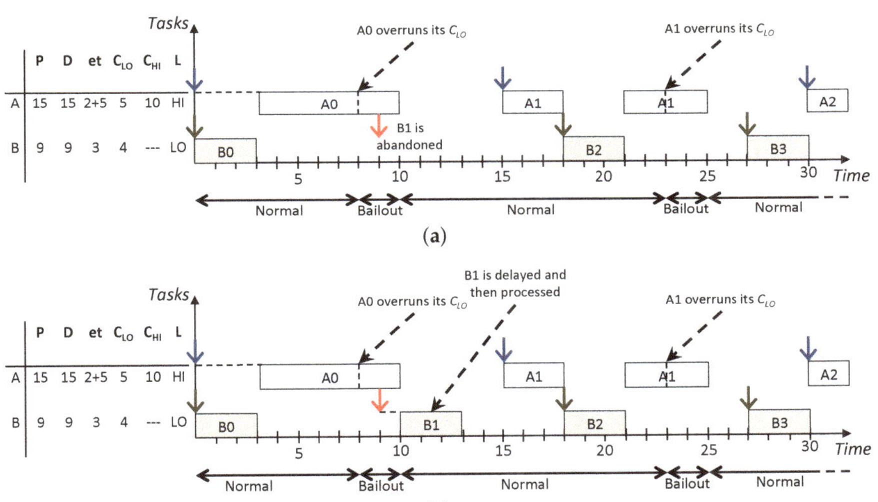

Figure 6. (Proof of Theorem 5, Part 2) LBPS always schedules more LO jobs than BPS: (**a**) BPS abandons job B_1; and (**b**) LBPS schedules all LO jobs.

5.4. Comparison between BPSG and LBPSG

With Corollary 2 and Theorem 5, we proved, respectively that the LBPG always outperforms LBP and LBPS always outperforms BPS. This is because the gain time collection made at runtime and the offline scaling up of the C_{LO} have the same benefits in the lazy Bailout method as in the standard Bailout protocol. On the other hand, from Corollary 1, we know that LBP always outperforms BP. It follows that:

Corollary 3. *LBPSG has a better mixed-criticality performance than BPSG, which can be formally written as:*

$$IsBetterMCS(LBPSG, BPSG)$$

In this section, we introduce the criterion `IsBetterMCS`(mtd_1, mtd_2) to compare the performance of mixed-criticality scheduling methods with priority given to HI jobs scheduled. Using this criterion, we show that LBP always performs better than BP. Moreover, we also show that the offline sensitivity analysis and online gain time collection always contribute to increase the amount of LO jobs scheduled and that they achieve better performances if used with LBP rather than BP. To conclude, the proposed LBPSG is consistently better than the existing BPSG [11].

Note that this result is strictly bound to our definition of $\texttt{IsBetterMCS}(mtd_1, mtd_2)$, which is motivated by systems where a sacrifice of HI jobs to increase performance of LO jobs is not acceptable.

6. Experimental Evaluation

In this section, we describe how we conducted our experiments and what were the final outcomes. We start by explaining how the experiments were structured. Section 6.2 explains how task sets were created, what were our application scenarios and what scheduling methods we compared. Next, Section 6.3 describes what type of performance metrics we considered to evaluate and compare different mixed-criticality scheduling methods. Finally, Section 6.4 contains the description and discussion of results according to what is shown within tables and charts.

6.1. Task Set Generation

The aim of the task set generation was to simulate situations in which the system easily switches to the Bailout execution mode in order to show the effectiveness of the LBP methods over BP and its derivatives in systems with high load (using a utilisation factor of 0.6 or more). The system load computed according to the pessimistic WCET of all HI tasks within each task set was always greater than that computed according to the optimistic WCET of all its tasks. Furthermore, the execution times of HI tasks almost always exceeded their C_{LO} in order to trigger the mode change.

As a result, task sets were generated to have an overall utilisation factor that varied randomly between 0.60% and 0.75% with respect to the optimistic WCET C_{LO} of all tasks while the utilisation factor computed according to the HI tasks was always 0.75%. The number of tasks per task set varied randomly between 4 and 20. Within each task set, the amount of HI tasks varied randomly between 20% and 70% of the task set. Moreover, the execution times of HI tasks varied between the 90% of their C_{LO} and their C_{HI}, while the execution times of LO tasks was between 40% of their C_{LO} and 10% more than the C_{LO}. Priorities were assigned to tasks according to the *Deadline Monotonic* (DM) strategy: task instances with shorter deadlines had higher priorities.

6.2. Description of Experiments

We conducted different experiments, each consisting of a group of 3000 task sets randomly generated. Tasks within every task set have implicit deadlines and their periods varied randomly between 3 and 22. Every task set group belonged to one of the three scenarios specified below:

HC-LP contains job sets where all HI jobs have larger deadlines than all LO jobs. More precisely, LO tasks' periods varied randomly between 3 and 10, while HI tasks' periods varied between 14 and 22. Therefore, all HI jobs had lower priority than all LO jobs:

$$\forall j \in J_{HI}. \ \forall j' \in J_{LO}. \ pr(j) < pr(j')$$

HC-MP contains job sets where HI jobs have deadlines that are smaller or larger than those of LO ones. In fact, periods of tasks, either HI or LO, varied randomly between 3 and 22. Therefore, HI and LO jobs had mixed priorities:

$$\forall j \in J_{HI}. \ \forall j' \in J_{LO}. \ pr(j) \leq pr(j') \ \lor \ pr(j) > pr(j')$$

HC-HP contains job sets where all HI jobs have smaller deadlines than all LO ones. More precisely, HI tasks' periods varied randomly between 3 and 10 while LO tasks' periods varied between 14 and 22. This implies that all HI jobs have higher priority than LO jobs:

$$\forall j \in J_{HI}. \ \forall j' \in J_{LO}. \ pr(j) > pr(j')$$

We compared the following scheduling protocols:

- The standard *Fixed-Priority Preemptive Scheduling* with DM as priority assignment (FPPS-DM).
- The standard *Bailout Protocol* (BP).
- The *Bailout Protocol with Gain Time* (BPG), where each job that finishes before its optimistic time threshold in Normal mode gives its gain time to increase the time budget of next job ready to be scheduled.
- The *Bailout Protocol-Slack* (BPS) and the *Bailout Protocol-Slack and Gain Time* (BPSG) that represent the execution of BP and BPG on task sets in which the C_{LO} of HI tasks is appropriately increased via sensitivity analysis [30,31] while the schedulability is guaranteed according to AMC-rtb [15].
- The *Lazy Bailout Protocol* (LBP).
- The *Lazy Bailout Protocol with Gain Time* (LBPG), the *Lazy Bailout Protocol-Slack* (LBPS) and the *Lazy Bailout Protocol-Slack and Gain Time* (LBPSG) that represent extensions of LBP made by using the offline scaling of C_{LO} of HI tasks with sensitivity analysis and the gain time collection at runtime.

We finally show the benefit of the lazy bailout approaches with respect to the former methods.

It is important to note that, if HI tasks all have higher priority than LO ones, then the scheduling problem so created becomes equivalent to the standard real-time scheduling problem since there is no criticality inversion. The same applies to those cases in which higher priority is assigned to the highest criticality tasks regardless of their periods or deadline as in *Criticality As Priority Assignment* (CAPA) [32].

Results of experiments are collected in Tables 1 and 2, which refer to the different scenarios described above. For each scenario, we show the results with different scheduling protocols.

Table 1. BP and LBP variants: comparison of task set schedulability (%).

| | HC-LP | | | HC-MP | | | HC-HP | | |
Method	TSSched	TSSchedHI	TSSchedLO	TSSched	TSSchedHI	TSSchedLO	TSSched	TSSchedHI	TSSchedLO
FPPS-DM	83.03	83.03	100.0	76.87	98.33	77.27	78.67	100.0	78.67
BP	2.20	100.0	2.20	0.97	100.0	0.97	0.87	100.0	0.87
BPG	4.87	100.0	4.87	1.17	100.0	1.17	0.93	100.0	0.93
BPS	7.23	100.0	7.23	11.17	100.0	11.17	12.30	100.0	12.30
BPSG	11.87	100.0	11.87	17.23	100.0	17.23	20.00	100.0	20.00
LBP	13.93	100.0	13.93	22.53	100.0	22.53	46.43	100.0	46.43
LBPG	21.17	100.0	21.17	23.57	100.0	23.57	46.63	100.0	46.63
LBPS	20.73	100.0	20.73	30.77	100.0	30.77	52.97	100.0	52.97
LBPSG	29.57	100.0	29.57	37.60	100.0	37.60	58.57	100.0	58.57

Table 2. BP and LBP variants: comparison of jobs scheduled within their deadline (%).

| | HC-LP | | | HC-MP | | | HC-HP | | |
Method	GJSched	GJSchedHI	GJSchedLO	GJSched	GJSchedHI	GJSchedLO	GJSched	GJSchedHI	GJSchedLO
FPPS-DM	99.19	88.64	100.0	98.51	99.55	97.91	99.11	100.0	98.18
BP	62.81	100.0	55.99	73.63	100.0	54.78	85.41	100.0	60.20
BPG	67.22	100.0	61.12	74.38	100.0	55.89	85.63	100.0	60.73
BPS	66.21	100.0	60.38	79.06	100.0	64.68	88.44	100.0	69.96
BPSG	71.03	100.0	66.07	81.64	100.0	69.41	90.05	100.0	75.13
LBP	83.64	100.0	80.94	92.71	100.0	88.71	97.87	100.0	95.16
LBPG	85.87	100.0	83.54	92.91	100.0	88.99	97.88	100.0	95.18
LBPS	85.23	100.0	82.92	93.24	100.0	89.54	97.99	100.0	95.51
LBPSG	87.57	100.0	85.66	93.77	100.0	90.48	98.08	100.0	95.78

6.3. Description of Performance Metrics

This subsection introduces the criteria we used to assess performances of scheduling protocols that process dual-criticality task sets. To evaluate the results, we defined two types of performance parameters, i.e., *task set schedulability* that is relative to the whole amount of task sets and the *global job set schedulability* that is relative to jobs within each individual task set.

We conducted our experiments on three sets of 3000 task sets *STS*, one per scenario (HC-LP, HC-MP and HC-HP). The task set schedulability formula *tsched* is defined as follows:

$$tsched(S, cat) = \frac{|STSsucc(S, cat)|}{|S|} \tag{4}$$

where S could be either a simple task set τ or set of task sets *STS* and the category $cat \in \{HI + LO, HI, LO\}$ represents the type of tasks within a set that is HI to indicate HI tasks, LO to indicate LO tasks and either when we use HI+LO. The function *STSsucc* depends on the scheduling protocol that is actually used and returns as output the set of task sets *STS* in which there are no jobs missed of category *cat*. The absolute values within the formula give the set cardinality. Equation (4) allows deriving the percentages of tasks set in *STS* that are successfully processed according to the category *cat* as follows:

TSSched is the amount of task sets scheduled with no jobs missing their deadlines.

$$TSSched = tsched(STS, HI + LO)$$

TSSchedHI is the amount of task sets scheduled with no HI jobs missing their deadlines.

$$TSSchedHI = tsched(STS, HI)$$

TSSchedLO is the amount of task sets scheduled with no LO jobs missing their deadlines.

$$TSSchedLO = tsched(STS, LO)$$

The task set schedulability permits showing the percentage of task sets in which no job of category *cat* misses its deadline. However, whenever a task set contains some jobs, either HI or LO, that miss their deadline, it is also useful to assess the level of service provided in terms of jobs completed within their deadlines and jobs abandoned or aborted. The job set completion rate method *jsched* returns the percentage of jobs of category *cat* generated by a specific task set that complete within their deadlines. The job set schedulability *jsched* is formally written as below:

$$jsched(\tau_{cat}) = \frac{|Jsucc(J(\tau_{cat}))|}{|J(\tau_{cat})|} \quad | \ cat \in \{LO, HI\} \tag{5}$$

The formula of the global job set schedulability *gjsched(STS, cat)* returns the average amount of jobs of category *cat* processed within their deadline that has been generated by a set of task sets *STS*:

$$gjsched(STS, cat) = \frac{\sum_{\tau \in STS} jsched(\tau_{cat})}{|STS|} \quad | \ cat \in \{LO, HI\} \tag{6}$$

As in the previous case, it is possible to filter the jobs meeting their deadline according to the category *cat* as below:

GJSched is the average number of jobs (either HI or LO) that is scheduled within a task set.

$$GJSsched = gjsched(STS, HI + LO)$$

GJSchedHI is the average number of HI jobs that is scheduled within a tasks set.

$$GJSchedHI = gjsched(STS, HI)$$

GJSchedLO is the average number of LO jobs that is scheduled within a task set.

$$GJSschedLO = gjsched(STS, LO)$$

Tables 1 and 2 contain, respectively, the results about task set and global job set schedulabilities. It is possible to comment on the data according to scenario or scheduling protocol. However, we use figures to describe graphically what is contained within the tables and to allow an easier and quicker comparison among the results.

Task set and global job set schedulabilities are averages and do not give information about how data are distributed and about outliers. Therefore, we use also boxplots charts to show the distribution of LO jobs scheduled per task set. The information shown in Table 1 is contained in Figures 7a, 8a and 9a. On the other hand, Figures 7b, 8b and 9b displays the average percentages of LO jobs completed within their deadlines.

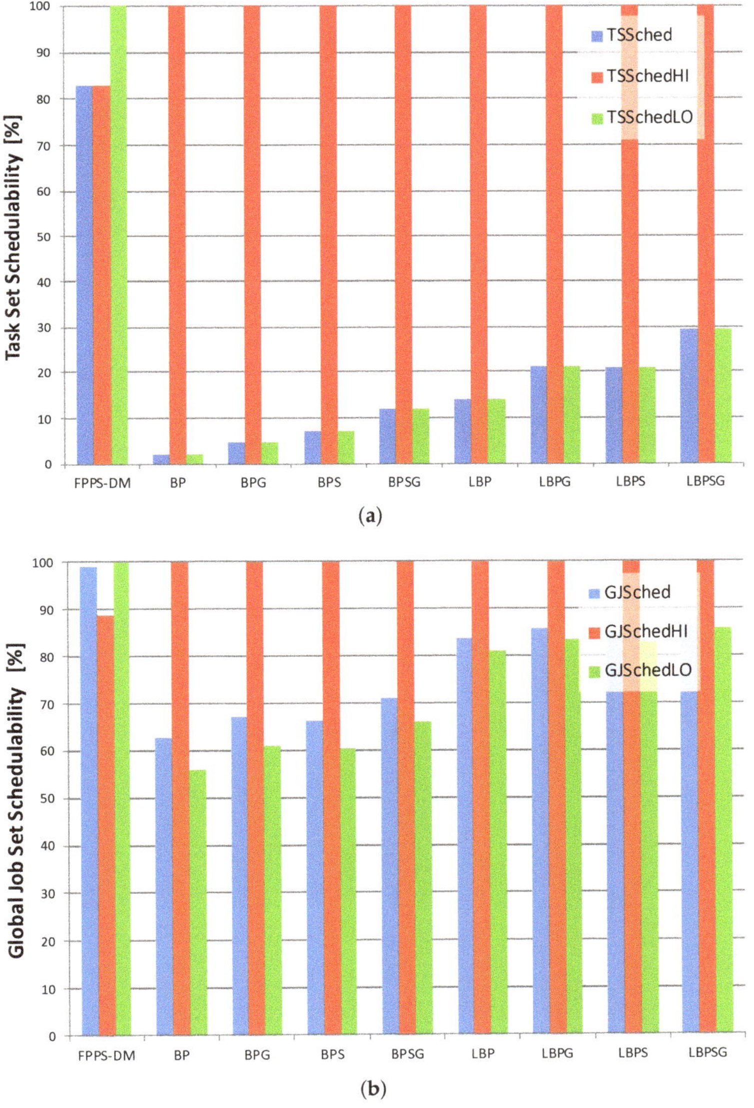

Figure 7. BP and LBP variants: schedulability in HC-LP scenario: (**a**) task sets with no jobs missed; and (**b**) average of jobs scheduled per task set.

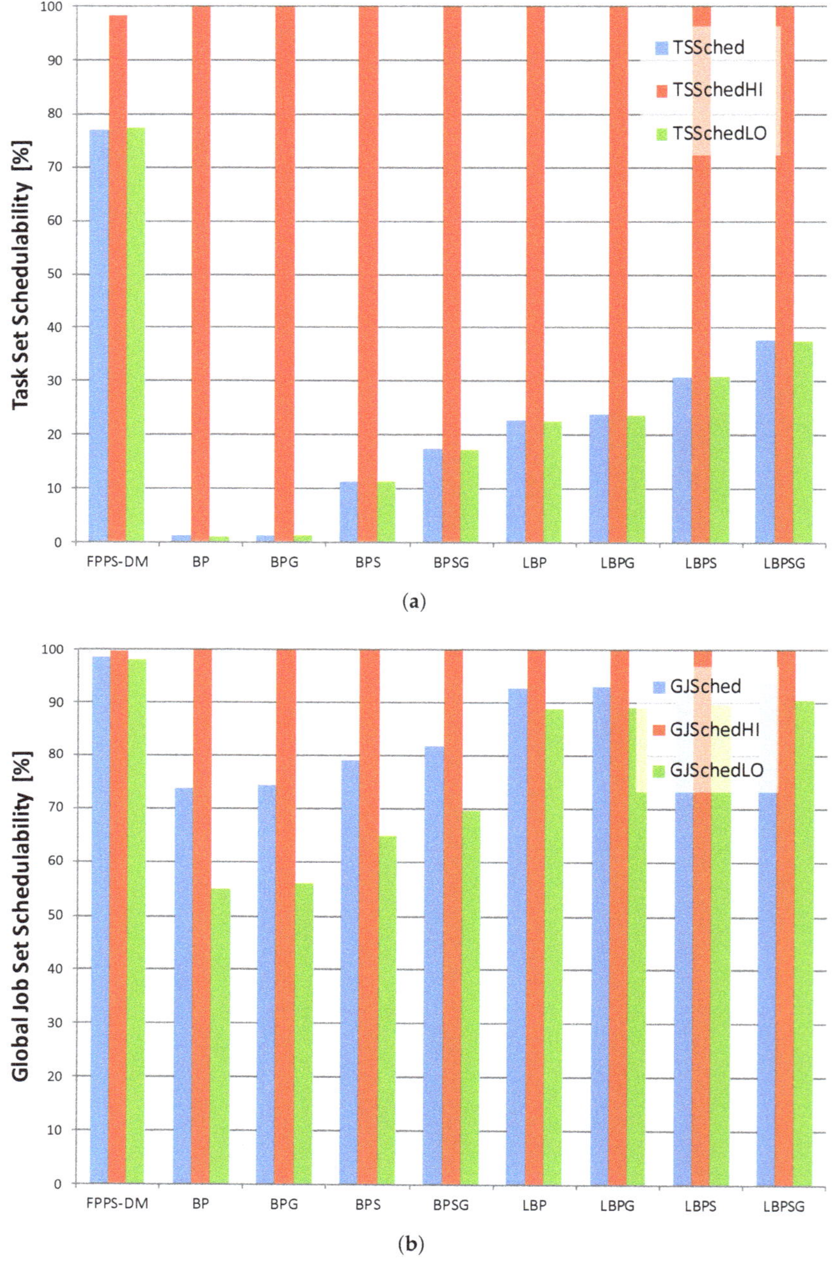

Figure 8. BP and LBP variants: schedulability in HC-MP scenario: (**a**) task sets with no jobs missed; and (**b**) average of jobs scheduled per task set.

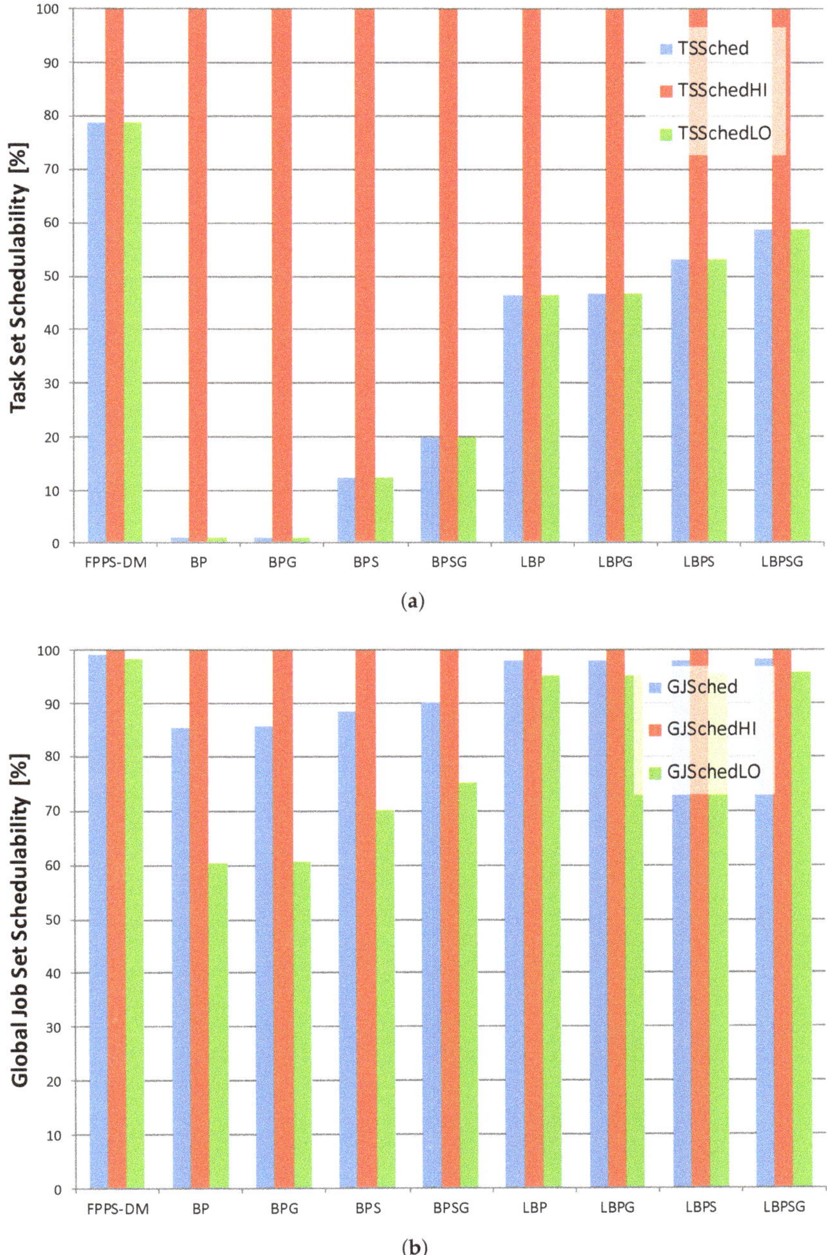

Figure 9. BP and LBP variants: schedulability in HC-HP scenario (as priority and criticality values have the same order, this is essentially a standard real-time scheduling problem): (**a**) task sets with no jobs missed; and (**b**) average of jobs scheduled per task set.

6.4. Discussion of Results

This subsection is dedicated to study the outcome of the experiments. It contains figures that summarise the results of our experiments with dual-criticality job sets. Figures 7 and 8 summarise the results in cases where there is criticality inversion. In these situations, if no HI job completes within its optimistic threshold estimate C_{LO}, then very likely there will be some new incoming higher priority LO jobs that will interfere with it. Then, Figure 9 contains information about cases in which all HI jobs have higher priority than LO jobs since all the critical jobs have smaller deadlines. This basically leads to having no interference between HI and LO jobs and thus no criticality inversion occurrence during the scheduling process.

Looking both at task set and job set schedulabilities results in Figures 7–9, it is possible to notice that, compared with mixed-criticality methods, the standard deadline monotonic approach always schedules jobs only according to priorities. In this case, the percentages of HI and LO jobs successfully scheduled mainly depend only on their priority, with all LO jobs always meeting their deadlines in HC-LP scenario and all HI jobs always meeting their deadlines in HC-HP scenario. Conversely, the mixed-criticality protocols always assure that there are no HI job missed regardless of job priorities. The experiments confirm what is stated in Section 5 with LBP in which the percentage of task set scheduled with no jobs missed is between 13% and 46% while BP schedules no more than 2.20% of task sets with no jobs missed.

Then, where the offline and online complementary techniques are used, there is an increase in LO jobs successfully processed. Furthermore, the usage of sensitivity analysis and the gain time mechanism have the same effects when applied both to the standard or to the lazy bailout method. A noticeable result is that each LBP-based approach always increases the amount of LO jobs completed within their deadlines compared with the corresponding standard BP-based protocol. Overall, according to the criteria defined in Section 5, LBPSG is the protocol that outperforms all other mixed-criticality scheduling methods with an amount of task set scheduled with no jobs missed that is between 29.57% and 58.57%.

Figures 10–12 display the distribution of the LO jobs percentages per task set that are completed within their deadlines. Each scheduling protocol is represented by a box-and-whisker diagram with the box itself representing the range in which at least 50% of results tend to be concentrated. The box also contains the indication of the median and the mathematical average of all the LO jobs scheduled by the related protocol. The results highlight how the LBP-based methods always increase the LO jobs success rate, as defined in Section 5, compared with the former BP ones.

In conclusion, the experiments confirm what is stated in Section 5 with lazy approaches increasing the amount of LO jobs successfully scheduled while guaranteeing the correct completion of all HI jobs. In other words, LBP has better mixed-criticality performances than BP, while LBPS, LBPG and LBPSG have, respectively, better mixed-criticality performances than BPS, BPG and BPSG. Finally, the usage of mixed-criticality protocols is recommended in HP-LP and HC-MP scenarios, i.e., when HI jobs could have lower priorities than LO jobs.

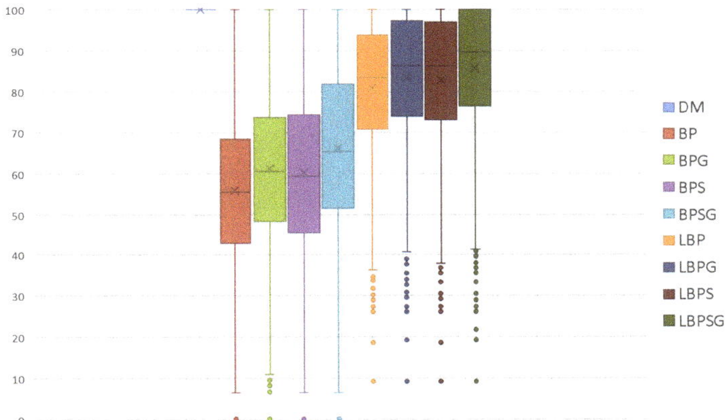

Figure 10. BP and LBP variants: LO jobs scheduled per task set in HC-LP scenario.

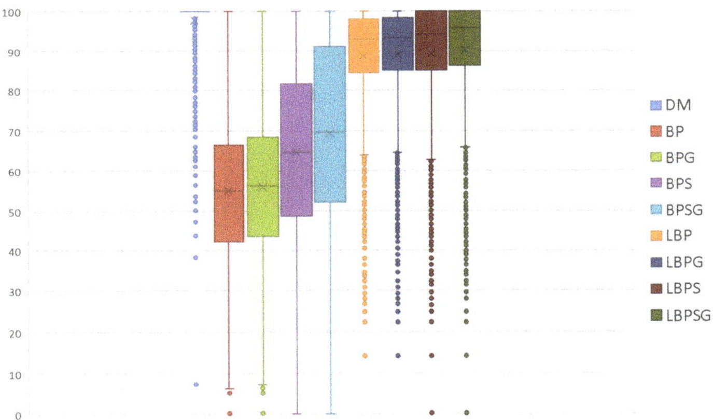

Figure 11. BP and LBP variants: LO jobs scheduled per task set in HC-MP scenario.

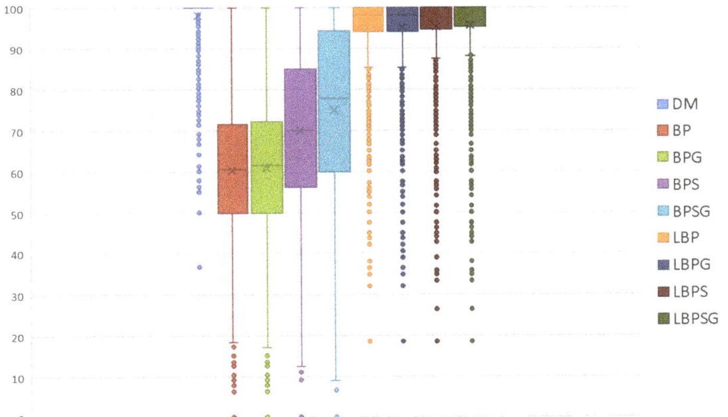

Figure 12. BP and LBP variants: LO jobs scheduled per task set in HC-HP scenario.

7. Summary and Conclusions

Mixed-criticality scheduling is important for cyber-physical systems to provide robustness against resource shortage. In this paper, we have introduced the mixed-criticality scheduling protocol *Lazy Bailout Protocol (LBP)*. LBP is a scheduling protocol for uni-processor platforms and is a refinement of Bailout Protocol (BP). We have also introduced a formal criterion to compare performances among mixed-criticality scheduling protocols. This criterion prioritises HI jobs against LO jobs, where HI indicates high-criticality and LO stands for low-criticality. Based on that criterion, we have proven that the complementary techniques used in [11] always contribute to increase the performances of the scheduling protocol. Similar to BP, LBP and its derivatives always guarantee the correct completion of HI jobs. Moreover, we have shown that LBP always schedules more LO jobs than BP and that each LBP derivative always outperforms the corresponding BP-based approach.

Besides these formal results, we have also presented experiments that give quantitative values of the comparisons between the different mixed-criticality scheduling protocols. LBP schedules between 13.93% and 46.63% of task sets with no jobs missed while BP at maximum schedules no more than 2.20% of task sets with no jobs missed. Finally, the experiments confirm that LBP is equivalent to BP in guaranteeing HI jobs and that the derivatives of LBP (LBPS, LBPG and LBPSG) outperform all the equivalent BP-based protocols by increasing the amount of LO jobs successfully scheduled. Overall, LBPSG has shown the best mixed-criticality performance with an amount of task sets processed with no jobs missed that is between 29.57% and 58.57%.

Future work will be on extending LBP towards support for many-core platforms.

Author Contributions: R.K. has contributed the majority of formal proofs, S.I. has done the implementation and experiments and refined some proofs.

Funding: The research leading to these results in its early stages has received funding from the FP7 ARTEMIS-JU research project *"ConstRaint and Application driven Framework for Tailoring Embedded Real-time Systems"* (CRAFTERS) under contract no 295371.

Acknowledgments: The authors would like to thank Olga Tveretina for general comments on how to improve proofs.

Conflicts of Interest: The authors declare no conflict of interest.

References

1. Liu, C.L.; Layland, J.W. Scheduling Algorithms for Multiprogramming in a Hard-Real-Time Environment. *J. ACM* **1973**, *20*, 46–61. [CrossRef]
2. Vestal, S. Preemptive Scheduling of Multi-criticality Systems with Varying Degrees of Execution Time Assurance. In Proceedings of the 28th IEEE International Real-Time Systems Symposium (RTSS'07), Tucson, AZ, USA, 3–6 December 2007; pp. 239–243. [CrossRef]
3. Burns, A.; Davis, R.I. *Mixed Criticality Systems—A Review*; Research Report V4-31/7/2014; Department of Computer Science, University of York: York, UK, 2014.
4. Wilhelm, R.; Engblom, J.; Ermedahl, A.; Holsti, N.; Thesing, S.; Whalley, D.; Bernat, G.; Ferdinand, C.; Heckman, R.; Mitra, T.; et al. The Worst-Case Execution Time Problem—Overview of Methods and Survey of Tools. *ACM Trans. Embed. Comput. Syst. (TECS)* **2008**, *7*. [CrossRef]
5. Kirner, R.; Iacovelli, S.; Zolda, M. Optimised Adaptation of Mixed-criticality Systems with Periodic Tasks on Uniform Multiprocessors in Case of Faults. In Proceedings of the 11th IEEE Workshop on Software Technologies for Future Embedded and Ubiquitous Systems (SEUS'15), Auckland, New Zealand, 13 April 2015.
6. RTCA SC-205. Software Considerations in Airborne Systems and Equipment Certification. RTCA/DO-178C. Available online: https://www.rtca.org (accessed on 20 December 2011).
7. International Organization for Standardization (ISO). *Road Vehicles—Functional Safety*; ISO Standard 26262; ISO: Geneva, Switzerland, 2011.
8. International Electrotechnical Commission. *Functional Safety of Electrical/Electronic/Programmable Electronic Safety-Related Systems*; IEC Standard 61508; International Electrotechnical Commission: Geneva, Switzerland, 1998.

9.	Esper, A.; Nelissen, G.; Nélis, V.; Tovar, E. How Realistic is the Mixed-criticality Real-time System Model? In Proceedings of the 23rd Int'l Conference on Real Time and Networks Systems (RTNS), Lille, France, 4–6 November 2015; ACM: New York, NY, USA, 2015; pp. 139–148. [CrossRef]

10.	Bate, I.; Burns, A.; Davis, R.I. A Bailout Protocol for Mixed Criticality Systems. In Proceedings of the 27th Euromicro Conference on Real-Time Systems, Lund, Sweden, 8–10 July 2015.

11.	Bate, I.; Burns, A.; Davis, R.I. An Enhanced Bailout Protocol for Mixed Criticality Embedded Software. *IEEE Trans. Softw. Eng.* **2017**, *43*, 298–320. [CrossRef]

12.	Crespo, A.; Alonso, A.; Marcos, M.; de la Puente, J.A.; Balbastre, P. Mixed Criticality in Control Systems. In Proceedings of the 19th World Congress of The International Federation of Automatic Control, Cape Town, South Africa, 24–29 August 2014.

13.	Ernst, R.; Natale, M.D. Mixed Criticality Systems—A History of Misconceptions. *IEEE Des. Test* **2016**, *33*, 65–74. [CrossRef]

14.	Burns, A.; Davis, R.I. A Survey of Research into Mixed Criticality Systems. *ACM Comput. Surv.* **2017**, *50*, 82:1–82:37. [CrossRef]

15.	Baruah, S.K.; Burns, A.; Davis, R.I. Response-Time Analysis for Mixed Criticality Systems. In Proceedings of the 2011 IEEE 32nd Real-Time Systems Symposium (RTSS '11), Vienna, Austria, 29 November–2 December 2011; IEEE Computer Society: Washington, DC, USA, 2011; pp. 34–43. [CrossRef]

16.	Burns, A.; Davis, R.I. Response Time Analysis for Mixed Criticality Systems with Arbitrary Deadlines. In Proceedings of the 5th International Workshop on Mixed Criticality Systems (WMC 2017), Paris, France, 5 December 2017.

17.	Fleming, T.; Burns, A. Extending Mixed Criticality Scheduling. In Proceedings of the 1st International Workshop on Mixed Criticality Systems (WMC), Vancouver, BC, Canada, 3–6 December 2013.

18.	Su, H.; Zhu, D. An Elastic Mixed-Criticality task model and its scheduling algorithm. In Proceedings of the Conference on Design, Automation and Test in Europe (DATE), Grenoble, France, 18–22 March 2013; pp. 147–152.

19.	Burns, A.; Baruah, S. Towards A More Practical Model for Mixed Criticality Systems. In Proceedings of the 1st International Workshop on Mixed Criticality Systems (WMC), Vancouver, BC, Canada, 3–6 December 2013; pp. 1–6.

20.	Baruah, S.; Burns, A.; Davis, R. An Extended Fixed Priority Scheme for Mixed Criticality Systems. In *Workshop on Real-Time Mixed Criticality Systems (ReTiMics)*; George, L., Lipari, G., Eds.; University of York: York, UK, 2013; pp. 18–24.

21.	Audsley, N.C.; Burns, A.; Richardson, M.F.; Wellings, A.J. Hard Real-Time Scheduling: The Deadline-Monotonic Approach. *IFAC Proc. Vol.* **1991**, *24*, 127–132.

22.	Zuhily, A.; Burns, A. Optimal (D-J)-monotonic priority assignment. *Inf. Process. Lett.* **2007**, *103*, 247–250. [CrossRef]

23.	Fleming, T.; Burns, A. Incorporating the Notion of Importance into Mixed Criticality Systems. In Proceedings of the 2nd International Workshop on Mixed Criticality Systems (WMC), Rome, Italy, 2 December 2014; pp. 33–38.

24.	Prasad, D.; Burns, A.; Atkins, M. The Valid Use of Utility in Adaptive Real-Time Systems. *Real-Time Syst.* **2003**, *25*, 277–296. [CrossRef]

25.	Audsley, N. *Optimal Priority Assignment and Feasibility of Static Priority Tasks With Arbitrary Start Times*; Technical Report YCS 164; Department of Computer Science, University of York: York, UK, 1991.

26.	Gettings, O.; Quinton, S.; Davis, R.I. Mixed criticality systems with weakly-hard constraints. In Proceedings of the 23rd International Conference on Real Time and Networks Systems (RTNS '15), Lille, France, 4–6 November 2015; ACM: New York, NY, USA, 2015; pp. 237–246.

27.	Bernat, G.; Burns, A.; Llamosi, A. Weakly Hard Real-Time Systems. *IEEE Trans. Comput.* **2001**, *50*, 308–321. [CrossRef]

28.	Frehse, G.; Hamann, A.; Quinton, S.; Woehrle, M. Formal Analysis of Timing Effects on Closed-Loop Properties of Control Software. In Proceedings of the 2011 IEEE 32nd Real-Time Systems Symposium. IEEE, 2014 (RTSS '14), Rome, Italy, 2–5 December 2014; pp. 53–62.

29.	Santy, F.; George, L.; Thierry, P.; Goossens, J. Relaxing Mixed-Criticality Scheduling Strictness for Task Sets Scheduled with FP. In Proceedings of the 24th Euromicro Conference on Real-Time Systems (ECRTS), Pisa, Italy, 11–13 July 2012; pp. 155–165.

30. Bini, E.; Natale, M.D.; Buttazzo, G.C. Sensitivity analysis for fixed-priority real-time systems. *Real-Time Syst.* **2008**, *39*, 5–30. [CrossRef]

31. Punnekkat, S.; Davis, R.; Burns, A. Sensitivity Analysis of Real-Time Task Sets. In *Advances in Computing Science—ASIAN'97: Third Asian Computing Science Conference Kathmandu, Nepal, December 9–11, 1997 Proceedings*; Springer: Berlin/Heidelberg, Germany, 1997; pp. 72–82. doi:10.1007/3-540-63875-X_44.

32. De Niz, D.; Lakshmanan, K.; Rajkumar, R.R. On the Scheduling of Mixed-Criticality Real-Time Task Sets. In Proceedings of the 2009 30th IEEE Real-Time Systems Symposium (RTSS '09), Washington, DC, USA, 1–4 December 2009; pp. 291–300.

Article

Fighting CPS Complexity by Component-Based Software Development of Multi-Mode Systems

Hang Yin [1,*] and Hans Hansson [2,†]

[1]   Zenuity, Lindholmspiren 2, 417 56 Gothenburg, Sweden
[2]   School of Innovation, Design and Engineering, Mälardalen University, Högskoleplan 1, 722 20 Västerås, Sweden; hans.hansson@mdh.se
*   Correspondence: hang.yin@zenuity.com; Tel.: +46-765-836-887
†   Current address: Lindholmspiren 2, 417 56 Gothenburg, Sweden.

Received: 7 October 2018; Accepted: 18 October 2018; Published: 22 October 2018

Abstract: Growing software complexity is an increasing challenge for the software development of modern cyber-physical systems. A classical strategy for taming this complexity is to partition system behaviors into different operational modes specified at design time. Such a multi-mode system can change behavior by switching between modes at run-time. A complementary approach for reducing software complexity is provided by component-based software engineering (CBSE), which reduces complexity by building systems from composable, reusable and independently developed software components. CBSE and the multi-mode approach are fundamentally conflicting in that component-based development conceptually is a bottom-up approach, whereas partitioning systems into operational modes is a top-down approach with its starting point from a system-wide perspective. In this article, we show that it is possible to combine and integrate these two fundamentally conflicting approaches. The key to simultaneously benefiting from the advantages of both approaches lies in the introduction of a hierarchical mode concept that provides a conceptual linkage between the bottom-up component-based approach and system level modes. As a result, systems including modes can be developed from reusable mode-aware components. The conceptual drawback of the approach—the need for extensive message exchange between components to coordinate mode-switches—is eliminated by an algorithm that collapses the component hierarchy and thereby eliminates the need for inter-component coordination. As this algorithm is used from the design to implementation level ("compilation"), the CBSE design flexibility can be combined with efficiently implemented mode handling, thereby providing the complexity reduction of both approaches, without inducing any additional design or run-time costs. At the more specific level, this article presents (1) a mode mapping mechanism that formally specifies the mode relation between composable multi-mode components and (2) a mode transformation technique that transforms component modes to system-wide modes to achieve efficient implementation.

Keywords: component-based software engineering; mode; mode-switch

1. Introduction

Growing software complexity is posing a challenge for the design of cyber-physical systems (CPS) [1]. Complexity related to CPS software is multifaceted, including specific requirements related to extra functional properties such as functional safety, resilience and timeliness. There is additionally a trade-off between the different aspects of complexity, e.g., added complexity at design-time can reduce complexity at run-time and vice versa. A key issue in making such trade-offs is the risk implied by different choices; and risks have to be balanced against benefits. For companies, this is a reality even for safety-critical systems regulated by safety standards. As a result, to maximize business benefits, it is standard practice for many companies to make a minimal, but sufficient, effort to comply with applicable regulation.

At the technological level, there are approaches developed to reduce complexity throughout the system life-cycle. Combining such approaches can potentially reduce the overall life-cycle complexity, but approaches are not always compatible, and each of them introduces both benefits and costs. This article presents the integration of two such approaches: multi-mode systems and component-based software engineering, which both provide composability, but are targeting different life-cycle phases. A specific challenge in combining the two is that they are conceptually incompatible, in the sense that component-based development is a bottom-up approach, whereas partitioning systems into operational modes is a top-down approach with its starting point from a system-wide perspective. Still, we are able to successfully integrate them in a single framework of multi-mode components, thereby providing the combined benefits of both, while being able to reduce costs to acceptable levels. These approaches, introduced below, are primarily focusing on the design, configuration and run-time phases of the system life-cycle, although they do have important implications also for the maintenance phase. Furthermore, other dimensions of complexity are affected by the considered technologies, including organizational complexity, as both component-based software engineering (CBSE) and the multi-mode approach provide a basis not only for system partitioning, but also for partitioning of the design activities, implying that the distribution of the design tasks to different departments or even different organizations are facilitated. A further implication of the enabled partitioning and inherent clearly-defined interfaces is that the approaches could scale also to systems-of-systems (SoS), e.g., different components or modes can correspond to different systems in an SoS.

1.1. Multi-Mode Systems

A common practice to manage software complexity is to partition system behaviors into different mutually-exclusive operational modes so that each mode corresponds to a specific system behavior. A multi-mode system [2] changes behavior by switching modes. A typical example is the control software of an airplane, which runs in different modes such as taking off, flight and landing. Multi-mode systems have also been motivated by many other reasons:

1. Faster development: system behavior for different modes can be designed and tested in parallel.
2. Diversified functionalities due to multiple modes.
3. Enable adaptivity by mode-switch.
4. Efficient resource usage: optimized resource reservation for each mode instead of fixed resource reservation.
5. Fault tolerance: safety-critical systems can switch to a safe mode in case of a fault.
6. Extensibility and scalability: it is flexible to add new modes and integrate them with an existing system.

1.2. Component-Based Software Engineering

As a complementary approach to multi-mode systems, CBSE [3] advocates the reuse of independently developed software components as a promising technique for the development of complex software systems. The success of CBSE has been evidenced by a variety of component models proposed both in industry and academia [4,5]. CBSE suggests software modularity and reusability to facilitate the development of high-quality software. For instance, different functional modules or even subsystems of the control software of an airplane can be encapsulated into reusable software components that can be reused multiple times for the same system or for other systems in the same software product line.

Applying CBSE in multi-mode systems or the other way around has been a largely unexplored research area, possibly because multi-mode systems and CBSE are fundamentally conflicting in the sense that the former traditionally is a top-down approach, whereas the latter is a bottom-up approach. Despite this apparent conflict, our research goal in this article is to combine these approaches and benefit from the advantages of both multi-mode systems and CBSE. Hence, we propose component-based software development of multi-mode systems, characterized by the independent development and reuse of multi-mode components (i.e., components that can run in multiple modes).

1.3. A Guiding Example

As a guiding example, consider a proof-of-concept healthcare monitoring system. The system consists of two subsystems: a data acquisition subsystem and a monitoring subsystem. The data acquisition subsystem uses cameras and microphones to collect video and audio data from a ward or a private home. Video and audio data are separately encoded and encrypted before transmission. The monitoring subsystem decrypts and decodes data that it receives and reports them to the health center. The monitoring subsystem also includes an alarm that is triggered when the person being monitored encounters a dangerous situation, such as falling or suffering from a heart attack.

Our focus is on the software architecture of the monitoring subsystem (MoS), which is composed of multi-mode components. Figure 1 illustrates the component hierarchy of the system on the left and component connections on the right. The system can be considered as a top-level component MoS with three subcomponents: DaD for data decryption, the multimedia decoder MuD and EvA for event analysis. Due to the tree structure of the component hierarchy, DaD, MuD and EvA are also called the children of MoS, which consequently is their parent. The system can run in two different modes: regular monitoring mode (denoted as *Rm*) and attention mode (denoted as *Att*).

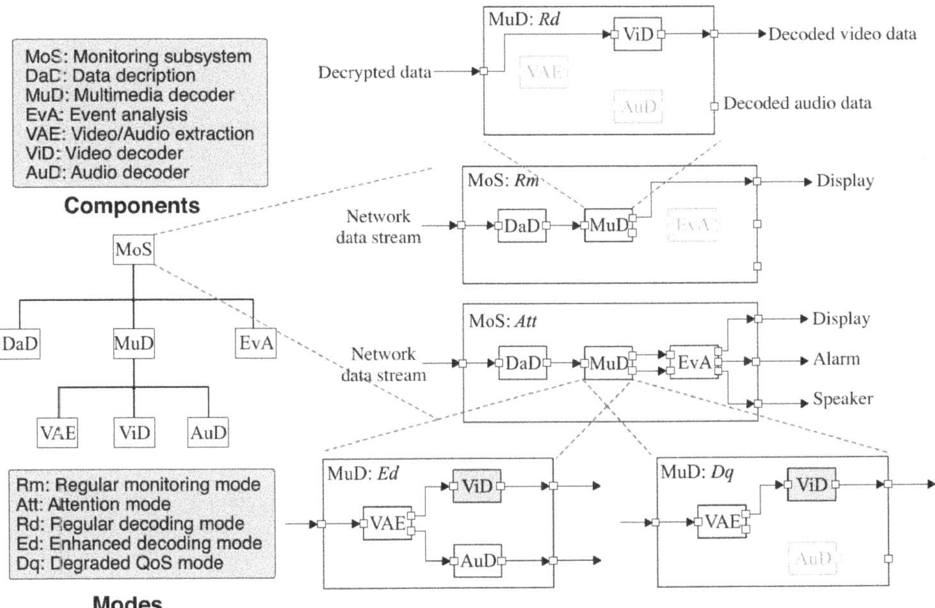

Figure 1. The architectural model of a multi-mode system built from multi-mode components.

The default mode of MoS is *Rm* when nothing special happens. To save resource in this mode, a fast video encoding/decoding algorithm can be selected and only low resolution video is transmitted to keep low CPU and bandwidth usage. Shown in Figure 1, small squares denote the input and output ports of a component, while each arrow that connects two ports denotes the data flow. Basically, each component has input data coming from its input ports, processes the data and provides output data at its output ports. Such a pipes and filters architectural style is fairly common for multimedia applications. The video data are first decrypted by DaD and subsequently decoded by MuD, which sends the decoded video data to the display in the health center. Represented by the dimmed color, EvA is deactivated when MoS runs in *Rm*. Meanwhile, DaD runs in a regular mode *R1* and MuD runs in a regular decoding mode *Rd*. MuD has three subcomponents: VAE for video/audio extraction, a video decoder, ViD, and an audio decoder, AuD. ViD is the only activated subcomponent running in a regular video decoding mode *Rvd* as MuD runs in *Rd*. VAE and AuD are deactivated as shown in Figure 1 because no audio data are transmitted.

When the data acquisition subsystem detects an accident, both subsystems will switch to an attention mode *Att* to raise the attention of the health center. The network load is increased due to transmission of video data with higher quality and audio data. Component EvA becomes activated running in a regular mode *R2*, to analyze the detected event and trigger an alarm when necessary. MuD starts to run in enhanced decoding mode *Ed*, where all its subcomponents (VAE, ViD, AuD) are activated and a different video encoding/decoding algorithm is used to support higher resolution video. VAE runs in a regular mode *R3* to separate decrypted video and audio data and send them to ViD and AuD, respectively. Represented by grey color in Figure 1, ViD runs in an enhanced video decoding mode *Evd* with a different video decoding algorithm for high quality video. AuD runs in a regular audio decoding mode *Rad*. In the case of poor network condition, MuD switches to a degraded QoS mode *Dq*, where the transmission of audio data is terminated to keep video quality, which is considered to be more important. Therefore, AuD becomes deactivated.

We distinguish two types of components in the monitoring subsystem: primitive components and composite components. DaD, EvA, VAE, ViD and AuD are primitive components, which are directly implemented by code, while MoS and MuD are composite components composed by other components. What makes this system distinctive compared with traditional component-based systems is its constitution of multi-mode components, i.e., components that can run in different modes at run-time. The system in Figure 1 indicates a clear mapping between the modes of different components. Such a mapping is summarized in Table 1, where modes in the same column are mapped to each other for the composite components MoS and MuD. For instance, when MoS runs in mode *Att*, DaD must run in *R1*, MuD can run in either *Ed* or *Dq* and EvA must run in *R2*. Even already, this simple example signifies the power of building multi-mode systems with multi-mode components, which has the potential to enrich system architectural variability at all levels. Since multi-mode components still comply with component-based development, the overall software design complexity is decoupled into building multi-mode components at different levels, thereby making the growing software complexity manageable.

Table 1. Mode mappings.

(a) Mode Mapping of MoS

Component	Modes		
MoS	*Rm*	*Att*	
DaD	*R1*		
MuD	*Rd*	*Ed*	*Dq*
EvA	*Deactivated*	*R2*	

(b) Mode Mapping of MuD

Component	Modes		
MuD	*Rd*	*Ed*	*Dq*
VAE	*Deactivated*		*R3*
ViD	*Rvd*		*Evd*
AuD	*Deactivated*	*Rad*	*Deactivated*

1.4. Contributions

In achieving component-based software development of multi-mode systems, this article includes two key contributions. First, we propose a formal mode mapping description in the form of mode mapping automata (MMA) that specifies how the modes of a composite component are mapped to the modes of its subcomponents. The MMA presented in this article partially builds on the MMA initially proposed in [6], which is here substantially refined and extended. Mode mapping elegantly links modes and software component reuse. The hierarchical composition of multi-mode components easily allows one to build multi-mode systems with multi-mode behaviors at various levels. A potential drawback of this approach is the run-time overhead due to inter-component communication for coordination of the mode-switches of different components. To eliminate this run-time drawback, while still being able to design systems from reusable multi-mode components, we introduce a mode transformation technique as our second contribution. This technique transforms component modes to system-wide modes to optimize the implementation. This is obtained by flattening the hierarchical structure of component modes mapped at different levels. Mode transformation can be included in the mapping from the design to implementation level ("compilation"), after the mode mappings of all composite components in a system have been specified. An initial version of the mode transformation technique is presented in [7].

The rest of this article is structured as follows: Section 2 elaborates on the composition of multi-mode components and the mode mapping mechanism. Section 3 presents the mode transformation technique. Related work is reviewed in Section 4. Finally, Section 5 concludes the article and discusses some future work.

2. The Composition of Multi-Mode Components

As an essential step in the composition of multi-mode components, mode mapping unambiguously specifies the mode relation between different multi-mode components at design time. This section highlights the essential properties of multi-mode components and the motivation of mode mapping, followed by a thorough explanation of MMA, i.e., a formal description of mode mapping.

2.1. Multi-Mode Components and Mode Mapping

A multi-mode component supports multiple modes and has a unique configuration defined for each mode. Figure 2 illustrates the key properties of a reusable multi-mode component. The configuration for each mode relies on various factors. For instance, a multi-mode primitive component may have different mode-specific behaviors for different modes; while a multi-mode composite component may have a different set of subcomponents activated depending on its mode.

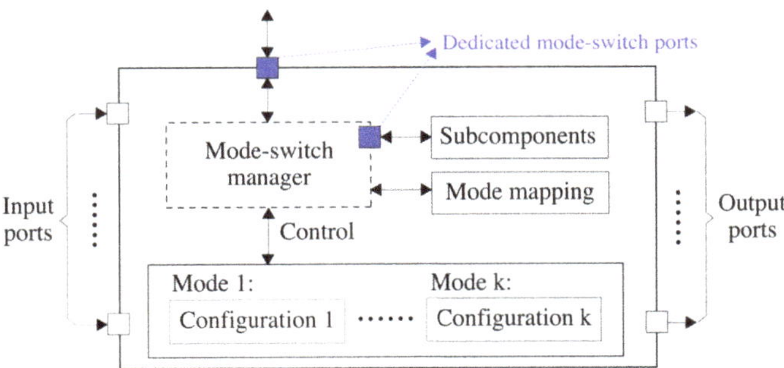

Figure 2. The illustration of a multi-mode component.

A multi-mode component can switch between certain modes at run-time, either on its own initiative or as the result of a request by another component. A mode-switch leads to the reconfiguration of the component by changing its configuration in the current mode to a new configuration in the target mode. A local mode-switch manager is used to handle the mode-switch of a multi-mode component. By having such a mode-switch manager in each component, a multi-mode component is able to exchange mode information with its parent and subcomponents via dedicated mode-switch ports (the blue ports in Figure 2) during a mode-switch, even without knowing the global mode information. These mode-switch ports do not deal with input or output data going through the component. Instead, they are only used for mode-switch coordination between a composite component and its subcomponents. Therefore, each mode-switch port is bidirectional, which allows mode-switch signals to be transmitted in both directions. For instance, in the guiding example presented in Section 1.3, when an accident is detected, MoS will switch from *Rm* to *Att*. Meanwhile, the mode-switch manager of MoS will send a signal to its subcomponents DaD, MuD and EvA, requesting them to switch mode based on the mode mapping defined in Table 1 (a). The mode-switch managers of different components are jointly responsible for propagating a mode-switch event to the affected components, keeping mode consistency between components and coordinating the mode-switches of different components. Designing the mode-switch manager is out of the scope of this article. We have previously developed distributed mode-switch algorithms [8,9] running in the mode-switch manager for the cooperative mode-switch of different components. Here, our focus is on mode mapping, also shown in Figure 2.

Since we assume that multi-mode components are independently developed, they typically support different numbers of modes and name them differently. It is necessary to specify the relation between the modes of different components at design-time without ambiguity. Such a specification is called mode mapping. To ensure reusability, the mode mapping must never violate the following principles:

- A primitive component only knows own mode's information such as supported modes, initial mode and the current mode of itself.
- A composite component knows the mode information of itself and its immediate subcomponents.

The principles imply that mode mapping should be managed by each composite component, not by its subcomponents. A primitive component requires no mode mapping. The mode mapping defined in Table 1 is simple and intuitive; however, it is incapable of showing the initial mode of each component, which component initiates a mode-switch and the exact mode-switch of each individual component due to a mode-switch event. For example, when MoS switches from *Rm* to *Att*, according to Table 1 (a), MuD may switch to either *Ed* or *Dq*. Such non-determinism can be eliminated by specifying either *Ed* or *Dq* as the default new mode of MuD for this particular mode-switch scenario. To be able to formally specify all types of mode mapping rules, we propose a more powerful representation: the mode mapping automata.

2.2. Mode Mapping Automata

Let c be a composite component with $\mathcal{SC}_c$ being the set of subcomponents of c and P_c being the parent of c. When c is running in one of its supported modes, it should always know its current mode and the current modes of all $c_i \in \mathcal{SC}_c$ by its mode mapping. Moreover, whenever the mode-switch manager of c notices the mode-switch of $c_i \in \mathcal{SC}_c \cup \{c\}$, it will refer to the mode mapping, which should tell which other components among $\mathcal{SC}_c \cup \{c\} \setminus \{c_i\}$ should also switch mode as a consequence, as well as the new modes of these components.

The complete mode mapping of c can be formally presented by a set of MMA, which consists of one mode mapping automaton of c (denoted as MMA_c^s) and one MMA of each subcomponent $c_i \in \mathcal{SC}_c$ (denoted as $MMA_{c_i}^c$). Here, we call MMA_c^s a self-MMA and $MMA_{c_i}^c$ a child MMA.

As an example, Figure 3 presents the set of MMA of the component MuD in Figure 1, including a self-MMA (MMA_{MuD}^s) and three child MMA (MMA_{VAE}^c, MMA_{ViD}^c and MMA_{AuD}^c). These MMA are hierarchically organized in the same way as the corresponding components. Each MMA can receive and emit internal or external signals. Internal signals are used to synchronize the pair of the self-MMA and its child MMA while external signals interact with its local mode-switch manager for requesting and returning mode mapping results.

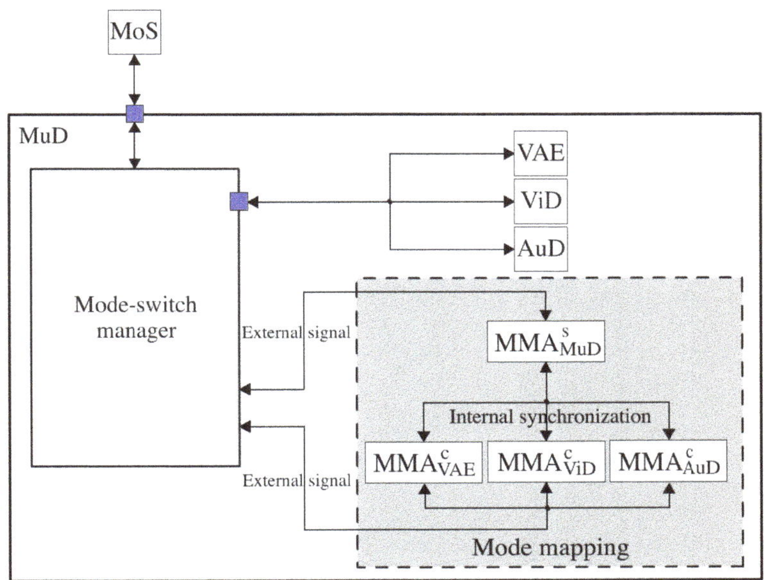

Figure 3. The role of the mode mapping of MuD at run-time. MMA, mode mapping automata.

An external signal indicates that a component is requested to switch to a particular mode. We use $x.E(y)$ to denote an external signal asking MMA_x, which is either a self-MMA or a child MMA, to switch to mode y. An internal signal is sent either from a self-MMA to a child MMA or from a child MMA to the self-MMA. A self-MMA only sends an internal signal to a child MMA if the current mode-switch event requires the mode-switch of the corresponding subcomponent. The self-MMA decides the new mode of this subcomponent. We use $x.I(y)$ to denote an internal signal emitted by a self-MMA to the child MMA MMA_x^c, asking the subcomponent x to switch to mode y. A child MMA can also send an internal signal to the self-MMA. This implies that the corresponding subcomponent is requesting a mode-switch. Since mode mapping is always determined by the self-MMA, the internal signal from a child MMA only needs to contain the current mode and new mode of the corresponding subcomponent. We use $x.I(z \rightarrow y)$ to denote an internal signal emitted by a child MMA MMA_x^c that requests to switch mode from z to y. Note that z must be present in this internal signal, as $x.I(z \rightarrow y)$ and $x.I(z' \rightarrow y)$ are two different mode-switch scenarios, which may lead to different mode mappings.

A self- or child MMA can be formally defined as follows:

Definition 1. *MMA: An MMA is defined as a tuple:*

$$\langle \mathcal{S}, s^0, \mathcal{SI}, \mathcal{T} \rangle$$

where $\mathcal{S}$ is a set of states; $s^0 \in \mathcal{S}$ is the initial state; $\mathcal{SI} = \mathcal{I} \cup \mathcal{E}$ ($\mathcal{I} \cap \mathcal{E} = \varnothing$) is a set of signals received or emitted during a state transition, with $\mathcal{I}$ as the set of internal signals and $\mathcal{E}$ as the set of external signals; $\mathcal{T} \subseteq \mathcal{S} \times \mathcal{SI} \times 2^{\mathcal{SI}} \times \mathcal{S}$ is a set of transitions of the MMA.

We use a state machine for the graphical representation of an MMA, where each state is one mode and each transition is a result of mode-switch. Ordinary states are marked by a circle, while the initial state is marked by a double circle. If the MMA is a self-MMA of c, then each state corresponds to a mode of c. If the MMA is a child MMA of c associated with $c_i \in \mathcal{SC}_c$, then each state corresponds to a mode of c_i or the deactivated status of c_i, if c_i can be deactivated, denoted as D. When a composite component is deactivated, then all its enclosed components must also be deactivated. A transition $t \in \mathcal{T}$ is represented by an arrow from a state s to a state s', denoted as $s \xrightarrow{In/Out} s'$, where In/Out is the label of the transition. "In" is an external/internal signal as the input that triggers the transition. "Out" is a set of external/internal signals as the output of the transition.

Figure 4 depicts MMA^s_{MuD}, i.e., the self-MMA of MuD of the guiding example. Three states are included in this MMA, implying that MuD can run in three modes. The state transitions of MMA^s_{MuD} and the corresponding child MMA MMA^c_{VAE}, MMA^c_{ViD} and MMA^c_{AuD} (Figure 5) are manually specified to determine the mode mapping of MuD. As an example for demonstrating MMA synchronization, the top left transition of MMA^s_{MuD}, $Rd \xrightarrow{MuD.E(Ed)/\{VAE.I(R3),ViD.I(Evd),AuD.I(Rad)\}} Ed$, implies that MuD requests a mode-switch to Ed, consequently requiring its subcomponents VAE, ViD and AuD to switch to modes $R3$, Evd and Rad, respectively. Figure 5 shows that this transition of MMA^s_{MuD} is synchronized with three transitions of the child MMA: $D \xrightarrow{VAE.I(R3)/\{VAE.E(R3)\}} R3$, $Rvd \xrightarrow{ViD.I(Evd)/\{ViD.E(Evd)\}} Evd$, $D \xrightarrow{AuD.I(Rad)/\{AuD.E(Rad)\}} Rad$.

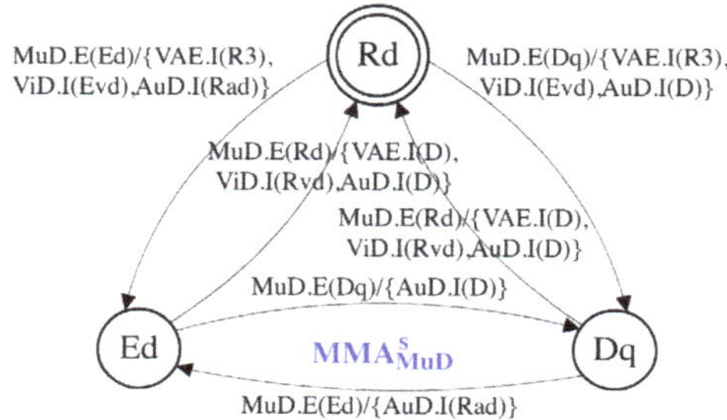

Figure 4. The self-mode mapping automaton of MuD.

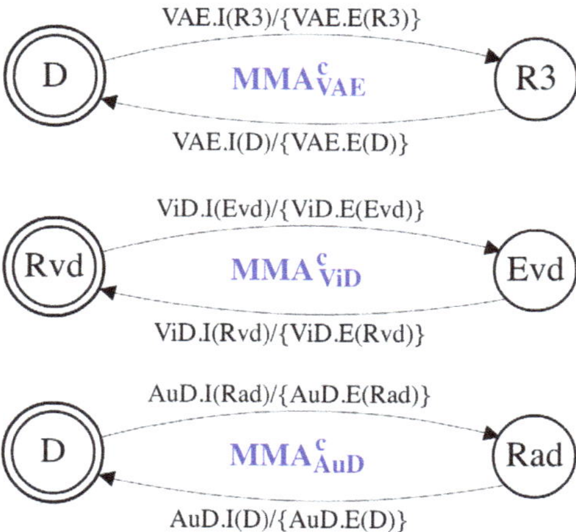

Figure 5. The child mode mapping automata of MuD.

2.3. MMA Composition

The internal synchronization between a set of MMA of a composite component is actually invisible to the mode-switch manager of the composite component. What the mode-switch manager sees is the composition of these MMA. MMA composition is achieved by merging a set of MMA into a single MMA without internal signals. For instance, based on the set of MMA of MuD depicted in Figures 4 and 5, the composed MMA is illustrated in Figure 6. After MMA composition, the mode mapping of MuD is defined by a single MMA, where each state represents a mode combination between MuD and its subcomponents, i.e., $s_1 = (Rd, D, Rvd, D)$, $s_2 = (Ed, R3, Evd, Rad)$ and $s_3 = (Dq, R3, Evd, D)$. All internal signals are eliminated. This composed MMA is the actual MMA referenced by the mode-switch manager of MuD, since the mode-switch manager does not care about the internal synchronization of a set of MMA. However, the composed MMA can be much more complex than any single MMA before the composition. Instead of specifying mode-switch directly with the composed MMA, it is much easier to design mode mapping with a self-MMA and the child MMA first and then compose them. The synchronization semantics of a set of MMA and the formal definition of MMA composition can be found in the extended technical report [10].

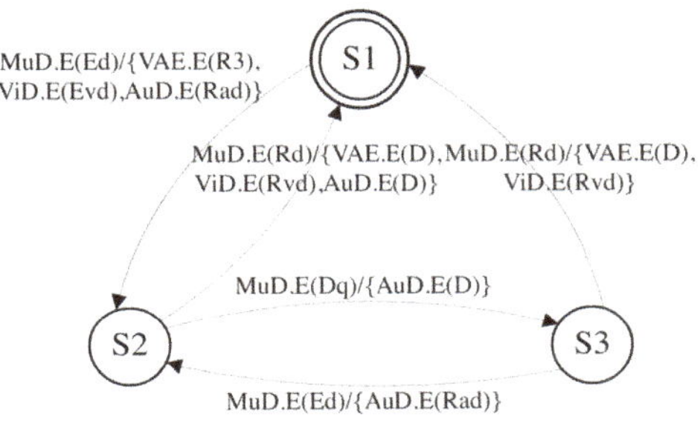

Figure 6. MMA composition for MuD.

2.4. Mode Mapping Verification

A crucial issue in designing mode mapping with MMA is ensuring the correctness of the mode mapping, i.e., for each input external signal from the mode-switch manager, the set of MMA should produce the expected set of external signals as the output back to the mode-switch manager. For instance, failing to synchronize an internal signal will never yield a mode mapping result.

The mode mapping of a composite component specified by MMA can be easily verified by model checking. We use the model checker UPPAAL [11] for mode mapping verification. Since UPPAAL is a convenient tool for modeling and verifying concurrent state transition systems, it is fairly straightforward to graphically model a set of MMA in UPPAAL. Using UPPAAL, we have modeled (the UPPAAL model is available at http://mdh.diva-portal.org/smash/record.jsf?pid=diva2%3A1244506& dswid=1426) the mode mapping of MuD specified by the set of MMA in Figures 4 and 5.

The behaviors of the local mode-switch manager of MuD, the self-MMA of MuD and the child MMA of its subcomponents are modeled as separate automata in UPPAAL. For instance, Figure 7 showcases three typical UPPAAL models for the mode-switch manager of MuD, MMA^c_{VAE} and MMA^s_{MuD}. Each mode of a component is represented by a state in these UPPAAL models (e.g., *mode_R3* represents mode $R3$ in Figure 7b). These models also contain committed states marked with "C" in a circle, which are intermediate states during a mode-switch. External and internal signals are simulated as channels synchronized between multiple UPPAAL models. For example, *VAE_I[R3]!* denotes the internal signal VAE.I(R3) emitted by MMA^s_{MuD}, while *VAE_I[R3]?* denotes the same signal VAE.I(R3)received by MMA^c_{VAE}. The UPPAAL model of MMA^s_{MuD} in Figure 7c is consistent with MMA^s_{MuD} in Figure 4. The reason why the UPPAAL model contains one or more intermediate states for each mode-switch is that receiving and sending each signal needs to be modeled sequentially in UPPAAL. This essentially does not change the execution semantics, as all intermediate states are committed states, whose incoming and outgoing transitions are performed as a single atomic transaction. In addition, shown in Figure 7a, the mode-switch manager of MuD consists of two states. *InitialS* is the initial state, where the mode-switch manager can send an external signal to MMA^s_{MuD} and switch to the state *ModeSwitching*. Meanwhile, a Boolean variable *switching* is set to true, indicating an ongoing mode-switch. Depending on the current mode of MuD and the new mode indicated by the external signal from the mode-switch manager, there are four possible events, leading to different transitions among these components: (1) k_1: MuD requests to switch from *Rd* to *Ed*; (2) k_2: MuD requests to switch from *Ed* to *Rd*; (3) k_3: MuD requests to switch from *Ed* to *Dq*; (4) k_4: MuD requests to switch from *Dq* to *Ed*. Each event ID is assigned to a variable *eventID* as shown in Figure 7c.

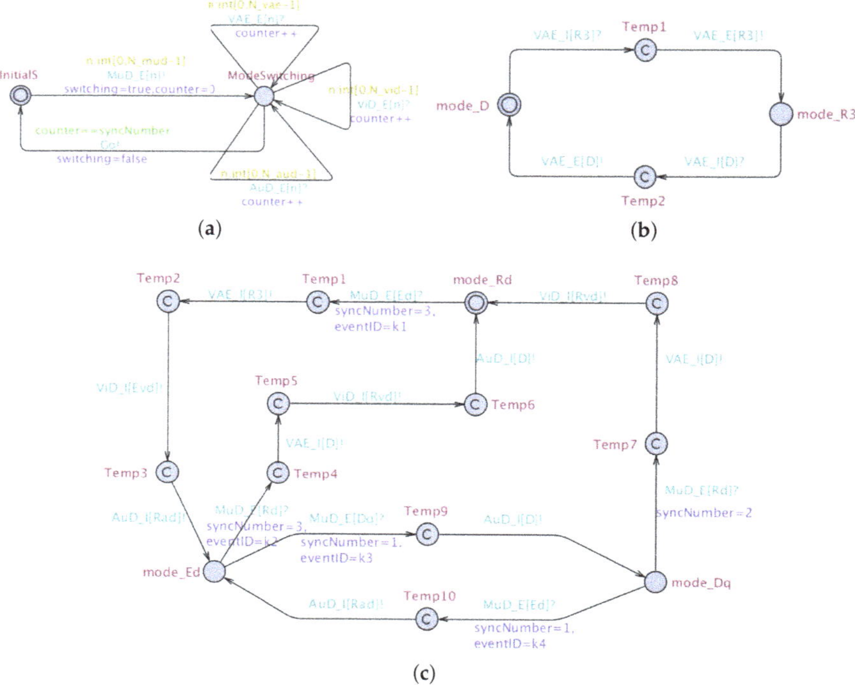

Figure 7. UPPAALmodels of the mode mapping of MuD. (**a**) UPPAAL model for the mode-switch manager of MuD. (**b**) UPPAAL model for the child MMA of VAE. (**c**) UPPAAL model for the self-MMA of MuD.

Based on the UPPAAL models, we can verify that the set of MMA of MuD satisfies the expected constraints by checking properties formulated in the UPPAAL query language, which is a subset of timed computation tree logic [12]. The following are four types of properties addressing different constraints:

P1. *A[] not deadlock*: The complete set of UPPAAL models is deadlock-free. This is not directly related to mode mapping, but it is a fundamental property that we expect the model to satisfy.
P2. *E<> sMMA_MuD.mode_Ed*: It is possible for MuD to run in mode *Ed*. This property should be verified for all the modes of MuD and its subcomponents.
P3. *A[] (sMMA_MuD.mode_Rd and !ModeSwitchManager.switching) imply (cMMA_VAE.mode_D and cMMA_ViD.mode_Rvd and cMMA_AuD.mode_D)*: When MuD runs in *Rd*, its subcomponents VAE and AuD must be deactivated, while the other subcomponent ViD must run in *Rvd*. This property should be verified for all possible mode combinations between MuD and its subcomponents according to the mode mapping table in Table 1.
P4. *(ModeSwitchManager.switching and eventID==k1)–>(sMMA_MuD.mode_Ed and cMMA_VAE.mode_R3 and cMMA_ViD.mode_Evd and cMMA_AuD.mode_Rad)*: An external signal requesting MuD to

switch from *Rd* to *Ed* will make VAE, ViD and AuD switch to *R3*, *Evd* and *Rad*, respectively. This property should be verified for all possible events from k_1–k_4.

All these properties are satisfied with verification time less than 4 ms. Furthermore, our UPPAAL models can be used as a common template for modeling any other mode mapping specified by MMA. Due to the graphical resemblance between an MMA and the corresponding UPPAAL model, it is possible to generate UPPAAL models from MMA described by a graphical or textual domain-specific language.

3. Mode Transformation

Our previous research results [9] show that the mode-switch of a multi-mode component may lead to mode-switches of other multi-mode components in the same system, and it is not trivial to coordinate the mode-switches of different components at run-time. The local mode-switch manager of each component needs to run delicate algorithms to communicate with the parent and subcomponents of the component via dedicated mode-switch ports to switch mode cooperatively. Such inter-component communication incurs run-time computation overhead and mode-switch latency. For instance, when MoS in the healthcare monitoring system triggers a mode-switch from *Rm* to *Att*, the mode-switch event is first propagated from MoS to MuD and EvA, and MuD subsequently propagates the mode-switch event to VAE, ViD and AuD. Further, more handshake messages are exchanged between these components to keep mode consistency. The communication overhead grows as the component hierarchy becomes more complex.

The purpose of mode transformation is to eliminate the need for the mode-switch coordination among different multi-mode components by a centralized mode management, and thereby achieve better run-time performance, provided that (1) all components are deployed on the same hardware platform and (2) the mode information of each component is globally accessible. Illustrated in Figure 8, mode transformation transfers the responsibility of mode-switch handling from the local mode-switch manager of each component to a single global mode-switch manager. As a result of mode transformation, each multi-mode component becomes unaware of modes. Instead, a global mode transition graph is generated for the global mode-switch manager to handle mode-switch at the system level.

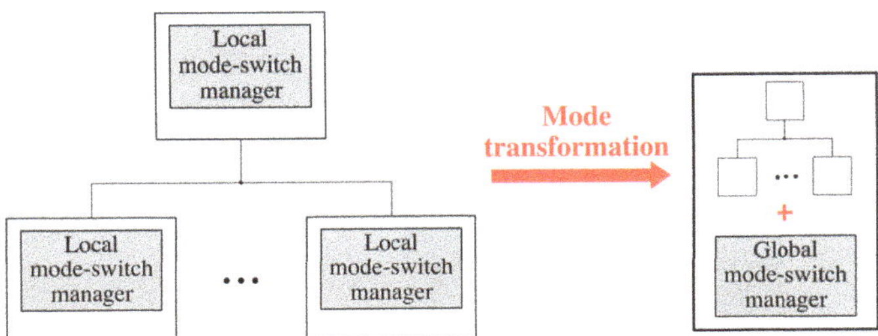

Figure 8. The overview of mode transformation.

Our mode transformation process includes two sequential steps. First, given the mode mappings of all composite components, we construct an intermediate representation, a mode combination tree (MCT), where all the possible system modes are identified. In the second step, based on a list of possible mode-switch events defined in the system, we add transitions between the identified system modes to construct the mode transition graph. The two steps are further explained in the following subsections separately.

3.1. Construction of the Mode Combination Tree

The aim of constructing the MCT is to identify all the system modes. Let $\mathcal{M}_c$ denote the set of supported modes of a component c and D denote the current mode of a deactivated component. Then, we define system modes as follows:

Definition 2. *System modes based on component modes: For a system composed by a set of components* $\mathcal{C} = \{c_1, c_2, \cdots, c_n\}$ $(n \in \mathbb{N})$, *the set of system modes is defined as* $\mathcal{M}_s \subseteq \underset{i \in [1,n]}{\times} \{\mathcal{M}_{c_i} \cup \{D\}\}$. *Each system mode* $m \in \mathcal{M}_s$ *is a mode combination of all components.*

By Definition 2, each system mode $m = (m_{c_1}, m_{c_2}, \cdots, m_{c_n})$, where $m_{c_i} \in \mathcal{M}_{c_i} \cup \{D\}$ for $i \in [1, n]$. In order to indicate more explicitly the relationship between c_i and m_{c_i}, we shall hereafter use an alternative expression to represent a system mode: $m = \{(c_i, m_{c_i}) | i \in [1, n]\}$, where $m_{c_i} \in \mathcal{M}_{c_i} \cup \{D\}$. Using the same formalism, an MCT is defined as follows:

Definition 3. *Mode combination tree: An MCT is a tree with a set of nodes* $\mathcal{N} = \{\mathcal{N}_0, \mathcal{N}_1, \cdots, \mathcal{N}_n\}$ $(n \in \mathbb{N})$, *where* $\mathcal{N}_0 = \varnothing$ *is the root node, and each other node* $\mathcal{N}_i = \{(c_j, m_{c_j}) | j \in [1, k], k \in \mathbb{N}\}$ $(i \in [1, n])$, *where for all* j, $m_{c_j} \in \mathcal{M}_{c_j} \cup \{D\}$ *and all* c_j *have the same depth level in the system component hierarchy.*

By Definition 3, each non-root node of an MCT provides a mode combination of components with the same depth level. A typical outlook of MCT is displayed in Figure 9, while the construction of the MCT will be further explained later.

A few more notations and concepts need to be introduced before the formal description of the MCT construction process. First, we introduce the valid local mode combination (LMC) of a composite component c, which is a feasible combination of a mode of c and the modes of all its subcomponents as per the local mode mapping of c. To define the valid LMC of a composite component formally, let $\mathcal{PC}$ and $\mathcal{CC}$ be the set of primitive components and composite components in a system, respectively. Let Top be the component at the top of the component hierarchy in a system. For each $c \in \mathcal{CC}$, a valid LMC of c is formally defined as follows:

Definition 4. *Valid local mode combination: For* $c \in \mathcal{CC}$ *with* $\mathcal{SC}_c = \{c_i^1, \cdots, c_i^n\}$ $(n \in \mathbb{N})$, *we call the set* $\mathcal{V}_c = \{(c, m_c), (c_i^1, m_{c_i^1}), \cdots, (c_i^n, m_{c_i^n})\}$ *a valid LMC of* c, *where* $m_c \in \mathcal{M}_c \cup \{D\}$ *and* $\forall k \in [1, n]$, $m_{c_i}^k \in \mathcal{M}_{c_i^k} \cup \{D\}$, *if* $(m_c, m_{c_i^1}, \cdots, m_{c_i^n})$ *is a state of the composed MMA of* c.

Note that each element in $\mathcal{V}_c$ is a pair (x, y), where $x \in \mathcal{SC}_c \cup \{c\}$ and $y \in \mathcal{M}_x \cup \{D\}$. For instance, the mode mapping of MoS in Table 1 (a) implies three valid LMCs of MoS: (1) $\{(\text{MoS}, Rm), (\text{DaD}, R1), (\text{MuD}, Rd), (\text{EvA}, D)\}$; (2) $\{(\text{MoS}, Att), (\text{DaD}, R1), (\text{MuD}, Ed), (\text{EvA}, R2)\}$; (3) $\{(\text{MoS}, Att), (\text{DaD}, R1), (\text{MuD}, Dq), (\text{EvA}, R2)\}$.

Based on Definition 4, we further introduce the valid LMC concerning a specific mode of a composite component c, which is a feasible combination of the modes of all subcomponents of c as per the local mode mapping of c when c is running in a particular mode. A formal definition is given as follows:

Definition 5. *Valid LMC concerning a specific mode: For* $c \in \mathcal{CC}$ *with* $\mathcal{SC}_c = \{c_i^1, \cdots, c_i^n\}$ $(n \in \mathbb{N})$, *if when* c *is running in* m_c, *and* $\forall c_i^k \in \mathcal{SC}_c$ $(k \in [1, n])$, $\exists m_{c_i^k}$ *such that* $\{(c, m_c), (c_i^1, m_{c_i^1}), \cdots, (c_i^n, m_{c_i^n})\}$ *is a valid LMC of* c, *then the set* $\mathcal{V}_{c, m_c} = \{(c_i^1, m_{c_i^1}), \cdots, (c_i^n, m_{c_i^n})\}$ *is a valid LMC of* c *for* m_c.

Depending on the mode mapping of c, multiple valid LMCs of c may exist for m_c. Let $\mathcal{W}_{c, m_c}$ be the set of all valid LMCs of $c \in \mathcal{CC}$ for m_c. Each element in $\mathcal{W}_{c, m_c}$ is a set $\mathcal{V}_{c, m_c}$. The total number of all valid LMCs of c for m_c is $|\mathcal{W}_{c, m_c}|$. For instance, according to Table 1 (a),

$\mathcal{W}_{\text{MoS},Att} = \{\mathcal{V}^1_{\text{MoS},Att}, \mathcal{V}^2_{\text{MoS},Att}\}$, where $\mathcal{V}^1_{\text{MoS},Att} = \{(DaD, R1), (MuD, Ed), (EvA, R2)\}$ and $\mathcal{V}^2_{\text{MoS},Att} = \{(DaD, R1), (MuD, Dq), (EvA, R2)\}$. By traversing the states of the composed MMA of c containing m_c, it is easy to automatically generate $\mathcal{W}_{c,m_c}$.

Next, we introduce an important operator for combining different valid LMCs:

Definition 6. *Valid LMC operation: Consider two sets of valid LMCs* $\mathcal{W}_1 = \{\mathcal{V}_1, \mathcal{V}_2, \cdots, \mathcal{V}_m\}$ *and* $\mathcal{W}_2 = \{\mathcal{V}_{k+1}, \mathcal{V}_{k+2}, \cdots, \mathcal{V}_{k+n}\}$, *where* $m, n, k \in \mathbb{N}$ *and* $k \geq m$. *Let* $\oplus$ *be an operator such that* $\mathcal{W}_1 \oplus \mathcal{W}_2 = \{\mathcal{V}_i \cup \mathcal{V}_{k+j} | i \in [1, m], j \in [1, n]\}$. *In addition, for each* $l \in \mathbb{N}$, $\mathcal{W}_1 \oplus \mathcal{W}_2 \oplus \cdots \oplus \mathcal{W}_l$ *can be represented as* $\bigoplus\limits_{o \in [1,l]} \mathcal{W}_o$.

For the sake of clarity, let us clarify the $\oplus$ operator using a small example. Suppose $\mathcal{W}_1 = \{\mathcal{V}_1, \mathcal{V}_2\}$ where $\mathcal{V}_1 = \{(a, m_a^1), (b, m_b^1)\}$ and $\mathcal{V}_2 = \{(a, m_a^2), (b, m_b^2)\}$; and $\mathcal{W}_2 = \{\mathcal{V}_3, \mathcal{V}_4\}$ where $\mathcal{V}_3 = \{(c, m_c^1), (d, m_d^1)\}$ and $\mathcal{V}_4 = \{(c, m_c^2), (d, m_d^2)\}$. Then,

$$\begin{aligned}
\mathcal{W}_1 \oplus \mathcal{W}_2 =& \{\mathcal{V}_1 \cup \mathcal{V}_3, \mathcal{V}_1 \cup \mathcal{V}_4, \mathcal{V}_2 \cup \mathcal{V}_3, \mathcal{V}_2 \cup \mathcal{V}_4\} \\
=& \{\{(a, m_a^1), (b, m_b^1), (c, m_c^1), (d, m_d^1)\}, \\
& \{(a, m_a^1), (b, m_b^1), (c, m_c^2), (d, m_d^2)\}, \\
& \{(a, m_a^2), (b, m_b^2), (c, m_c^1), (d, m_d^1)\}, \\
& \{(a, m_a^2), (b, m_b^2), (c, m_c^2), (d, m_d^2)\}\}
\end{aligned}$$

Given the mode mappings of all composite components, the MCT of the system can be constructed by creating nodes top-down from the root node. For each node $\mathcal{N}$ of an MCT, let $d_{\mathcal{N}}$ be its depth level and $\lambda_{\mathcal{N}}$ be the number of new nodes created from this node. We use $\mathcal{N}_i \succ \mathcal{N}_j$ to denote that a new node $\mathcal{N}_i$ is created from on old node $\mathcal{N}_j$. Moreover, let $\mathcal{M}_{Top} = \{m_T^1, m_T^2, \cdots, m_T^{|\mathcal{M}_{Top}|}\}$ be the set of supported modes of *Top*. The MCT is constructed by the following steps:

1. From $\mathcal{N}_0$, create $\lambda_{\mathcal{N}_0} = |\mathcal{M}_{Top}|$ new nodes, such that for each new node $\mathcal{N}_i \succ \mathcal{N}_0$, $\mathcal{N}_i = \{(Top, m_T^i)\}$ $(i \in [1, |\mathcal{M}_{Top}|])$.
2. From each $\mathcal{N}_i = \{(Top, m_T^i)\}$ $(i \in [1, |\mathcal{M}_{Top}|])$, create $\lambda_{\mathcal{N}_i} = |\mathcal{W}_{Top,m_T^i}|$ new nodes, such that for each $\mathcal{N}' \succ \mathcal{N}_i, \mathcal{N}' \in \mathcal{W}_{Top,m_T^i}$. Moreover, if $\lambda_{\mathcal{N}_i} > 1$, then for each $\mathcal{N}', \mathcal{N}'' \succ \mathcal{N}_i$, we have $\mathcal{N}' \neq \mathcal{N}''$.
3. For each node $\mathcal{N} = \{(c_1, m_{c_1}), (c_2, m_{c_2}), \cdots, (c_n, m_{c_n})\}$ $(n \in \mathbb{N})$ with $d_{\mathcal{N}} \geq 2$, if $\forall i \in [1, n]$, $c_i \in \mathcal{PC}$, then $\mathcal{N}$ is marked as a leaf node, and no new node is created from $\mathcal{N}$. Otherwise, if $\exists i \in [1, n]$ such that $c_i \in \mathcal{CC}$, then create $\lambda_{\mathcal{N}} = \prod\limits_{\substack{i \in [1,n], \\ c_i \in \mathcal{CC}}} |\mathcal{W}_{c_i,m_{c_i}}|$ new nodes, such that for each

   $\mathcal{N}' \succ \mathcal{N}, \mathcal{N}' \in \bigoplus\limits_{\substack{i \in [1,n], \\ c_i \in \mathcal{CC}}} \mathcal{W}_{c_i,m_{c_i}}$. Moreover, if $\lambda_{\mathcal{N}} > 1$, then for each $\mathcal{N}', \mathcal{N}'' \succ \mathcal{N}$, we have $\mathcal{N}' \neq \mathcal{N}''$.
4. Repeat Step 3 until all branches of the MCT have reached the leaf node.

The MCT construction process is implemented as Algorithm 1, which is a recursive function *constructMCT*$(\mathcal{N}, d_{\mathcal{N}})$ that has two input parameters: $\mathcal{N}$ is the node currently being explored and $d_{\mathcal{N}}$ is the depth level of $\mathcal{N}$. Initially, $\mathcal{N} = \varnothing$ and $d_{\mathcal{N}} = 0$. We assume that *Top* must have subcomponents. Otherwise, *Top* itself will be the entire system, and mode transformation will be meaningless. Moreover, for each component c running in mode m, we assume that $\mathcal{W}_{c,m}$ is an indexed set such that $\mathcal{W}_{c,m}[i]$ represents the i-th element of $\mathcal{W}_{c,m}$.

Algorithm 1 *constructMCT*$(\mathcal{N}, d_{\mathcal{N}})$.

1: **if** $d_{\mathcal{N}} = 0$ **then**

2:     $\lambda_{\mathcal{N}} := |\mathcal{M}_{Top}|$;
3:     **for** i *from* 1 *to* $\lambda_{\mathcal{N}}$ **do**

4:        $\mathcal{N}_i := \{(Top, m^i_T)\}$;
5:        *constructMCT*$(\mathcal{N}_i, 1)$;
6:     **end for**
7: **end if**
8: **if** $d_{\mathcal{N}} = 1$ **then**

9:     $\{(Top, m)\} := \mathcal{N}$;
10:     *Derive* $\mathcal{W}_{Top,m}$;
11:     $\lambda_{\mathcal{N}} := |\mathcal{W}_{Top,m}|$;
12:     **for** i *from* 1 *to* $\lambda_{\mathcal{N}}$ **do**

13:        *constructMCT*$(\mathcal{W}_{Top,m}[i], 2)$;
14:     **end for**
15: **end if**
16: **if** $d_{\mathcal{N}} \geq 2$ **then**

17:     $\{(c_1, m_{c_1}), (c_2, m_{c_2}), \cdots, (c_n, m_{c_n})\} := \mathcal{N}$;
18:     **if** $\forall i \in [1, n] : c_i \in \mathcal{PC}$ **then**

19:        **return** ;
20:     **else**

21:        *Derive* $\mathcal{W} := \bigoplus\limits_{\substack{i \in [1,n], \\ c_i \in \mathcal{CC}}} \mathcal{W}_{c_i, m_{c_i}}$;

22:        $\lambda_{\mathcal{N}} := \prod\limits_{\substack{i \in [1,n], \\ c_i \in \mathcal{CC}}} |\mathcal{W}_{c_i, m_{c_i}}|$;

23:        **for** i *from* 1 *to* $\lambda_{\mathcal{N}}$ **do**

24:           *constructMCT*$(\mathcal{W}[i], d_{\mathcal{N}} + 1)$;
25:        **end for**
26:     **end if**
27: **end if**

Once the MCT is constructed, the system modes can be derived as the set of paths from the root node to the leaf nodes of the MCT. The total number of system modes is equal to the total number of leaf nodes of the MCT. Among the system modes, the initial system mode can be recognized based on the specification of the initial modes of all components.

As an example, Figure 9 illustrates the MCT of the monitoring subsystem introduced in Section 1. The MCT consists of nine nodes $\mathcal{N}_0$–$\mathcal{N}_8$ with four depth levels. Represented by the respective paths of the MCT, one of the three identified system modes is:

$$m_1 = \mathcal{N}_0 \cup \mathcal{N}_1 \cup \mathcal{N}_3 \cup \mathcal{N}_6$$
$$= \{(MoS, Rm), (DaD, R1), (MuD, Rd), (EvA, D), (VAE, D), (ViD, Rvd), (AuD, D)\}$$
$$m_2 = \mathcal{N}_0 \cup \mathcal{N}_2 \cup \mathcal{N}_4 \cup \mathcal{N}_7$$
$$= \{(MoS, Att), (DaD, R1), (MuD, Ed), (EvA, R2), (VAE, R3), (ViD, Evd), (AuD, Rad)\}$$
$$m_3 = \mathcal{N}_0 \cup \mathcal{N}_2 \cup \mathcal{N}_5 \cup \mathcal{N}_8$$
$$= \{(MoS, Att), (DaD, R1), (MuD, Dq), (EvA, R2), (VAE, R3), (ViD, Evd), (AuD, D)\}$$

Assuming that the monitoring subsystem starts with mode Rm, m_1 is the initial system mode after mode transformation. Figure 10 shows the configurations of the three system modes based on the component connections in Figure 1.

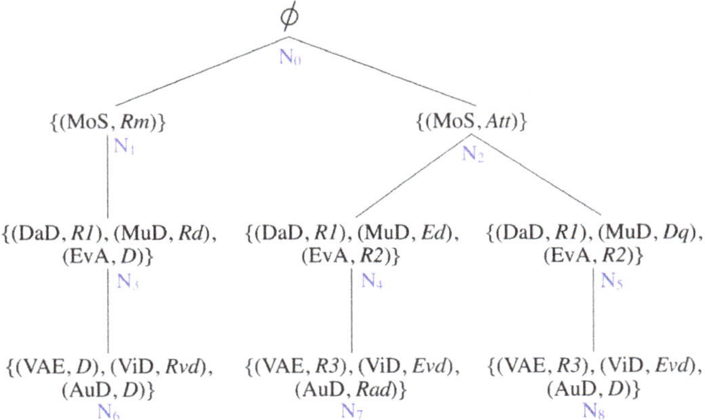

Figure 9. The mode combination tree of the monitoring subsystem.

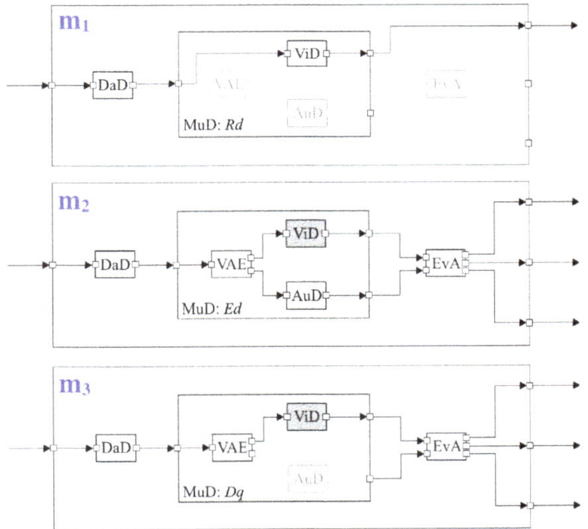

Figure 10. The configurations of different system modes after mode transformation.

The complexity of an MCT depends on the structure of the component hierarchy, the number of modes of each component and the mode mappings in the involved components. The worst-case combination of factors, such as the number of components and the number of component modes, may lead to a huge number of system modes, increasing the overhead exponentially. However, in practice, the expected number of system modes should be limited. If mode transformation becomes intractable due to extreme computation overhead, this would imply that the system is too complex to adopt centralized mode management. Then, it may be more suitable to go for distributed mode management without mode transformation, although the run-time overhead for the required message exchange may be substantial if the component hierarchy is deep. Alternatively, a better solution could be partial mode transformation, i.e., performing mode transformation within one or more composite components instead of the entire system. Our mode transformation technique is flexible enough to support partial mode transformation at any component level. Furthermore, we expect noticeably different behaviors in different modes. Depending on the application, it could be more efficient to merge several modes with similar global configurations into a single mode. The criteria for merging system modes are application-dependent and out of the scope of this article. Nevertheless, we believe that it is possible to partially automate the merging of system modes in a later optimization phase by certain application independent merging rules.

3.2. Deriving the Mode Transition Graph

The constructed MCT identifies system mode, which is subsequently used to derive the mode transition graph on top of these system modes based on the definition of mode-switch events. We assume that a mode-switch event is triggered by a component c requesting to switch mode from m_c^1 to m_c^2, denoted as $c : m_c^1 \rightarrow m_c^2$. The triggering of each mode-switch event may lead to the mode-switches of some other components in the same system. For a system with a set of identified system modes $\mathcal{M} = \{m_1, m_2, \cdots, m_n\}$ $(n \in \mathbb{N})$, a mode-switch is a transition from m_{old} to m_{new}, where $m_{old}, m_{new} \in \mathcal{M}$ and $m_{old} \neq m_{new}$. A mode transition graph contains all the possible transitions between these system modes and associates each transition with the corresponding mode-switch event. Similar to an MMA, each state of a mode transition graph can be graphically represented by a circle, with the initial state being marked by a double circle. A graphical illustration of the mode transition graph can be found in Figure 11.

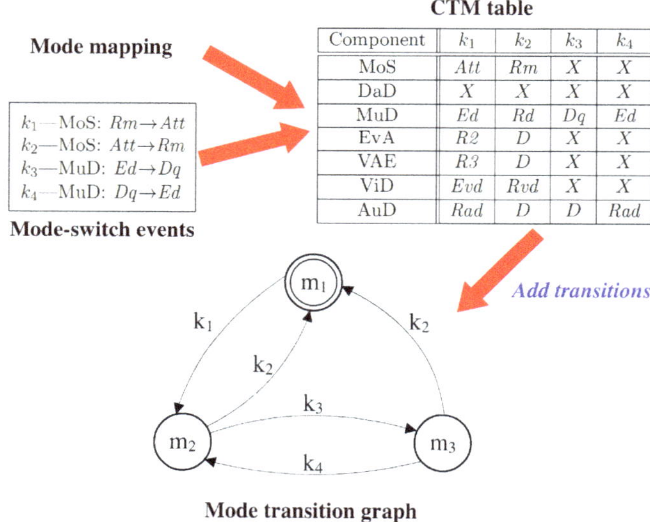

Component	k_1	k_2	k_3	k_4
MoS	Att	Rm	X	X
DaD	X	X	X	X
MuD	Ed	Rd	Dq	Ed
EvA	$R2$	D	X	X
VAE	$R3$	D	X	X
ViD	Evd	Rvd	X	X
AuD	Rad	D	D	Rad

Figure 11. Deriving the mode transition graph of the monitoring subsystem. CTM, component target mode.

The key issue of deriving the mode transition graph is to identify the system modes m_{old} and m_{new} for each mode-switch event for which a system mode-switch is possible. Consider a mode-switch event k identified as $c : m_c^1 \rightarrow m_c^2$. The only condition satisfying the triggering of k is that the triggering source c is currently running in mode m_c^1. For each k, m_{old} can be easily identified as long as $(c, m_c^1) \in m_{old}$. Note that more than one system mode could be identified as m_{old}. Depending on the current system mode, a mode-switch event may enable different transitions.

In contrast to m_{old}, only one system mode can be the m_{new} for each mode-switch event k. The identification of m_{new} for k is more difficult because it depends not only on m_c^2, but also on the target modes of the other components. We identify the m_{new} for each mode-switch event with the assistance of a component target mode (CTM) table. A CTM table is a table with n_1 rows and n_2

columns, where n_1 is the number of components of a system and n_2 is the number of mode-switch events. An example of a CTM table is shown above the mode transition graph in Figure 11. In the CTM table, each row is associated with a component, each column is associated with a mode-switch event and each cell contains the target mode m_c of the corresponding component c for the corresponding mode-switch event k. The cell with X indicates that m_c is independent of k, i.e., k does not lead to the mode-switch of c.

A CTM table can be automatically constructed offline based on the list of mode-switch events and the mode mapping of each composite component. Let m_c^k be the target mode of c for k in a CTM table. Taking advantage of the CTM table, the new system mode m_{new} for each mode-switch event k can be identified as follows: For each system mode $m = \{(c_i, m_{c_i}) | i \in [1, n], n \in \mathbb{N}\}$, if $\forall i$ where $m_{c_i}^k \neq X$ in the CTM table (i.e., k leads to the mode-switch of c_i to a new mode $m_{c_i}^k$), we have $m_{c_i} = m_{c_i}^k$, then m is the m_{new} for k. Algorithm 2 describes the process of building the mode transition graph, with a search space of $O(|\mathcal{M}| \cdot |\mathcal{K}|)$.

Algorithm 2 *constructMTG*$(\mathcal{C}, \mathcal{M}, \mathcal{K})$.

1:   $\mathcal{C} = \{c_1, \cdots, c_o\}$ $(o \in \mathbb{N})$; *{The set of all components}*
2:   $\mathcal{M} = \{m_1, \cdots, m_n\}$ $(n \in \mathbb{N})$; *{The set of identified system modes}*
3:   $\mathcal{K} = \{k_1, \cdots, k_l\}$ $(l \in \mathbb{N})$; *{The set of all mode-switch events}*
4:   **for all** $k_i \in \mathcal{K}$ *where* $k \in [1, l]$ *and* $k_i = c : m_c^1 \to m_c^2$ **do**

5:      **if** $\exists m_j \in \mathcal{M}$ *s.t.* $(\forall c_p \in \mathcal{C}$ *and* $m_{c_i}^{k_i} \neq X, (c_p, m_{c_p}^{k_i}) \in m_j)$ **then**

6:        $m_{new} = m_j$;
7:        **for all** $m_j \in \mathcal{M}$ **do**

8:          **if** $(c, m_c^1) \in m_j$ **then**

9:            *addTransition*(m_j, m_{new}, k_i); *{Add a transition from m_j to m_{new} labeled with k_i}*
10:          **end if**
11:        **end for**
12:      **end if**
13: **end for**

Figure 11 presents the workflow for deriving the mode transition graph of the monitoring subsystem. The CTM table is derived based on two inputs: (1) the mode mapping of composite components MoS and MuD specified by MMA and (2) the possible mode-switch events. For this example, four mode-switch events, from k_1–k_4, are specified at design time. k_1 and k_2 are triggered by MoS for switching between modes Rm and Att, while k_3 and k_4 are triggered by MuD for switching between modes Ed and Dq. The target modes of all components in the monitoring subsystem for all mode-switch events are listed in the CTM table. Previously, the MCT in Figure 9 has identified three system modes m_1 (the initial mode), m_2 and m_3. The CTM table additionally adds transitions between the system modes based on each possible mode-switch event, thereby yielding the mode transition graph. The mode transition graph helps the global mode-switch manager to keep track of the current system mode and makes the system switch to the right target mode when a mode-switch is triggered.

After mode transformation, local mode-switch managers are replaced with a single global mode-switch manager, whose complexity is much lower than each local mode-switch manager. A local mode-switch manager needs complex algorithms [9] to coordinate mode-switches between a composite component and its subcomponents, such as checking component states before mode-switch is performed, handling multiple concurrent mode-switch events triggered by different components and handling emergency mode-switch events, which are more critical than regular mode-switch events. By contrast, the global mode-switch manager only takes care of system mode based on a single CTM table.

Mode transformation assumes no dynamic change of modes or mode mappings at any level. If there is a need to change the mode of a component or its mode mapping (e.g., adding new modes, removing modes, changing mode names), then mode transformation must be applied again from scratch. The chain effect of such a change must be considered when it is propagated to other components. The change of mode mapping of one component may entail the change of mode mapping of its parent component and subcomponents. The architecture designer should decide how to update the MMA of components impacted by a change. Mode transformation can always be automated in the same way once all MMA are in place.

A potential drawback of mode transformation is the loss of potential concurrency between local mode managers. If multiple mode-switch events are triggered concurrently and affect disjoint sets of components, distributed mode management before mode transformation allows that these mode-switch events can be handled concurrently, whereas different mode-switch events have to be sequentially handled by the global mode-switch manager. Nonetheless, the centralized mode management after mode transformation eliminates inter-component communication, which is a complex process [9]. Hence, mode transformation is still more likely to yield a faster mode-switch.

The correctness of the two steps of mode transformation has been verified by manual theorem proving. All the detailed theorems and proofs can be found in the extended technical report [10].

3.3. Concrete Implementation of Mode Transformation

A prototype tool MCORE [13], the Multi-mode COmponent Reuse Environment, has been developed to support the modeling of multi-mode systems with multi-mode components by integrating mode mapping and mode transformation. Compared with other component-based development tools, a distinguishing feature of MCORE is the reuse of multi-mode software components. As far as we know, MCORE is the first (and possibly only) tool for building multi-mode systems with multi-mode components. MCORE can be potentially used as a preprocessor for Rubus ICE [14], which is an IDEfor the Rubus component model [15] developed by Arcticus Systems (http://www.arcticus-systems.com/). As an industrial component model, Rubus is targeting the component-based development of vehicular systems. Rubus supports multi-mode systems; however, modes can only be specified at the system level, and the reuse of multi-mode components is not supported. This limitation can be alleviated by MCORE. The system model built by multi-mode components in MCORE is in compliance with the Rubus component model after mode transformation. Hence, the system model designed in MCORE can be imported to Rubus ICE for further analysis, test and code generation.

4. Related Work

The extended MechatronicUML [16,17] (EUML) allows the hierarchical composition of reconfigurable components, which are comparable to our multi-mode components. EUML introduces an additional reconfiguration port for each component, which resembles the dedicated mode-switch ports of a multi-mode component. In EUML, the reconfiguration of a composite component is handled by two dedicated subcomponents, which play similar roles as the local mode-switch manager of a multi-mode component. Unlike our approach, EUML does not pre-define component configurations at design-time, thus allowing more flexible reconfiguration at run-time. Compared with such reconfigurable systems, multi-mode systems built by multi-mode components are more predictable due to static configurations specified at design-time.

Pop et al. proposed an Oracle-based approach [18] that also supports the reuse of multi-mode components. Component behaviors are abstracted into a global property network. Component mode is treated as a property dependent on other property values. The change of one property is propagated throughout the property network, potentially leading to the change of other properties. At the end of propagation, component modes are updated top-down. Similar to our mode transformation, a finite-state machine called Oracle is offline constructed to guarantee a predictable update time of the property network. The mapping between component modes is however not systematically specified in the Oracle-based approach.

Weimer et al. proposed a set of input-output blocks for building multi-mode systems [19]. Each multi-mode component contains a set of mode blocks (MBs), while each MB includes all the components used for the corresponding mode. The mode-switch of a component is achieved by switching the currently selected MB controlled by a supervisor block (SB). These blocks were implemented in Simulink [20]. Another work similar to this is the mode-oriented design [21] in Gaspard2 [22]. A multi-mode component is represented by a macro component, which consists of a state graph component and one or more mode-switch components. A mode-switch component plays the same role as the MB in [19]. Both approaches in [19,21] use completely different components for different modes, whereas in our approach, it is possible to share some components and connections in different modes. Hence, our approach is more suitable for the reuse of multi-mode components.

Mode-switch has been addressed in a number of component models, e.g., SaveCCM [23], COMDES-II [24] and MyCCM-HI [25], to name a few. There are also some other component models that have been commercialized, e.g., Koala [26] (targeting consumer electronics) and Rubus [15] (targeting ground vehicles). These component models have different notions of mode-switch handling. Koala and SaveCCM both use a special connector *switch* to achieve the structural diversity of a component. *switch* selects outgoing connections based on input data. In COMDES-II, a state-machine component switches component configurations in different modes. Rubus only considers system-level mode, which is in line with our system mode after mode transformation. MyCCM-HI supports mode-aware components whose mode-switch is controlled by a mode automaton associated with each component. Another component model supporting component reconfiguration is Fractal [27]. Each Fractal component has a membrane (a container for local controllers) that is able to control the reconfiguration of the component.

Mode-switch has also been covered by some programming and specification languages, such as AADL [28], Giotto [29], TDL [30], the extended Darwin [31] and mode-automata [32]. In AADL, component mode-switch is represented by a state machine, including states, transitions and input/output event ports used for mode-switch triggering. Both Giotto and TDL are time-triggered languages for embedded programming, which require periodic checking of conditions to decide whether to trigger a mode-switch or not. The extended Darwin [31] extends the existing Architecture Description Language Darwin [33] by incorporating the notion of mode. The mode of a composite component is directly related to the modes of its subcomponents. Yet, the mapping between modes is unclear in [31]. Mode-automata is a programming model supporting the description of running modes of reactive systems. The behavior of a system is a sequence of modes, each of which corresponds to a collection of execution states. Our MMA differs from mode-automata in the sense that mode-automata specifies the hierarchical structure of system-wide modes, whereas MMA specifies the local mode mapping within composite components.

Dynamic software product lines (DSPL) [34], which originates from the conventional software product lines (SPL) [35] for producing a family of software systems, is an emerging technique for developing adaptive systems. Different systems configured from the same SPL share certain common features, whereas the SPL uses variation points to distinguish the unique features of each system. DSPL allows the binding of variation points at run-time so that a system can dynamically change configurations on the fly to accommodate the changing environment. DSPL is becoming more adaptive [36]; however, to the best of our knowledge, DSPL only considers global system configurations without considering reuse of adaptive software components.

Different types of automata have been proposed for component-based systems and multi-mode systems. For instance, constraint automata [37] is used to model the functional coordination of components, thereby enabling the formal verification of coordination mechanisms. Besides, multi-mode automata [38] is intended for compositional analysis of multi-mode real-time systems. The MMA presented in this article serves as a formalism for a unique and dedicated purpose: mode mapping, which to our knowledge has not been addressed by other existing automata.

Criado et al. [39] proposed a method for an adaptive component-based architecture using model transformation. Software architecture can be dynamically constructed based on transformation rules defined in a repository. Their proposal was applied to component-based GUIs for web applications. Compared to our approach, their adaptation runs at the system level only.

5. Conclusions and Future Work

Partitioning system behaviors into modes and component-based software engineering are both successful software development methods to tame the growing software complexity of modern cyber-physical systems (CPS). It is still an under-researched area to combine both methods, due to their conflicting natures: multi-mode systems are built top-down, while component-based systems are built bottom-up. In this article, we combine the advantages of both methods and propose the component-based software development of multi-mode systems, characterized by the reuse of multi-mode components, i.e., components that can run in different modes and switch mode guided by a local mode-switch manager. We specify the local mode mapping of each composite component by mode mapping automata. Mode mapping is then complemented by a mode transformation technique that transforms component modes to system modes for centralized mode management to improve the mode-switch performance, since the transformation eliminates the need for inter-component communication to coordinate a mode-switch at run-time, thereby reducing mode-switch overhead and shortening mode-switch time. Mode transformation is an optional and flexible process that can be taken for the entire system if the mode information of all components is globally accessible and all software components are deployed on the same hardware platform, or within certain composite components instead of the entire system. It can even be performed iteratively. For instance, in scenarios where systems are built from composite components provided by different vendors that do not want to reveal the internal structure of their components to the integrator, each vendor could apply mode transformation on the level of their respective composite component, and the integrator could then compose the resulting mode-mappings to a system-wide centralized mode management.

The healthcare monitoring system introduced in this article is only a proof-of-concept guiding example. In future work, our software development approach should be further evaluated, and before being deployed, its applicability and concrete implementation should be explored in more substantial real-world systems. Moreover, some remaining efforts need to be invested to complete the development of our prototype tool MCORE fully and its integration in the commercial tool Rubus ICE developed by Arcticus Systems. This will allow us to develop reusable multi-mode software components in MCORE as a preprocessor of Rubus ICE, perform mode transformation therein and then export the system model with global system modes to Rubus ICE for further analysis, test and code generation. Still, the actual effects in terms of resulting improvements, additional and/or reduced efforts, improved quality, etc., throughout the life-cycle of a CPS require empirical evidence much beyond what is presented in this article.

At a more general level, this article presents essential bricks and related glue for a small part of the wall needed to tame the complexity of CPS-related development and life-cycle challenges to a level that allows future deployment of the many technical solutions required to address several of the key challenges of modern society successfully. Many more bricks are however needed, as well as the glue that enables their successful composition.

Author Contributions: methodology, H.H. (Hang Yin); validation, H.H. (Hang Yin); formal analysis, H.H. (Hang Yin); writing—original draft preparation, H.H. (Hang Yin); writing—review and editing, H.H. (Hans Hansson); supervision, H.H. (Hans Hansson); project administration, H.H. (Hans Hansson).

Funding: This work was funded by the Swedish Research Council via the framework project ARROWS (ref. 90447401) and Mälardalen University

Acknowledgments: The authors would like to thank Arcticus Systems for discussions and support.

Conflicts of Interest: The authors declare no conflict of interest.

References

1. Rajkumar, R.; Lee, I.; Sha, L.; Stankovic, J. Cyber-physical systems: The next computing revolution. In Proceedings of the Design Automation Conference, Anaheim, CA, USA, 13–18 June 2010; pp. 731–736.

2.   Degani, A.; Kirlik, A. Modes in human-automation interaction: Initial observations about a modeling approach. In Proceedings of the IEEE International Conference on Systems, Man, and Cybernetics, Vancover, BC, Canada, 22–25 October 1995; pp. 3443–3450.

3.   Crnković, I.; Larsson, M. *Building Reliable Component-Based Software Systems*; Artech House: Norwood, MA, USA, 2002.

4.   Crnković, I.; Sentilles, S.; Vulgarakis, A.; Chaudron, M.R.V. A Classification Framework for Software Component Models. *IEEE Trans. Softw. Eng.* **2011**, *37*, 593–615. [CrossRef]

5.   Pop, T.; Hnětynka, P.; Hošek, P.; Malohlava, M.; Bureš, T. Comparison of component frameworks for real-time embedded systems. *Knowl. Inf. Syst.* **2013**, 1–44. [CrossRef]

6.   Yin, H.; Hansson, H. A mode mapping mechanism for component-based multi-mode systems. In Proceedings of the 4th Workshop on Compositional Theory and Technology for Real-Time Embedded Systems, Vienna, Austria, 29 November–2 December 2011; pp. 38–45.

7.   Yin, H.; Hansson, H. Flexible and efficient reuse of multi-mode components for building multi-mode systems. In Proceedings of the 14th International Conference on Software Reuse, Miami, FL, USA, 4–6 January 2015; pp. 237–252.

8.   Yin, H.; Hansson, H. Handling multiple mode-switch scenarios in component-based multi-mode systems. In Proceedings of the 20th Asia-Pacific Software Engineering Conference, Ratchathewi, Bangkok, Thailand, 2–5 December 2013; pp. 404–413.

9.   Yin, H.; Hansson, H. Handling emergency mode-switch for component-based systems. In Proceedings of the 21st Asia-Pacific Software Engineering Conference, Jeju, Korea, 1–4 December 2014; pp. 158–165.

10.   Yin, H.; Hansson, H.; Orlando, D.; Miscia, F.; Marco, S.D. *Component-Based Software Development of Multi-Mode Systems—An Extended Report*; Technical Report MDH-MRTC-312/2016-1-SE; Mälardalen University: Västerås, Sweden, 2016.

11.   Larsen, K.G.; Pettersson, P.; Yi, W. UPPAAL in a nutshell. *Int. J. Softw. Tools Technol. Transf.* **1997**, *1*, 134–152. [CrossRef]

12.   Alur, R.; Courcoubetis, C.; Dill, D. Model-checking for real-time systems. In Proceedings of the 5th Annual IEEE Symposium on Logic in Computer Science, Philadelphia, PA, USA, 4–7 June 1990; pp. 414–425.

13.   Miscia, F. Design and Implementation of the MCORE IDE: A Multi-Mode COmponent Reuse Environment. Master's Thesis, University of L'Aquila, L'Aquila, Italy, 2015.

14.   Systems, A. Rubus ICE. Available online: https://www.arcticus-systems.com/products/ (accessed on 20 October 2018).

15.   Hänninen, K.; Mäki-Turja, J.; Nolin, M.; Lindberg, M.; Lundbäck, J.; Lundbäck, K. The Rubus component model for resource constrained real-time systems. In Proceedings of the 3rd International Symposium on Industrial Embedded Systems, La Grande Motte, France, 11–13 June 2008; pp. 177–183.

16.   Schubert, D.; Heinzemann, C.; Gerking, C. Towards Safe Execution of Reconfigurations in Cyber-Physical Systems. In Proceedings of the 2016 19th International ACM SIGSOFT Symposium on Component-Based Software Engineering (CBSE), Venice, Italy, 5–8 April 2016; pp. 33–38.

17.   Heinzemann, C.; Becker, S.; Volk, A. Transactional Execution of Hierarchical Reconfigurations in Cyber-Physical Systems. *Softw. Syst. Model.* **2017**. [CrossRef]

18.   Pop, T.; Plasil, F.; Outly, M.; Malohlava, M.; Bures, T. Property networks allowing oracle-based mode-change propagation in hierarchical components. In Proceedings of the 15th International ACM SIGSOFT Symposium on Component Based Software Engineering, Bertinoro, Italy, 25–28 June 2012; pp. 93–102.

19.   Weimer, J.E.; Krogh, B.H. Hierarchical Modeling of Mode-Switching Systems. In Proceedings of the 2007 Summer Computer Simulation Conference, San Diego, CA, USA, 15–18 July 2007; pp. 567–574.

20.   MathWorks. Simulink. Available online: http://se.mathworks.com/products/simulink/ (accessed on 20 October 2018).

21.   Quadri, I.R.; Gamatié, A.; Boulet, P.; Dekeyser, J.L. Modeling of Configurations for Embedded System Implementations in MARTE. In Proceedings of the 1st Workshop on Model Based Engineering for Embedded Systems Design, Dresden, Germany, 12 March 2010.

22.   Gamatié, A.; Beux, S.L.; Piel, E.; Etien, A.; Atitallah, R.B.; Marquet, P.; Dekeyser, J.L. *A Model Driven Design Framework for High Performance Embedded Systems*; Technical Report RR-6614; Institut National de Recherche en Informatique et Automatique: Rocquencourt, France, 2008.

23. Hansson, H.; Åkerholm, M.; Crnković, I.; Törngren, M. SaveCCM—A component model for safety-critical real-time systems. In Proceedings of the Euromicro Conference, Special Session on Component Models for Dependable Systems, Rennes, France, 31 August–3 September 2004; pp. 627–635.

24. Ke, X.; Sierszecki, K.; Angelov, C. COMDES-II: A Component-Based Framework for Generative Development of Distributed Real-Time Control Systems. In Proceedings of the 13th IEEE International Conference on Embedded and Real-Time Computing Systems and Applications, Daegu, Korea, 21–24 August 2007; pp. 199–208.

25. Borde, E.; Haïk, G.; Pautet, L. Mode-based reconfiguration of critical software component architectures. In Proceedings of the Conference on Design, Automation and Test in Europe, Nice, France, 20–24 April 2009; pp. 1160–1165.

26. Ommering, R.V.; Linden, F.V.D.; Kramer, J.; Magee, J. The Koala component model for consumer electronics software. *Computer* **2000**, *33*, 78–85. [CrossRef]

27. Bennour, B.; Henrio, L.; Rivera, M. A reconfiguration framework for distributed components. In Proceedings of the 2009 ESEC/FSE Workshop on Software Integration and Evolution, Amsterdam, The Netherlands, 25 August 2009; pp. 49–56.

28. Feiler, P.H.; Gluch, D.P.; Hudak, J.J. *The Architecture Analysis & Design Language (AADL): An Introduction*; Technical Report CMU/SEI-2006-TN-011; Software Engineering Institute: Pittsburgh, PA, USA, 2006.

29. Henzinger, T.A.; Horowitz, B.; Kirsch, C.M. Giotto: A time-triggered language for embedded programming. *Proc. IEEE* **2003**, *91*, 84–99. [CrossRef]

30. Templ, J. *TDL Specification and Report*; Technical Report; Department of Computer Science, University of Salzburg: Salzburg, Austria, 2003.

31. Hirsch, D.; Kramer, J.; Magee, J.; Uchitel, S. Modes for software architectures. In Proceedings of the 3rd European Conference on Software Architecture, Nantes, France, 4–5 September 2006; pp. 113–126.

32. Maraninchi, F.; Rémond, Y. Mode-Automata: About Modes and States for Reactive Systems. In Proceedings of the European Symposium on Programming, Lisbon, Portugal, 28 March–4 April 998; pp. 185–199.

33. Magee, J.; Dulay, N.; Eisenbach, S.; Kramer, J. Specifying Distributed Software Architectures. In Proceedings of the 5th European Software Engineering Conference, Sitges, Spain, 25–28 September 1995; pp. 137–153.

34. Capilla, R.; Bosch, J.; Trinidad, P.; Ruiz-Cortés, A.; Hinchey, M. An overview of Dynamic Software Product Line architectures and techniques: Observations from research and industry. *J. Syst. Softw.* **2014**, *91*, 3–23. [CrossRef]

35. Clements, P.; Northrop, L. *Software Product Lines: Practices and Patterns*; Addison-Wesley: Boston, MA, USA, 2001.

36. Sharifloo, A.M.; Metzger, A.; Quinton, C.; Baresi, L.; Pohl, K. Learning and Evolution in Dynamic Software Product Lines. In Proceedings of the 11th International Symposium on Software Engineering for Adaptive and Self-Managing Systems, Austin, TX, USA, 14–22 May 2016; pp. 158–164.

37. Baier, C.; Sirjani, M.; Arbab, F.; Rutten, J. Modeling component connectors in Reo by constraint automata. *Sci. Comput. Program.* **2006**, *61*, 75–113. [CrossRef]

38. Phan, L.T.X.; Lee, I.; Sokolsky, O. Compositional Analysis of Multi-mode Systems. In Proceedings of the 22nd Euromicro Conference on Real-Time Systems, Brussels, Belgium, 6–9 July 2010; pp. 197–206.

39. Criado, J.; Rodríguez-Gracia, D.; Iribarne, L.; Padilla, N. Toward the adaptation of component-based architectures by model transformation: Behind smart user interfaces. *Softw. Pract. Exp.* **2015**, *45*, 1677–1718. [CrossRef]

Article

Adaptive Time-Triggered Multi-Core Architecture

Roman Obermaisser *, Hamidreza Ahmadian, Adele Maleki, Yosab Bebawy, Alina Lenz and Babak Sorkhpour

Department of Electrical Engineering and Computer Science, University of Siegen, 57068 Siegen, Germany; hamidreza.ahmadian@uni-siegen.de (H.A.); Adele.Maleki@uni-siegen.de (A.M.); Yosab.Bebawy@uni-siegen.de (Y.B.); alina.lenz@uni-siegen.de (A.L.); Babak.Sorkhpour@uni-siegen.de (B.S.)
* Correspondence: roman.obermaisser@uni-siegen.de; Tel.: +49-271-740-3332

Received: 27 September 2018; Accepted: 18 January 2019; Published: 22 January 2019

Abstract: The static resource allocation in time-triggered systems offers significant benefits for the safety arguments of dependable systems. However, adaptation is a key factor for energy efficiency and fault recovery in Cyber-Physical System (CPS). This paper introduces the Adaptive Time-Triggered Multi-Core Architecture (ATMA), which supports adaptation using multi-schedule graphs while preserving the key properties of time-triggered systems including implicit synchronization, temporal predictability and avoidance of resource conflicts. ATMA is an overall architecture for safety-critical CPS based on a network-on-a-chip with building blocks for context agreement and adaptation. Context information is established in a globally consistent manner, providing the foundation for the temporally aligned switching of schedules in the network interfaces. A meta-scheduling algorithm computes schedule graphs and avoids state explosion with reconvergence horizons for events. For each tile, the relevant part of the schedule graph is efficiently stored using difference encodings and interpreted by the adaptation logic. The architecture was evaluated using an FPGA-based implementation and example scenarios employing adaptation for improved energy efficiency. The evaluation demonstrated the benefits of adaptation while showing the overhead and the trade-off between the degree of adaptation and the memory consumption for multi-schedule graphs.

Keywords: time-triggered system; real-time; cyber-physical systems; adaptation; scheduling; multi-core

---

1. Introduction

Safety-critical CPS demand assures services under all considered load and fault assumptions in order to minimize the risk for people, property and the environment. Therefore, electronic systems are subject to domain-specific certification processes (e.g., ISO26262, IEC61508, and DO178/254), which provide documented evidence and safety arguments.

Time-triggered systems facilitate the establishment of safety arguments and have thus become prevalent in many safety-critical applications. Schedule tables are defined at development time and determine global points in time for the initiation of activities, such as sending a message or starting a computation. Major benefits are temporal predictability, temporal composability and support for fault containment [1]. Time-triggered systems simplify the system design and reduce the probability of design faults by offering implicit synchronization, implicit flow control and transparent fault tolerance. By deriving all control signals from the global time base, there is no control flow between application components, which can be independently developed and seamlessly integrated. Furthermore, a priori knowledge about the permitted temporal behavior can be used by network guardians or operating systems for isolating faulty messages or tasks, thereby preventing fault propagation via shared resources. This fault containment is a prerequisite for active redundancy as well as modular and incremental certification [2,3].

Time-triggered operation has been realized at different levels. Many safety-critical distributed systems are deployed with time-triggered communication networks such as Time-Triggered Ethernet (TTE), Time-Triggered Protocol (TTP), FlexRay and IEEE 802.1Qbv/TSN [1,4]. Time-triggered operating systems and hypervisors (e.g., ARINC653 [5]) adopt scheduling tables for cyclic time-based executions of partitions to virtualize the processor. Time-triggered multi-core architectures (e.g., TTMPSoC [6] and COMPSoC [7]) use time-triggered Network-on-Chips (NoCs) in analogy to the time-triggered networks in distributed systems.

At the same time, adaptive system behaviors upon relevant events are desirable to improve energy efficiency, reliability and context awareness. For example, execution slack enables energy management, such as voltage/frequency scaling and clock gating. Information about faults can serve for fault recovery by redistributing application services on the system's remaining resources. Changing environmental conditions or operational modes may demand for different application services (e.g., take-off vs. in-flight of an airplane).

Time-triggered systems can support this adaptation through the deployment of precomputed schedules for the relevant events [8]. However, the major challenge is preserving the properties of time-triggered systems, such as temporal predictability, fault containment and implicit synchronization. Therefore, all building blocks must consistently switch between schedules. This requires system-wide consistent information about the context events determining the adaptation. In addition, the adaptation must work correctly in the presence of faults to prevent the introduction of vulnerabilities. Further requirements are bounded times of adaptation for fault recovery and the efficient storage of large numbers of schedules.

This paper introduces the *Adaptive Time-triggered Multi-core Architecture (ATMA)* that fulfills these requirements. The architecture establishes a system-wide consistent agreement on context events as well as robust and efficient switching between schedules.

Prior work has addressed computing schedule graphs for time-triggered systems (e.g., [9,10]). However, the combination of agreement, adaptation and meta-scheduling as part of a time-triggered multi-core architecture supporting implicit synchronization, fault containment and timeliness is an open research problem. Furthermore, the presented techniques for minimizing the overhead of adaptation (e.g., difference encoding of multi-schedule graphs, and adjustable reconvergence horizons for events) enable scalability and the deployment in resource-constrained applications.

The paper builds on previous work of the authors where a non-adaptive time-triggered multi-core architecture [6] was introduced as well as individual components for agreement [11] and adaptation [12]. This paper introduces the overall architecture of an adaptive time-triggered multi-core architecture along with the interplay of agreement, adaptation and meta scheduling. The paper provides experimental results for the overall architecture and shows the suitability for improved energy efficiency and fault tolerance. In addition, the paper introduces adaptation concepts for time-triggered systems and describes the services and system properties, which are essential to preserve implicit synchronization, temporal predictability and avoidance of resource conflicts.

The remainder of the paper is structured as follows. Section 2 analyzes the challenges and requirements for adaptation in time-triggered systems. The ATMA is the focus of Section 3. Section 4 describes the computation of multiple schedules, each serving for certain context events. Section 5 introduces agreement services, which establish chip-wide consistent context information. The context information is used for adaptive communication in Section 6. Section 7 presents example scenarios and the experimental evaluation.

2. Adaptation in Time-Triggered Systems

Adaptation in time-triggered systems is motivated by higher energy efficiency, fault recovery and the adjustment to changing environmental conditions. However, the fundamental properties and strengths of time-triggered systems must be preserved in order to obtain suitability for safety-critical systems.

2.1. Properties of Time-Triggered Systems

Time-triggered systems exhibit specific properties, which result from the dispatching of cyclic activities driven by the global time base and precomputed schedule tables:

- **Avoidance of resource contention without dynamic resource arbitration:** The precomputed schedules ensure that each resource is used by at most one user at any particular point in time. Thus, dynamic resource arbitration is not necessary.
- **Implicit synchronization:** The precomputed schedules satisfy synchronization requirements based on the global time including precedence constraints and avoidance of race conditions.
- **Guaranteeing of timing constraints:** The computation of schedules at development time ensures that deadlines are met without further runtime efforts beyond time-triggered dispatching.
- **Implicit flow control:** The computation of schedules considers the receivers' ability for handling received messages. Therefore, an overload of receivers is prevented without acknowledgments or flow control protocols.
- **Fault containment:** A priori knowledge about the permitted behavior of components allows to block faulty messages, thereby preventing fault propagation via shared resources.

These properties are essential characteristics of a time-triggered system and must be preserved despite adaptation. Henceforth, we call these properties the *ATMA properties*. In the ATMA, these properties are determined by the schedules computed offline, as well as the correct dispatching and schedule switching at runtime.

In order to realize these properties, a schedule encompasses three dimensions as depicted in Figure 1. The schedule defines in the temporal dimension when activities need to be dispatched with respect to the global time base. In the spatial dimension, the schedule defines the triggers for the different resources of the multi-core architecture such as Network Interfaces (NIs), routers and communication links. The third dimension corresponds to the contextual dimension, where different plans for the context events are distinguished. Overall, the time-triggered schedule thus defines what activities shall be dispatched for each resource and each relevant context event at which global points in time.

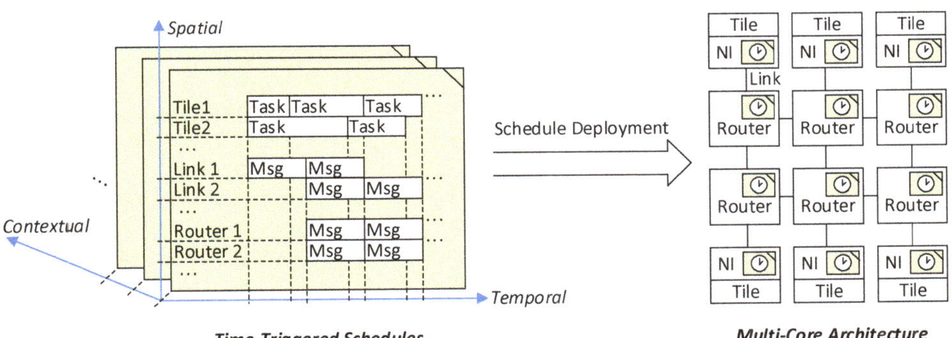

Figure 1. Deployment of time-triggered schedules (**left**) on the building blocks of the ATMA, such as NIs and routers (**right**).

In general, the schedules must be deployed in multiple building blocks of the architecture as depicted on the right hand side of Figure 1. Each schedule is fragmented along the spatial dimension and the resulting parts are assigned to the resources. NIs are the target in case of NoCs with source-based routing and routers in case of distributed routing. Furthermore, in order to acquire the ATMA properties, consistency of the schedules in temporal and spatial dimension is essential. For instance, messages must be passed without resource conflicts along the communication links in the spatial and temporal dimensions, while satisfying the deadlines and precedence constraints.

2.2. Need for Adaptation

In the following, the motivation for adaptation in time-triggered systems is detailed.

2.2.1. Energy Efficiency

Energy efficiency is relevant for many safety-critical systems. Examples are battery-operated devices (e.g., medical equipment, wireless sensors for railway systems [13]) and systems with thermal constraints (e.g., avionics [14]). While techniques for energy management such as Dynamic Voltage and Frequency Scaling (DVFS) and clock gating are common in many application areas (e.g., consumer electronics), the applicability for safety-critical systems is often limited. Certification and the computation of Worst Case Execution Times (WCETs) for multi-core processors is challenging by itself [15] and further complicated by energy management. If dynamic slack of one task is exploited for dynamic modifications of frequencies or clock gating, then unpredictable timing effects and fault propagation can occur for other tasks due to shared resources such as caches and I/O.

In safety-critical systems, we can distinguish between two types of energy sensitivity with respect to safety certification:

1. **Energy efficiency without detrimental effects on safety.** Energy management is desirable for availability or economic reasons (e.g., energy cost) but not required for safety. For example, a safe state is reached when energy resources are depleted. Therefore, energy management must not affect the system's safety functions.
2. **Energy efficiency as part of the safety argument.** Energy efficiency is part of the safety argument, e.g., because battery capacity must suffice for the mission time or thermal constraints must be satisfied under all load and fault conditions.

Both types can be effectively supported by adaptable time-triggered systems. Adaptable time-triggered systems promise to improve energy efficiency without detrimental effects on temporal predictability and fault containment. Different schedules serve for potential system conditions like various dynamic-slack values. Each schedule can be analyzed in isolation as in a fully static time-triggered system.

2.2.2. Lower Cost for Fault-Tolerance by Fault Recovery and Reduced Redundancy Degrees

In safety-critical applications, an embedded computer system has to provide its services with a dependability that is better than the dependability of any of its constituent components. Considering the failure-rate data of available electronic components, the required level of dependability can only be achieved if the system supports fault tolerance.

N-modular redundancy is a widely deployed technique, where the assumptions concerning failure modes and failure rates determine the necessary replication degrees. For example, a triple-triple redundant primary fight computer is deployed in the Boeing 777 aircraft [16]. However, emerging application areas with safety-critical embedded systems and stringent cost-pressure for electronic equipment cannot afford the high cost associated with excessive replication. For example, fully autonomous vehicles depend on ASIL-D [17] functions for environmental sensing and vehicle control. At the same time, the extreme cost pressure of the automotive industry precludes the massive deployment of redundant components.

Fault recovery is a viable alternative by switching to configurations that do not use failed resources. The following fault-recovery approaches can be deployed to achieve fault-tolerance:

1. **Modifying allocation of services to resources.** The system design involves scheduling and allocations decisions in order to perform a spatial and temporal mapping of the application (e.g., computational service, message-based communication) to the resources (e.g., routers, communication links, processing cores). Fault recovery by reconfiguration activates a configuration with a changed mapping, thereby avoiding the use of failed resources.

2. **Substitution of failed services.** In many cases applications can be reconfigured in order to provide services using alternative computational paths or different sensory inputs. For example, a failure of the sensor measuring the steering angle in a car can be compensated by determining the curve radius of the vehicle via rotational sensors of the wheels [18].

3. **Degraded service modes.** If remaining resources are insufficient for providing all services, then a degraded service mode should ensure a minimal level of service to guarantee safety.

ATMA supports this goal by precomputing configurations for different types of faults. However, in order to comply with the requirements of safety-critical systems and to have a consistent fault-recovery, a number of challenges need to be addressed. The key challenges are the completeness of considering all potential faults, while also analyzing each configuration individually to ensure correct system states (e.g., no resource collisions, satisfaction of precedence constraints, timeliness). Another major challenge is the fault-tolerance of the fault recovery mechanism itself. A fault affecting the reconfiguration must not lead to an incorrect system state such as the partitioning of the system into subsystems with the old and the new configuration.

2.3. Challenges for Adaptation

The main challenges for adopting an adaptive system behavior in a time-triggered system are as follows:

2.3.1. System-Wide Consistent State After Adaptation

A fundamental requirement for a time-triggered multi-core architecture is a consistent state of all building blocks at any point in time (e.g., tiles, NIs, routers). System-wide consistency of the active schedules with respect to the temporal and contextual dimensions is a prerequisite to maintain the ATMA properties. The consistency upon adaptation depends on:

- **Consistent context information:** At runtime, each tile must be provided with consistent information about the context events triggering the adaptation.
- **Computation of aligned schedules:** Schedules must be precomputed at development time for each context event and each building block of the ATMA. The schedules must be temporally and spatially aligned according to the required ATMA properties.
- **Consistent and robust switching:** At runtime, the execution of consistent distributed switching actions is a prerequisite for a consistent new state.

2.3.2. Bounded Time for Adaptation

The delay of reacting to context events often determines the utility of adaptation (e.g., exploitation of slack for energy efficiency). In particular, adaptation within bounded time is essential for fault recovery where the dynamics of the controlled object determine the permitted time intervals without service provision (e.g., maximum actuator freezing time).

2.3.3. Fault-Tolerant Adaptation

The adaptation must be fault tolerant to ensure that a hardware or software fault does not bring the system into an erroneous state through faulty schedule switching.

2.3.4. Avoidance of State Explosion

The scheduling algorithm for computing the schedules needs to avoid state explosion. For example, enforcing reconvergence horizons for context events prevents an exponential growth of the number of schedules with increasing numbers of context events [19].

Furthermore, a memory-efficient representation and storage of the schedules is required. Since schedules for different context events will typically differ only in small parts, differential representations are most suitable.

3. Adaptive Time-Triggered Multi-Core Architecture

3.1. Architectural Building Blocks

Figure 2 gives an overview of the *Adaptive Time-triggered Multi-core Architecture (ATMA)*. The architecture encompasses tiles, which are interconnected by a NoC. The NoC consists of routers, each of which is connected to other routers and to tiles using communication links. A tile includes three parts: cores for the application services (cf. green area in Figure 2), adaptation logic (cf. orange area in Figure 2) and the NI for accessing the NoC (cf. blue area in Figure 2).

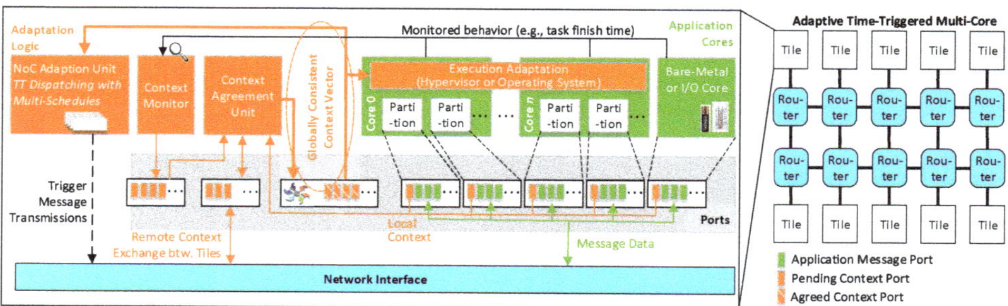

Figure 2. Adaptive time-triggered multi-core architecture.

The cores of a tile can be heterogeneous encompassing processors that are managed by an operating system, processors with bare-metal application software or state machines implemented in hardware. Regardless of the implementation, message-based ports provide the interface of the cores towards the adaptation logic and the NI.

The adaptation logic is the key element for the consistent and timely processing of context events. It includes the following building blocks:

- The *context monitor* is responsible for observing the behavior of the cores and generating local context information. An example is a slack event when a task finishes before its WCET. Another example of a context event is a relevant change of state observed via input/output channels (e.g., low battery state and environmental conditions).

- The *context agreement unit* establishes unanimous context information, which is globally consistent at a chip-wide level. Despite multiple clock domains and the occurrence of faults, the context agreement unit ensures that all operational tiles possess identical context information.

- The *NoC adaptation unit* contains multiple schedules and the next schedule is chosen based on the context information provided by the context agreement unit. The potential switching between schedules is triggered in a time-triggered manner and has its own schedule. Since the context information is chip-wide consistent, it is guaranteed that all tiles perform aligned schedule changes.

- In analogy to the adaptation at the network level, adaptation for the execution resources is supported by the *execution adaptation*. The consistent context information serves for switching between cyclic schedules of time-triggered operating systems or hypervisors.

The NI serves as an interface to the NoC for the processing cores by injecting the messages into the NoC as well as delivering the received messages from the NoC to the cores. The NI generates the message segments (known as packets and flits), in the case a message needs to be injected into the interconnect. At the opposite direction, i.e., for the incoming messages, the NI assembles each message from the received segments and stores it to be read by the connected core.

3.2. Local and Global Adaptation

We distinguish between local and global adaptations. *Local adaptations* are those changes in a subsystem, which do not introduce any changes in the use of shared resources. For example, dynamic slack in a processor core would result in the early completion of a task and the generation of a message before its transmission is due according to the time-triggered schedule. DVFS can be used locally within the processor to complete the task and produce the message just before the scheduled transmission time. In this case, energy is saved locally without any implications on the rest of the system and the time-triggered schedule.

Global adaptations, in contrast, are those changes in a subsystem, which result in a temporal or spatial change in the usage of the shared resources. For example, dynamic slack of a sender can be used to transmit a message earlier. The receivers can then start their computations earlier with more time until their deadlines. The longer time budget can be used for DVFS in the receivers, thereby saving more energy because several receivers are clocked down instead of a single sender. However, a new schedule is required for global adaptation in order to preserve the ATMA properties. Such global adaptations and schedule changes must not introduce an inconsistency in the system and thus need to be harmonized over the entire chip.

To establish a chip-wide aligned adaptation of subsystems, the operation of the dispatchers of different tiles shall be harmonized by a *global time base* to have a common understanding of the time, despite different clock domains. The global time base is a low-frequency digital clock, which is based on the concept of a sparse time base [20]. The global time base provides a system-wide *notion of the time* between different components.

3.3. Fault Tolerance

ATMA offers inherent temporal predictability and fault containment for processing cores. The adaptation unit is deployed with a schedule that provides a priori knowledge about the permitted temporal behavior of messages. Using this knowledge, the adaptation unit blocks untimely messages from the processing cores, thereby preventing resource contention in the NoC. Likewise, the adaptation units and the network interfaces autonomously insert the source-based routing information into messages, thus a faulty processing core cannot influence the correct routing of messages in the NoC.

While faults of the processing cores are rigorously contained, faults affecting the adaptation units, context-agreement units, network interfaces and routers have the potential to cause the failure of the entire multi-core chip. In previous work on time-triggered multi-core architectures, the network interfaces and routers have thus been denoted as a *trusted subsystem* [6]. The risk of a failure due to design faults can be minimized through a rigorous design of these building blocks.

Two strategies can be distinguished for dealing with random faults of the trusted subsystem:

- **Fault-tolerant trusted system**: Fault-tolerance techniques can be deployed for the adaptation units, context-agreement units, network interfaces and routers. For example, these building blocks can be synthesized as design units with triple modular redundancy [21].
- **Consideration in reliability models and off-chip fault-tolerance:** Alternatively, the designer can accept the risk from random faults and consider the fault propagation probabilities from adaptation units, context-agreement units, network interfaces and routers, which potentially represent single points of failures. Evidently, this strategy is only reasonable if the chip area consumed by the processor cores is dominant. In many safety-critical systems (e.g., ultra-reliable fail-operational systems such as ADAS systems with highest SAE levels [22]), on-chip fault-tolerance mechanisms are complemented by off-chip fault-tolerance techniques because of non-negligible probabilities of chip failures (e.g., due to shared power supply, common clock source, spatial proximity [23] (p. 155)). In these systems, the fault-tolerance mechanisms of ATMA serve for increasing the reliability, but do not replace off-chip redundancy.

Another prerequisite for correct adaptation is the correctness of the schedules. The state of the art offers algorithms and tools supporting the verification of time-triggered schedules (e.g., TTPVerify [1] (p. 489)), which can be applied on each node of the multi-schedule graph.

4. Meta Scheduling

The meta scheduler [24] is an offline tool for computing time-triggered schedules considering the system's contextual, spatial and temporal dimensions. Different schedules are computed for all relevant context events and are deployed in the NoC adaptation units and the execution adaptation. For a given context, the corresponding schedules determine the dispatching of each resource at any point in time within the system's period.

Two types of schedules are taken into consideration: The computational schedules are computed for the tasks and define their allocation to cores and their start times. The communication schedules define the paths and injection times of the messages on the NoC. The computational and communication schedules must be synchronized as the messages to be sent are computed by the tasks.

4.1. Input Models

The meta scheduler requires the following three types of models as input (cf. Figure 3):

- **Application Model:** The application model describes the computational tasks with their deadlines, WCETs and resource requirements (e.g., memory, I/O). The assumed WCET of a task is based on its criticality [25]. High-criticality tasks must be considered with a more pessimistic WCET in order for the schedule to assure that applications have enough time to finish the task within the assigned time frame. In addition, the application model describes each message with its sender task and receiver tasks, thereby reflecting the precedence constraints between the tasks and the messages.
- **Platform Model:** The platform model informs the meta scheduler about the available hardware resources, such as cores, memories, I/Os, routers and the interconnection pattern of routers.
- **Context Model:** The context model covers all context events that are relevant for the adaptation including faults, dynamic slack, resource alerts and environmental changes. Examples of faults are permanent failures of cores and routers, which require a recovery action such as the reallocation of tasks and messages. Dynamic slack of a task occurs if the execution time of the task is shorter than its WCET. It can be exploited by DVFS or by clock/power-gating of resources to save energy. Alternatively, dynamic slack can be passed to subsequent tasks or messages. An example of a resource alert is a high thermal level of passively-cooled electronics that demands for a reduction of the computational load to avoid thermal damage. Likewise, certain battery levels motivate degradation concepts such as disabling comfort functions in an electric vehicle. Environmental changes are context events that originate from outside the computer system, e.g., entering of the takeoff phase in an airplane.

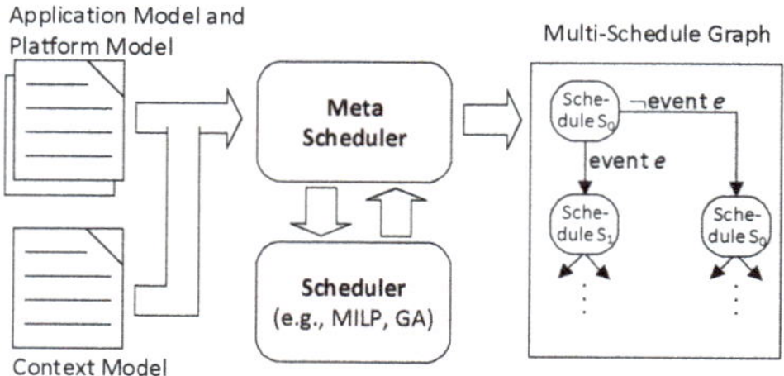

Figure 3. Meta scheduler overview.

4.2. Meta Scheduling

The meta scheduler computes a Multi-schedule Graph (MG) using the application, platform and context models. The MG is a directed acyclic graph of time-triggered schedules, which are precomputed at development time. At any instant during runtime, the time-triggered system is in one of the nodes of the multi-schedule graph. This node defines the temporal and spatial allocation of all computational and communication resources. Upon a relevant event from the context model, the node is left and the system traverses to another node of the multi-schedule graph.

To compute the MG, the meta scheduler repeatedly invokes a scheduler (cf. Figure 3). The scheduler takes an application model and a platform model as inputs and computes a time-triggered schedule fulfilling the scheduling constraints (e.g., collision avoidance, precedence constraints, and deadlines). The decision variables are the allocations of tasks to cores, start times of tasks, messages paths along routers, message injection times and fetch times for the adaptation manager to read the latest instance of the context vector.

Decision variables also include parameters for improving energy efficiency such as the time intervals for clock gating and DVFS. The time-triggered schedule specifies the frequencies of cores and routers in different time intervals of the time-triggered schedule [24,26]. The resulting overhead such as the time needed to adjust the voltage needs to be considered in the constraints of the optimization problem.

The state of the art encompasses a broad spectrum of algorithms (e.g., genetic algorithms, SAT/SMT, and MILP) to solve this static scheduling problem.

The algorithm of the meta scheduler is depicted in Figure 4. The meta scheduler starts in an initial state S_0 assuming the absence of faults, slack events and resource alerts. It invokes the scheduler to obtain a schedule for S_0. The meta scheduler then performs a time step until one of the events from the context model can occur.

Some events such as the occurrence of a particular dynamic-slack value occur at specific points in time (e.g., termination of a task after 50% of its WCET). For events that can occur at any time such as faults and resource alerts, predefined sampling points for fault detection or resource levels are assumed.

The meta scheduler applies the earliest context event, which results in a changed application or platform model. For example, a fault event results in the removal of a resource from the platform model. A dynamic slack event results in a shorter execution time of a task. Thereafter, the meta scheduler invokes the scheduler again to compute a new schedule S_1 for the updated application and platform models. Some decision variables are fixed in the new scheduling problem, namely those that correspond to actions before the time of the processed context event.

Subsequently, the meta scheduler continues to perform time steps, each time applying the context event, invoking the scheduler and adding the new schedule to the MG. In this process, the meta scheduler considers the different potential state traces. For example, the second potential context event will be applied in both schedule S_0 as well as in schedule S_1. Therefore, each context event results in a branching point of the MG, since the context event may occur or it can remain inactive.

The meta scheduler considers mutually exclusive events. For example, a certain dynamic slack value for a task precludes another dynamic slack event for the same task.

A major challenge in meta scheduling is the avoidance of state explosion in the MG. The meta scheduling addresses this challenge using the following techniques:

- **Reconvergence of paths in MG.** Whenever the meta scheduler computes a new node for the MG, it is checked whether this node was generated before. In this case, the meta scheduler connects the predecessor node to the existing node and terminates the further exploration of the current path (see Line 11 in Figure 4).
- **Reconvergence horizon.** For a given context event, the new schedule may only differ from the previous schedule within a limited time interval after the occurrence of the context event. All

decision variables after this horizon are fixed in analogy to the decision variables corresponding to actions before the time of the context event (see Line 21 in Figure 4). Thereby, reconvergence of paths in the MG is ensured.

```
01  initial application model AM
02  initial platform model PM
03  initial context model CM
04  initial multi-schedule graph SG={}
05  initially fixed decision variables FIX={}

06  proc meta-scheduler(AM, PM, CM, FIX, prev)
07    invoke scheduler(AM, PM, FIX) to obtain schedule S
08    S={<d,t(d)>}// decision variable d with action time t(d)
09    n = <S, CM>  // new node for schedule graph
10    if n ∈ SG   // dejavu
11       connect previous node prev to existing node n in SG
12    else
13       add n to SG
14       if (prev≠NULL) connect node prev to new node n in SG
15       while CM≠{}
16          e=earliest context event from CM with event time t(e)
17          EX=context events that are mutually exclusive with e
18          CM'=CM \ (EX ∪ {e})
19          AM'=result of applying e to AM
20          PM'=result of applying e to PM
21          FIX={<d,t(d)>∈S | t(d) ≤ t(e) ∨ t(d) ≥ t(e)+HORIZON)

22          recursively invoke meta-scheduler(AM', PM', CM', FIX, n)
23       end
24    endif
25  end
26  meta-scheduling: invoke meta-scheduler(AM, PM, FIX, NULL)
```

Figure 4. Meta-scheduling algorithm.

The size of the MG depends on the application, platform and context models. There is a linear relationship between the size of the MG and the number of tasks and messages in the application model, since the schedule must provide a resource allocation for each of these elements from the application model. In addition, the size of the MG depends linearly on the number of resources, which are deployed with dedicated schedules such as cores in case of source-based routing. For those resources, which are only referred to with indices by the schedules, there is a logarithmic relationship between the schedule size and the number of resources (e.g., routers in case of source-based routing). Without reconvergence, there would be an exponential relationship between the number of events in the context model and the schedule size. With a constant reconvergence horizon, there is a polynomial dependency between the number of events and the size of the MG [19].

4.3. Tile-Specific Schedule Extraction and Difference Encoding

Even with reconvergence, the meta scheduler can still generate MGs with hundreds or sometimes thousands of nodes. Storing those schedules in the memory space of the tiles would consume significant chip resources. Therefore, we extract an individual MG for each tile and we introduce difference encoding. This transformation of MGs is performed by the *MG compressor* in Figure 5.

The meta scheduler computes a MG where each node provides a schedule for the entire system with all tiles. However, an event will typically lead to changes in message injections at only a small subset of the tiles. Therefore, the graph compressor extracts graphs for the individual tiles from the MG. Each of the resulting tile-specific graphs contains a subset of the original nodes and the nodes contain a subset of the decision variables. Consequently, the resulting graphs are significantly smaller than the original MG.

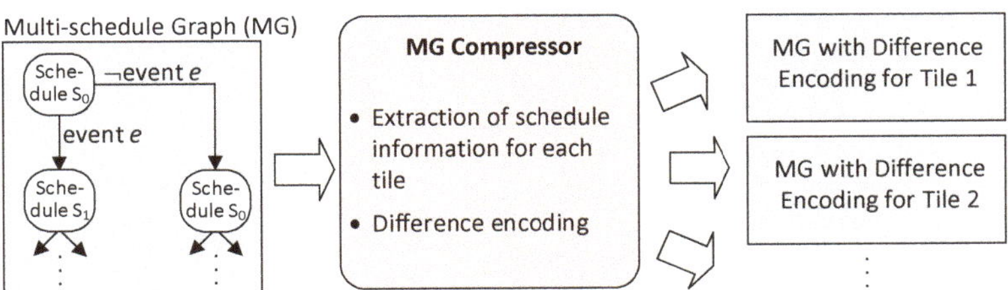

Figure 5. Multi-schedule graph compressor.

Each tile is deployed with its complete base schedule, but it stores only the differences of the other schedules in the graph-based structure. As shown in Figure 5, the MG generator extracts for each tile the schedule information and their changes and stores them in a format that is suitable for the NoC adaptation unit. The NoC adaptation unit, in turn, uses the schedule information to control the source-based NoC and to change the NI behavior at runtime based on the global context information that is provided by the context agreement unit.

The effectiveness of the difference encoding depends on the number of tasks and messages with changed temporal/spatial resource allocations after an event. The ratio between the reconvergence horizon and the makespan provides a lower bound for the compression ratio, if the number of messages and tasks per unit of time is uniform throughout the makespan.

The effectiveness of the schedule extraction is determined by the number of resources (e.g., tiles in case of source-based routing) with changed scheduling information after an event. In the worst-case, all resources are effected by an event, thus yielding no benefit from the schedule extraction. However, typically only a small fraction of the resources requires updated schedules after an event. For example, the extraction results in an average reduction of the schedule size by 71% in the three experimental scenarios described in Section 7.

The compression also has an impact on the fault-tolerance of the multi-core architecture. On the one hand, the reduced size of the schedule information decreases the susceptibility to Single Event Upsets (SEUs). On the other hand, an SEUs affecting the schedule information for a task or a message can have a more severe effect, potentially corrupting also the parameters of subsequent tasks and messages.

5. Context Monitor and Agreement Unit

All decisions taken by the adaptation unit are taken in a distributed manner at each tile. This helps to avoid having a central tile with the role of a managing device that collects the information of all tiles and pushes the new schedule to the entire system. Otherwise, such a tile would represent a single point of failure and could degrade scalability as well as performance. However, in the proposed distributed manner for decision making, we must ensure that all tiles are aware of the same global context to achieve a coherent distributed decision process. Inconsistencies in the global view can lead to inconsistent schedule changes, which in turn can cause collisions on the NoC and deadline misses.

As shown in Figure 6, the context monitors and the context-agreement units in ATMA establish a global view on the context using the following three steps: (1) context reporting, (2) context distribution; and (3) context convergence. At the beginning of the process each tile has only a local view on the context and is not aware of the status of the other tiles. The three steps are explained in the following.

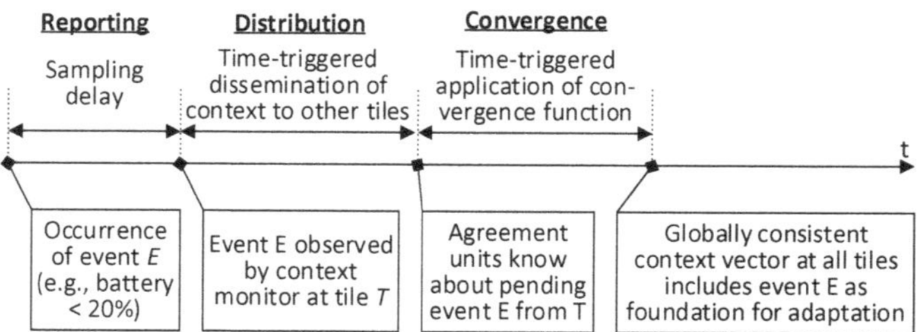

Figure 6. Timeline of agreement for an example context event.

5.1. Context Reporting

The context monitor focuses on the collection of the local context events within the associated context-agreement units.

We differentiate events based on their temporal properties and the safety implications:

- **Synchrony.** *Synchronous context events* are predictable in terms of their occurrence time and thus can be checked at particular points in time. Such events either happen at a particular time within the period or they will never happen. This kind of events allows for predictable changes. For instance, if slack happens, it can be used for clock gating or for applying DVFS. *Asynchronous context events*, on the other hand, happen at random and unconstrained points in time throughout the system lifetime. They indicate a degradation of the system health or resource availability which needs to be reacted to (e.g., fault or low battery level). These events are detected at runtime through periodic sampling performed by the respective context monitors.
- **Urgency.** The urgency of a context event indicates how long the information about its occurrence can be used before it loses its worth for the adaptation. Slack for example is highly urgent, as it expires at the latest by the corresponding task's WCET. The synchronous events have a predictable offset between the potential occurrence and the resulting schedule change, therefore we can schedule the agreement to fit to the event's urgency. Asynchronous events, on the other hand, are unpredictable in terms of their occurrence. These events can only be sampled and agreed on regularly, which can lead to protocol overhead, as the events may not occur at all during runtime. The more often the asynchronous events are monitored and agreed on, the faster the system can adapt to them. Depending on the safety requirements and the dynamics of the environment, a frequent observation of the context can be necessary for fault tolerance, as the adaptation of the network must be performed quickly after the fault event is detected. Hence, the frequency for the sampling of asynchronous events is a trade-off between the overhead caused by the agreement and the gain that a short delay for reconfiguration can provide for the system.
- **Safety.** Safety-critical events and non-safety-critical events can be distinguished depending on the impact with respect to functional safety. For example, fault recovery through adaptation can be necessary to avoid critical failure modes of the system after fault events. In contrast, slack events are often relevant for energy-efficiency and availability without direct safety implications.

The context reporting is performed by context monitors in ATMA. These monitors observe the system state with respect to relevant indicators serving as context. The context typically comprises both state information from within the computer system and from the environment. Therefore, monitors and sensors are realized in hardware and software. Software monitors are driver functions that can be used by the application to give direct feedback to the agreement unit. Hardware monitors contain sensors to evaluate the system or the environment. The output of the monitors consists of a bit for every potential event, indicating whether the event actually occurred. For example, an event can represent the exceeding of a threshold value at a sensor (e.g., battery level and temperature).

This information is encoded in a bit string where each event is mapped to a single bit within the string. The event bit thus indicates if the event occurred. This context string is prepared by each tile locally and thus encodes only the locally observed events, therefore we refer to this bit string as the local context. This local context bitstring is the information that is distributed in the following phase.

The monitors write the context information into dedicated ports. Each context event has a dedicated bit in one of the ports.

5.2. Context Distribution

The agreement on the context is realized by a broadcast protocol, which sends the messages with the context using a ring relay between all tiles, meaning that a message is sent by each tile to its neighbor and gets incrementally relayed until reaching the original sender [11].

The protocol is triggered periodically and executed by each context-agreement unit at the same time. Figure 7 shows which events are agreed upon in an example scenario. The start instant of the agreement process serves as a deadline d_1 for the event reporting of period 1 and all events that happen before d_1 are taken into consideration. Events happening after d_1 are considered for the next period, even if they occur during the context distribution phase of the protocol. Upon the trigger, the context-agreement unit reads the context ports and assembles a local context vector by concatenating the context information for different events. Once the information is gathered, the local context vector is sent to the other tiles within an agreement message that identifies the context string using the source tile id [11].

This can be done via the NoC or via a dedicated network [11]. The use of the NoC involves no hardware overhead for the implementation, but the agreement messages need to be added to the scheduling problem. This extension can render the communication more difficult to be scheduled. In such cases, a dedicated second network can be used. For the evaluation in Section 7, we implemented a FIFO ring structure where the agreement messages can be sent at arbitrary times without consideration of the application communication at the NoC. This implementation has shown to be able to save more energy than a NoC implementation in a benchmark setup [11].

Each tile in the FIFO ring sends its local context vector to its direct neighbors. This way no collisions can occur as the links that are used are predefined. Once the local context is sent, the tile also receives local context from its neighbors. The tile extracts the new information and saves it locally. Afterwards, it relays the received context to the next neighbors. This way, the local context is transmitted within a ring-like structure between all neighbors until it returns back to its original sender. There the sender knows that it has received all context information in the tile and it can proceed to build the global context vector. This exchange takes n transmission hops, with n being the number of tiles in the network, as the ring needs to be passed completely by all messages. Therefore, the execution time increases with the number of tiles.

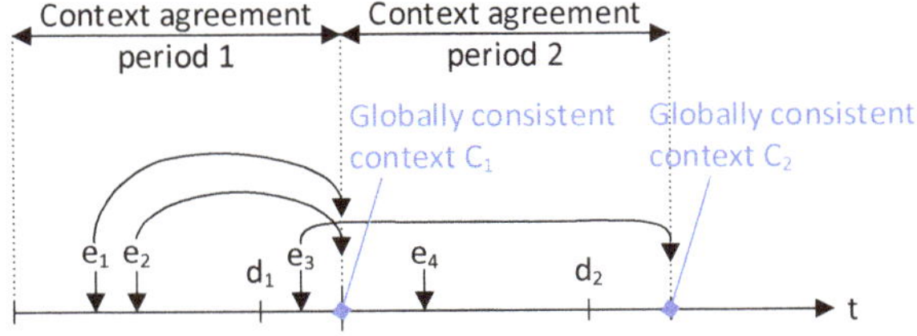

Figure 7. Events before the context reporting deadline d_n of the respective period n are considered for the agreed context C_n. Events happening between the agreement start and the agreed global context are considered for the next period.

5.3. Context Convergence

Once all messages with context information are received, the global context vector is produced. In this phase, the global context can also be converged from redundant information using majority voting. This is important if the same context event is observed redundantly by multiple tiles.

If the context events are observed only by one dedicated tile, the local context information is simply concatenated according to the predefined global context-vector layout in which each tile has a predefined interval where its local information is to be placed.

6. NoC Adaptation Unit

Robust and chip-wide aligned switching of schedules at all tiles are major requirements to ensure consistency and to preserve the properties of time-triggered systems. The NoC adaptation unit supports this requirement based on the assumption of a consistent context vector and correctness of the MG.

The NoC adaptation unit performs time-triggered dispatching of messages by deploying precomputed schedules, each of which is mapped to a particular set of context events. This unit receives the global context vector from the context agreement unit and triggers the ports for transmission of messages to support adaptive time-triggered communication. The global context vector is saved in a dedicated register within the NoC adaptation unit.

The operation of the adaptation unit occurs using the following four steps:

1. **Fetching step:** Fetch the global context vector from the dedicated register at the scheduled time.
2. **Compare mask step:** Compare the global context vector with a context bit mask to determine the occurrence of specific combinations of context events.
3. **Selection step:** Select the new schedule based on the masked signal.
4. **Triggering step:** Trigger the message injection for the respective ports based on the new schedule.

Figure 8 represents the internal structure of the NoC adaptation unit, which is composed of the following three internal building blocks: (1) the context register; (2) the Linked-List Multi-schedule Graph (LLMG) that stores the MG as a linked list; and (3) the adaptation manager.

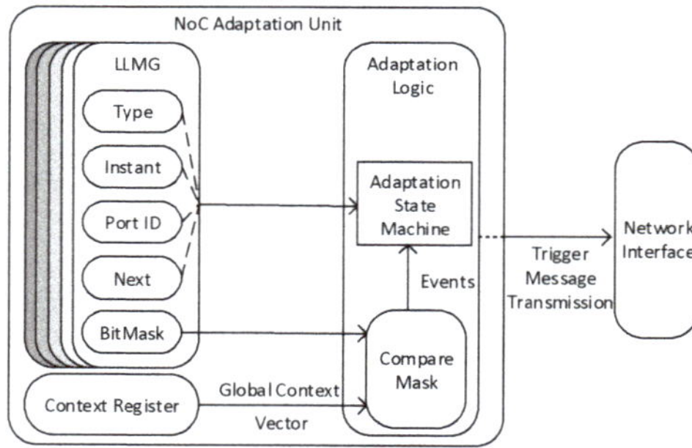

Figure 8. Internal structure of the NoC adaptation unit.

The global context vector is stored in the *context register* to be read by the compare-mask building block. The global context vector is fetched at a scheduled time, which is defined by the LLMG.

Figure 9 shows an example of a LLMG, which is stored as a circular linked list of instant entries. Each entry in the linked list is associated with an instant of time and address in the schedule file.

Two types of instant entries are provided: message entries and branching entries. Table 1 presents the content of the two types of entries. The message entries contain Type, Instant, PortID and Next values. The branching entries contain Type, Instant, BitMask, NextTaken and NextNotTaken values:

- *Type* defines the entry type (i.e., message or branching).
- *Instant* represents the injection time of the message or the branching time.
- *PortID* shows the ID of the port, from which the message is injected.
- *BitMask* is a mask to detect the simultaneous occurrence of several events from the global context vector.

- *Next* is a pointer to the next entry of the schedule.
- *NextTaken* is a pointer to the next entry when a specified event has occurred.
- *NextNotTaken* is a pointer to the next entry when the specified event has not occurred.

Table 1. Structure of the schedule entries.

Type					
Message	Type	Instant	PortID	Next	
Branching	Type	Instant	BitMask	NextTaken	NextNotTaken

Figure 9 presents an example LLMG, in which two different events E1 and E2 occur. The type of the entry is distinguished by the color, where message entries are shown in white and the branching entries are shown in gray. The number in each entry represents the address of the entry in the schedule file and the context events are shown on the links between the entries.

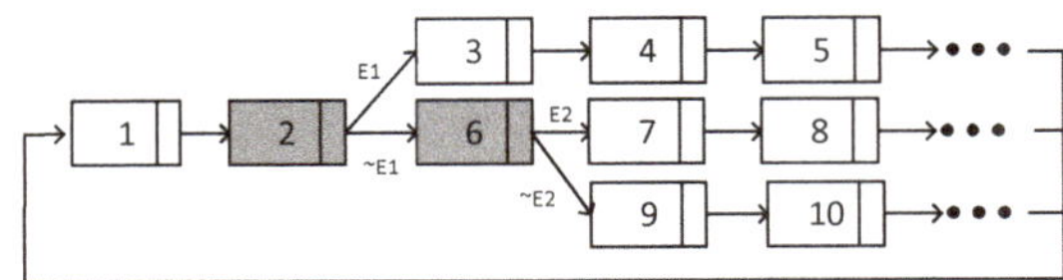

Figure 9. Linked-List Multi-schedule Graph (LLMG).

The process of selecting the correct schedule is started by tracing the entries of the LLMG, based on the address and next pointers. The adaptation manager is triggered at the instant, which is given by the entries. In the example in Figure 9, entry 1 is a message entry so the message of the corresponding *Port ID* is transmitted at the *Instant* specified by the entry. In this entry type, the *Next* value points to the address of the next entry. Entry 2 is a branching entry, which points to two different entries. The selection of the next schedule depends on the occurrence of the specified context event. The event occurrence is determined by the global context vector and MaskedBit. In the case of E1 occurrence, entry 3 is selected as the next entry by *NextTaken*. If E1 does not occur, then entry 6 is selected by *NextNotTaken*. The same operation is applied for the other entries of the LLMG. After applying the operations for all entries, the next pointer of the last entry is followed, which points back to the first entry of the period.

The adaptation manager serves the ports by triggering the message transmissions. The selection of the schedules is fully dependent on the received global context vector and is implicitly consistent with the other tiles, because the context vector is globally consistent and the actions of the different tiles within each schedule are temporally aligned by the meta scheduler.

The adaptation manager consists of a state machine and a compare mask. The state machine reads the input from the context vector register and the LLMG for switching schedules. Figure 10 presents the states and transitions of the state machine.

The state machine wakes up at an instant of time, which is defined by the LLMG. At this instant of time, the type of the LLMG entry is checked. In the message-type entries, the state machine reads the portID and injects the message of the specified port at the instant of time. After the injection, the new entry is fetched by the Next value.

In the case of a branching entry, the global context vector is fetched from the context register at the specified instant of time. In parallel, the BitMask is read from the LLMG. The global context vector and the BitMask values are received by the compare mask to extract the specified event. The event occurrence value can be 0 or 1, indicating whether the event has occurred. Therefore, the NextTaken entry is selected. An event bit 1 means the event has not occurred, so the NextNotTaken entry is selected. After the selection of the next address, the state machine waits again for the dispatching time of the next entry.

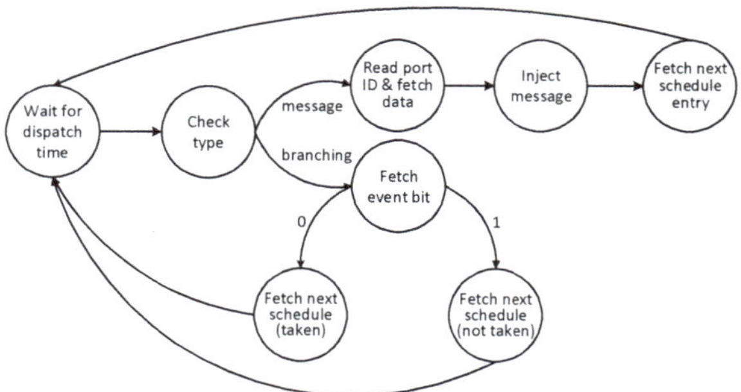

Figure 10. Adaptation state machine.

7. Results and Discussion

The introduced architecture was instantiated for example scenarios in order to validate the adaptation services. In addition, the scenarios served as an evaluation of the improvements with respect to energy efficiency and investigate the overhead with respect to memory and logic.

7.1. Zynq Prototype

The architecture was instantiated on a Xilinx Zynq-7000 SoC ZC706 FPGA board. The System-on-Chip (SoC) of this board consisted of an ARM-based processing system and a programmable-logic unit on a single die.

The hardware platform comprised four tiles interconnected by a time-triggered NoC [27]. Each tile was deployed with a network interface, a NoC adaptation unit and a context agreement unit. The Nostrum NoC [28] served as the basis for the implementation of the adaptable NoC. One tile was located in the processing system and contained two ARM Cortex-A9 processor cores. The other tiles were implemented in the programmable logic, where each tile contained a single core realized as a soft-core MicroBlaze-processor. The resource consumption is shown in Table 2.

Table 2. FPGA resource utilization.

Hardware	LUTs
Soft-core processors (3 Microblazes)	5673
Nostrum NoC with 4 routers	6225
NoC Adaptation Units (at 4 tiles)	8179
Agreement Units (at 4 tiles)	254

The meta scheduler and the MG compressor served for the generation of the schedules of the tiles, which were loaded to the dedicated memory of each corresponding tile. The meta scheduler is an implementation of the pseudo code in Figure 4. An existing optimal scheduler [29] for time-triggered networks was extended to use energy efficiency as the objective function. This optimal scheduler was implemented with IBM CPLEX and repeatedly invoked by the meta scheduler.

7.2. Slack-Based Adaptation Scenarios

Figures 11–13 show the three scenarios for the evaluation. In each scenario, different tasks with different WCETs and precedence relationships were hosted by the four tiles. The tasks had a period of 2 ms and it was assumed that each task could be subjected to dynamic slack of 50%. Table 3 summarizes the input models (i.e., AM, PM and CM) for the meta scheduling in the three scenarios.

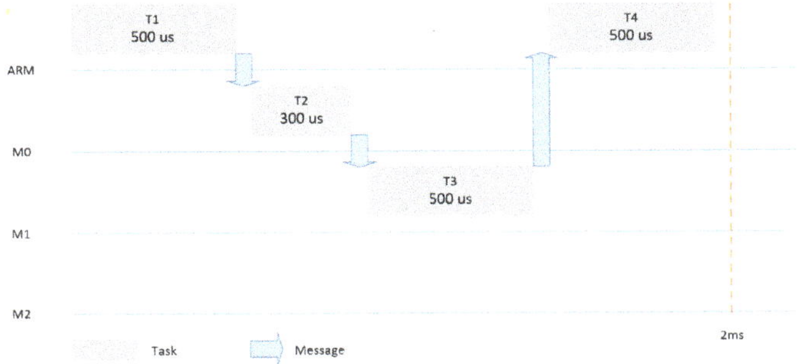

Figure 11. Scenario 1.

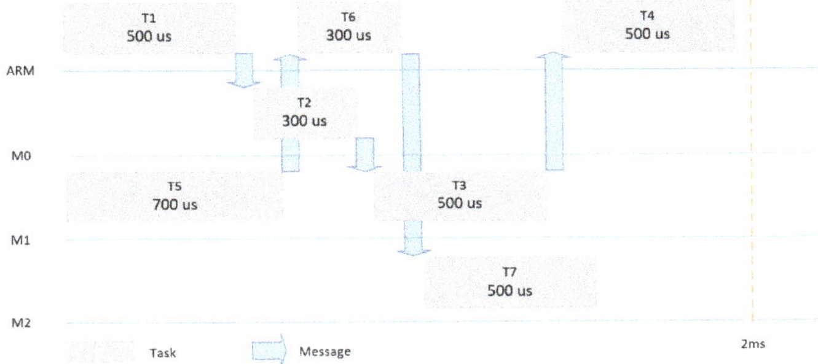

Figure 12. Scenario 2.

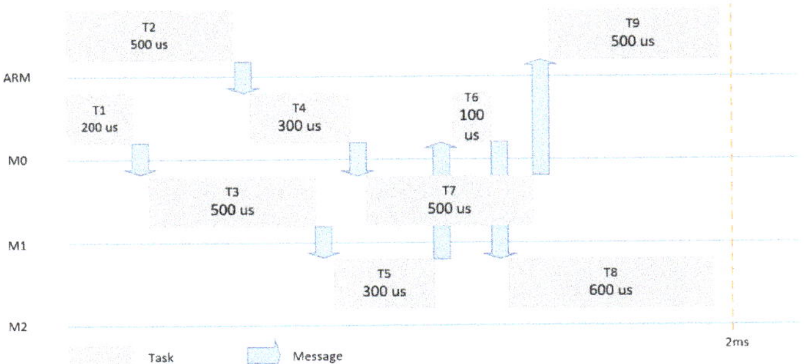

Figure 13. Scenario 3.

Table 3. Inputs models for scenarios.

	Input Model						
Scenarios	Application				Platform	Context	
	Jobs	WCETs	Deadlines	Msgs.	Tiles	# of Schedules	Dynamic Slack
1	4	300–500 µs	500–1850 µs	3	4	16	50% of WCET
2	7	300–700 µs	500–1850 µs	5	4	128	50% of WCET
3	9	200–600 µs	200–1850 µs	7	4	512	50% of WCET

Dynamic slack is the time difference between the WCET of the task and the actual point of time, at which the task ends. Slack can be used to save energy, as, in each execution, some or all tasks can be finished either as planned or earlier. In the case no slack happens, only one schedule can be used, as shown for example in Figure 11. In the case of a slack of 50% (e.g., T1 finishes 250us earlier), we changed the communication schedule to make use of the remaining time of the execution of T1 to start T2 (and consequently T3 and T4) earlier and achieve a shorter makespan for the system, as shown in Figure 14.

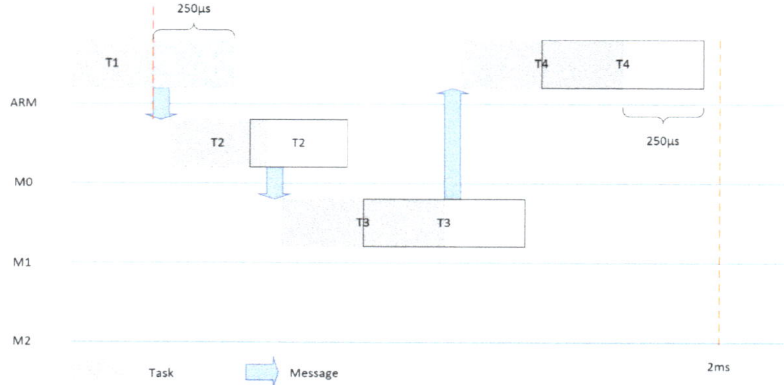

Figure 14. Example for slack in Task T1 of Scenario 1.

Energy reduction was achieved by clock-gating all tiles, their dedicated IPs and the interconnect. Clock gating aws performed between the completion of the makespan and the start of the subsequent period. In other words, all tiles as well as the NoC were in sleep mode after the termination of the last task, until the next period starts. This procedure was repeated every period.

7.3. Evaluation and Results

The evaluation was performed on the introduced prototype using the UCD90120A device, which is a 12-rail addressable power-supply sequencer and monitor. This chip was mounted on the evaluation board and it aws accessible by the processing system via the PMBus/I2C communication bus.

The experimental results encompass the power consumption for all combinations of slack events, thus showing the power savings depending on the completion times of tasks. Table 4 summarizes the numerical measurement results for the three scenarios. In each scenario, all possible slack combinations of different tasks were taken into consideration for the computation of the multi-schedule graph [30]. Each row of the table corresponds to an observed makespan value and it contains the number of schedules in the MG with this makespan, the corresponding power consumption in milliwatt and the power reduction percentage achieved by the proposed adaptation mechanisms. In addition, the table provides the average power reduction percentage under the assumption of a uniform probability distribution of the slack event combinations. This value is pessimistic, because in reality high slack values of tasks are common (e.g., as described by the literature on WCET analysis).

Table 4. Evaluation Results for the three scenarios.

Scenario 1

Duplications	Makespan (µs)	Consumption (mW)	Power Saving
1	950	451.2	32%
1	1100	482.2	28%
3	1200	502.8	25%
3	1350	533.8	20%
3	1450–1460	554.4–556.5	17%
3	1600–1610	585.4–587.5	12%
1	1710	608.1	9%
Average	1372	538.3	19%

Scenario 2

Duplications	Makespan (µs)	Consumption (mW)	Power Saving
1	1040	469.8	30%
2	1131	488.6	27%
12	1190–1200	500.7–502.8	25%
1	1210	504.9	24%
8	1281–1300	521.4–523.9	22%
15	1340–1350	531.7–533.8	20%
8	1381–1391	540.2–542.2	19%
27	1440–1460	552.4–556.5	17%
11	1531–1541	571.2–573.2	14%
10	1590–1595	583.3–584.4	13%
17	1600–1610	585.4–575.5	12%
8	1700–1710	606.1–608.1	9%
5	1850	637.0	5%
2	1860	639.1	4%
Average	1467.2	558.0	16%

Scenario 3

Duplications	Makespan (µs)	Consumption (mW)	Power Saving
1	930	447.0	33%
1	961	453.4	32 %
1	1001	461.7	31%
2	1051–1061	472.0–474.0	29%
5	1080–1090	478.0–480.1	28%
3	1111–1141	484.4–490.0	27%
3	1151–1160	492.6–494.5	26%
19	1180–1201	498.6–503.0	25%
9	1210–1211	504.8–505.1	24%
12	1240–1261	511.0–515.4	23%
8	1290–1302	521.3–523.9	22%
12	1310–1330	525.5–529.6	21%
54	1340–1361	531.71–536.0	20%
12	1371–1401	538.1–544.3	19%
10	1411–1431	546.4–55.5	18%
70	1440–1461	552.4–446.7	17%
6	1470–1491	558.6–562.9	16%
30	1500–1521	564.7–569.1	15%
22	1540–1561	573.0–577.3	14%
27	1570–1591	579.2–583.5	13%
75	1600–1621	585.4–589.7	12%
18	1640–1660	593.7–597.8	11%
11	1671–1691	600.1–604.2	10%
41	1700–1711	606.1–608.3	9%
10	1750–1752	616.4–616.8	8%
7	1761–1771	618.6–620.7	7%
2	1790–1800	624.6–626.7	6%
16	1840–1850	635.0–637.0	5%
6	1860	639.1	4%
7	1890–1900	645.3–647.4	3%
10	1940–1950	655.6–657.7	2%
Average	1517.4	568.3	15%

In addition, the memory size for the storage of the generated schedules is indicated in Table 5. In general, the number of schedules would increase exponentially with the number of tasks. However, the mechanisms for reconvergence, tile-specific schedule partitioning and difference encoding result in a significant reduction of the state space and the memory consumption. The baseline memory consumption and the memory consumption with difference encoding and tile-specific schedule partitioning are shown in Table 5 as well.

Table 5. Memory usage in different scenarios.

Results				
Scenario	Energy Reduction	Initial Memory Consumption	Mem. with Difference Encoding	Mem. after Tile-Based Extraction
1	19% (9%–32%)	768 B	44 B	11 B
2	16% (4%–30%)	10 kB	200 B	50 B
3	15% (2%–33%)	56 kB	896 B	224 B

The architectural building blocks for adaptation imposed delays, which added up to 20 clock cycles for a schedule change after a context event in the prototype implementation. The context monitor imposed an implementation-specific latency for the detection of context events. In addition, there was an additional delay of up to one sampling period for asynchronous context events. The prototype contained a hardware implementation of the context monitor in conjunction with synchronous context events involving a delay of two cycles.

The context agreement unit imposed a delay for distributing the context information among all tiles and establishing the globally consistent context vector. In the implementation, four clock cycles were needed for forwarding the context information between tiles using the FIFOs. Hence, 16 clock cycles were needed for the prototype system with four tiles.

The state machine of the adaptation unit in the prototype required two additional clock cycles for each message compared to the non-adaptive NoC in order to process the compressed schedule information and to traverse the linked lists.

8. Conclusions

The presented time-triggered multi-core architecture supports adaptation with multi-schedule graphs, while preserving significant properties of time-triggered systems including freedom from interference at shared resources, implicit synchronization, timeliness, implicit flow control and fault containment.

Architectural building blocks for agreement establish a consistent chip-wide view on context events, which is used by the adaptation unit in each tile for temporally aligned changes of schedules.

The meta-scheduler of the architecture introduces techniques for avoiding state explosion such as reconvergence of adaptation paths with bounded horizons for context events. In addition, memory consumption is minimized using difference-encoding and the tile-specific extraction of schedule information.

The meta-scheduler of the architecture computes Multi-schedule Graphs (MGs), which incorporate for each combination of context events the corresponding fixed scheduling decisions. These decisions include the start times of jobs, message injection times, messages paths and parameters for energy management such as time intervals with different frequency values for cores and routers.

For each schedule of the MG, deadlines, precedence constraints, resource contention and the adaptation overheads (e.g., delays for DVFS and delays for establishment of globally consistent context vectors) are considered. Consequently, correctness can be verified at design time and the presented time-triggered multi-core architecture enables adaptation for safety-critical embedded systems, where significant improvements with respect to energy efficiency and fault recovery can be obtained.

Plans for future work include the experimental evaluation of the fault tolerance using fault-injection experiments and the extension of ATMA towards a hierarchical architecture for the interconnection of adaptable multi-core chips via reconfigurable off-chip communication networks.

Author Contributions: R.O. contributed to the overall conceptualization, the adaptive multi-core architecture and the scheduling algorithms. H.A. contributed to the overall conceptualization, the adaptive multi-core architecture and the experimental evaluation. A.M. contributed to the conceptualization, implementation of the adaptation services and the experimental evaluation. Y.B. contributed to the prototype implementation and the experimental evaluation. A.L. contributed to the conceptualization and implementation of the agreement services. B.S. contributed to the scheduling algorithms.

Funding: This work was supported by the European project SAFEPOWER under the Grant Agreement No. 687902.

Conflicts of Interest: The authors declare no conflict of interest.

Abbreviations

The following abbreviations are used in this manuscript:

CPS	Cyber-Physical System
ATMA	Adaptive Time-triggered Multi-core Architecture
DAG	Directed Acyclic Graph
SoC	System-on-Chip
NoC	Network-on-Chip
NI	Network Interface
TTNoC	Time-Triggered Network-on-Chip
MPSoC	Multiprocessor System-on-Chip
MCNoC	Mixed-Criticality Network-on-Chip
MCS	Mixed-Criticality System
WPMC	Wormhole NoC Protocol for Mixed Criticality Systems
RT	Real-Time
WCAT	Worst-Case Arrival Time
SEU	Single Event Upset
VL	Virtual Link
TTEthernet	Time-Triggered Ethernet
VC	Virtual Channel
VN	Virtual Network
DVFS	Dynamic Voltage and Frequency Scaling
WCET	Worst Case Execution Time
PE	Periodic
SP	Sporadic
AP	Aperiodic
TT	Time-Triggered
ET	Event-Triggered
RC	Rate-Constrained
MINT	Minimum Inter-Arrival Times
BE	Best-Effort
GS	Guaranteed Services
ASM	Adaptation State Machine
MT	Message Transmission
BP	Branching Point
TTEL	Time-Triggered Extension Layer
PQ	Priority Queue
MG	Multi-schedule Graph
LLMG	Linked-List Multi-schedule Graph
TDM	Time Division Multiplexing
TDMA	Time Division Multiple Access

References

1. Obermaisser, R. *Time-Triggered Communication*; Embedded Systems; CRC Press: Boca Raton, FL, USA, 2012.
2. Gatti, S.; Aimé, F.; Treuchot, S.; Jourdan, J. Incremental functional certification for avionic functions reuse & evolution. In Proceedings of the 31st IEEE/AIAA Digital Avionics Systems Conference (DASC), Williamsburg, VA, USA, 14–18 October 2012; pp. 7A5–1–7A5–16. [CrossRef]
3. Spanoudakis, G.; Damiani, E.; Maña, A. Certifying Services in Cloud: The Case for a Hybrid, Incremental and Multi-layer Approach. In Proceedings of the 2012 IEEE 14th International Symposium on High-Assurance Systems Engineering, Omaha, NE, USA, 25–27 October 2012; pp. 175–176. [CrossRef]
4. Wollschlaeger, M.; Sauter, T.; Jasperneite, J. The Future of Industrial Communication: Automation Networks in the Era of the Internet of Things and Industry 4.0. *IEEE Ind. Electron. Mag.* **2017**, *11*, 17–27. [CrossRef]
5. Windsor, J.; Hjortnaes, K. Time and Space Partitioning in Spacecraft Avionics. In Proceedings of the 3rd IEEE International Conference on Space Mission Challenges for Information Technology, Pasadena, CA, USA, 19–23 July 2009; pp. 13–20. [CrossRef]
6. Obermaisser, R.; Kopetz, H.; Paukovits, C. A Cross-Domain Multiprocessor System-on-a-Chip for Embedded Real-Time Systems. *IEEE Trans. Ind. Inform.* **2010**, *6*, 548–567. [CrossRef]
7. Bensalem, S.; Goossens, K.; Kirsch, C.M.; Obermaisser, R.; Lee, E.A.; Sifakis, J. Time-predictable and composable architectures for dependable embedded systems. In Proceedings of the 9th ACM International Conference on Embedded Software (EMSOFT), Taipei, Taiwan, 9–14 October 2011; pp. 351–352. [CrossRef]
8. Heilmann, F.; Syed, A.; Fohler, G. Mode-changes in COTS Time-triggered Network Hardware Without Online Reconfiguration. *SIGBED Rev.* **2016**, *13*, 55–60. [CrossRef]

9. Fohler, G.; Gala, G.; Perez, D.G.; Pagetti, C. Evaluation of DREAMS resource management solutions on a mixed-critical demonstrator. In Proceedings of the 9th European Congress on Embedded Real Time Software and Systems, Toulouse, France, 31 January–2 February 2018.

10. Theis, J.; Fohler, G.; Baruah, S. Schedule Table Generation for Time-Triggered Mixed Criticality Systems. In Proceedings of the 1st Workshop on Mixed Criticality Systems at IEEE Real-Time Systems Symposium, Vancouver, BC, Canada, 3 December 2013.

11. Lenz, A.; Obermaisser, R. Global Adaptation Controlled by an Interactive Consistency Protocol. *J. Low Power Electron. Appl.* **2017**, *7*, 13. [CrossRef]

12. Maleki, A.; Ahmadian, H.; Obermaisser, R. Fault-Tolerant and Energy-Efficient Communication in Mixed-Criticality Networks-on-Chips. In Proceedings of the IEEE Nordic Circuits and Systems Conference (NorCAS), Tallinn, Estonia, 30–31 October 2018.

13. Hodge, V.J.; O'Keefe, S.; Weeks, M.; Moulds, A. Wireless Sensor Networks for Condition Monitoring in the Railway Industry: A Survey. *IEEE Trans. Intell. Transp. Syst.* **2015**, *16*, 1088–1106. [CrossRef]

14. Schoon, H.; Marni, A.; Musiol, R.; Nandagiri, N. In-depth lessons learned: Review of an avionics thermal analysis project. In Proceedings of the Semiconductor Thermal Measurement and Management Symposium, San Jose, CA, USA, 9–13 March 2014.

15. *Position Paper–Multi-core Processors, CAST-32A*; Technical report; Certification Authorities Software Team (CAST), Federal Aviation Administration: Washington, DC, USA, 2016.

16. Yeh, Y.C. Safety critical avionics for the 777 primary flight controls system. In Proceedings of the 20th Digital Avionics Systems Conference, Daytona Beach, FL, USA, 14–18 October 2001.

17. ISO. *ISO26262, Road Vehicles—Functional Safety*; International Organization for Standardization: Geneva, Switzerland, 2011.

18. Höftberger, O.; Obermaisser, R. Ontology-based runtime reconfiguration of distributed embedded real-time systems. In Proceedings of the 16th IEEE International Symposium on Object/component/service-oriented Real-time distributed Computing, Paderborn, Germany, 19–21 June 2013. [CrossRef]

19. Lenz, A.; Pieper, T.; Obermaisser, R. Global Adaptation for Energy Efficiency in Multicore Architectures. In Proceedings of the 25th Euromicro International Conference on Parallel, Distributed and Network-based Processing (PDP), IEEE, St. Petersburg, Russia, 6–8 March 2017. [CrossRef]

20. Kopetz, H. Sparse time versus dense time in distributed real-time systems. In Proceedings of the 12th International Conference on Distributed Computing Systems, Yokohama, Japan, 9–12 June 1992; pp. 460–467. [CrossRef]

21. Xilinx. *TMRTool User Guide. TMRTool Software Version 13.2*; Xilinx: San Jose, CA, USA, 2017.

22. Sari, B.; Reuss, H.C. Fail-Operational Safety Architecture for ADAS Systems Considering Domain ECUs. In Proceedings of the WCX World Congress. SAE International, Detroit, MI, USA, 10–12 April 2018. [CrossRef]

23. Kopetz, H. *Real-Time Systems: Design Principles for Distributed Embedded Applications*, 2nd ed.; Real-Time Systems Series; Springer: Berlin, Germany, 2011.

24. Sorkhpour, B.; Murshed, A.; Obermaisser, R. Meta-scheduling techniques for energy-efficient robust and adaptive time-triggered systems. In Proceedings of the 4th IEEE International Conference on Knowledge-Based Engineering and Innovation (KBEI), Tehran, Iran, 22 December 2017; pp. 143–150.

25. Vestal, S. Preemptive Scheduling of Multi-criticality Systems with Varying Degrees of Execution Time Assurance. In Proceedings of the 28th IEEE International Real-Time Systems Symposium, Tucson, AZ, USA, 3–6 December 2007; pp. 239–243. [CrossRef]

26. Sorkhpour, B.; Obermaisser, R.; Murshed, A. Optimization of Frequency-Scaling in Time-Triggered Multi-Core Architectures using Scenario-Based Meta-Scheduling. In Proceedings of the 10th VDE/GMM-Symposium Automotive meets Electronics, Dortmund, Germany, 12–13 March 2019.

27. Ahmadian, H.; Obermaisser, R.; Abuteir, M. Time-Triggered and Rate-Constrained On-chip Communication in Mixed-Criticality Systems. In Proceedings of the 10th IEEE International Symposium on Embedded Multicore/Many-core Systems-on-Chip (MCSoC), Lyon, France, 21–23 September 2016; pp. 117–124.

28. Millberg, M.; Nilsson, E.; Thid, R.; Kumar, S.; Jantsch, A. The Nostrum backbone-a communication protocol stack for Networks on Chip. In Proceedings of the 17th International Conference on VLSI Design, Mumbai, India, 5–9 January 2004; pp. 693–696. [CrossRef]

29. Murshed, A.; Obermaisser, R. Scheduler for reliable distributed systems with time-triggered networks. In Proceedings of the 15th IEEE International Conference on Industrial Informatics (INDIN), Emden, Germany, 24–26 July 2017; pp. 425–430.

30. Sorkhpour, B.; Obermaisser, R. MeSViz: Visualizing Scenario-based Meta-Schedules for Adaptive Time-Triggered Systems. In Proceedings of the AmE 2018—Automotive meets Electronics, 9th GMM-Symposium, St. Louis, MO, USA, 7–8 Marh 2018; pp. 80–85.

Article

A Two-Layer Component-Based Allocation for Embedded Systems with GPUs

Gabriel Campeanu [1,*] and Mehrdad Saadatmand [2]

[1]   Bombardier Transportation, 722 14 Västerås, Sweden
[2]   RISE Research Institutes of Sweden, RISE SICS Västerås, 722 12 Västerås, Sweden;
       mehrdad.saadatmand@ri.se
*   Correspondence: gabriel.campeanu@rail.bombardier.com

Received: 14 December 2018; Accepted: 16 January 2019; Published: 19 January 2019

Abstract: Component-based development is a software engineering paradigm that can facilitate the construction of embedded systems and tackle its complexities. The modern embedded systems have more and more demanding requirements. One way to cope with such a versatile and growing set of requirements is to employ heterogeneous processing power, i.e., CPU–GPU architectures. The new CPU–GPU embedded boards deliver an increased performance but also introduce additional complexity and challenges. In this work, we address the component-to-hardware allocation for CPU–GPU embedded systems. The allocation for such systems is much complex due to the increased amount of GPU-related information. For example, while in traditional embedded systems the allocation mechanism may consider only the CPU memory usage of components to find an appropriate allocation scheme, in heterogeneous systems, the GPU memory usage needs also to be taken into account in the allocation process. This paper aims at decreasing the component-to-hardware allocation complexity by introducing a two-layer component-based architecture for heterogeneous embedded systems. The detailed CPU–GPU information of the system is abstracted at a high-layer by compacting connected components into single units that behave as regular components. The allocator, based on the compacted information received from the high-level layer, computes, with a decreased complexity, feasible allocation schemes. In the last part of the paper, the two-layer allocation method is evaluated using an existing embedded system demonstrator; namely, an underwater robot.

Keywords: embedded systems; software component; component-based development; CBD; GPU; GPU component; allocation; component allocation; architecture layer

---

1. Introduction

Nowadays, embedded systems become more and more common in the daily life. Modern embedded systems, characterized by new and demanding functionalities, deal with huge amount of information resulted from the interaction with the environment. For instance, the Google autonomous car (i.e., the Waymo project) handles 750 MB of data per second that is produced by its sensors (e.g., LIDAR). The huge amount of information needs to be processed with a particular performance, in order to satisfy the system requirements. For example, the autonomous Google car needs to process its captured data in real-time in order to detect various objects and pedestrians, to avoid accidents. One solution to enhance the processing capacity of embedded systems comes from the usage of embedded boards with Graphics Processing Units (GPUs). A GPU is a processing unit that is equipped with hundreds of computation threads, excelling in parallel data-processing.

Although, on one side, the use of GPU increases the system (parallel-processing) performance, on the other side it increases the complexity of the system design. In particular, the software-to-hardware allocation is already not an easy task: when having several processing units of different kinds and with different capabilities, a major design challenge will then be in finding an optimal allocation of software artifacts (e.g., components) onto the processing units in a way that system constraints are also met and not violated. For allocating a set of n software artifacts onto m processing units, a total number of m^n combinations are to be considered [1]. The challenge, as mentioned, is then to find, from all combinations, a single permutation as an optimal allocation scheme with respect to the constraints and characteristics of the system. With the GPU in the landscape, the allocation becomes even more complicated and challenging. The software is characterized now, besides the properties regarding the CPU resources, with properties that refer to GPU such as the GPU memory usage or the execution performance on the GPU. Similarly, the platform also has characteristics regarding the CPU but also the GPU hardware. Hence, the allocation challenge is increased due to the extra application properties and hardware characteristics that must be considered. In short, the challenge of finding optimal allocations of software artifacts to hardware has increasingly attracted attention, especially with the advent and growing prevalence of heterogeneous hardware platforms and increasing use of software in mission-critical applications. In [2], for instance, Baruah has discussed the challenge of allocating a set of recurring tasks in real-time systems onto processing units of different kinds while respecting all timing constraints, and has identified this to be an NP-hard problem. One way to relieve the allocation challenge and cope with its complexity is by managing the amount of information that is fed to the allocator to make allocation decisions. This is the main topic and solution that we introduce and investigate in this paper. In other words, in this current work we do not focus mainly on how to derive and what will be an optimal allocation scheme for a system (which is the main focus in [1,3]), but rather how the burden on the allocator can be relieved to relax the complexity of the allocation process in general.

In this work, we use the component-based development (CBD) to construct embedded systems with GPUs. In general, this software engineering methodology promotes the construction of applications by composing existing software units called software components. CBD is used and promoted in industry to construct embedded systems; such as in AUTOSAR [4] which is now the de-facto standard in automotive industry, and IEC 61131 [5] used to develop programmable logic controllers (PLCs). In the context of component-based embedded systems with GPUs, we focus on the component-to-hardware allocation, proposing a semi-automatic allocation method. When using platforms with GPUs, the allocation challenge increases even more due to the (higher) complexity of the software and hardware. The software in such systems is composed of: (i) (traditional) components that have requirements on common resources (e.g., CPU load, RAM memory usage), and (ii) (GPU-specialized) components that have GPU requirements (e.g., number of GPU threads usage). These GPU-specialized components, although they contain small CPU functionality (e.g., activities to trigger execution on GPU), are seen as components with only GPU computation. Having a pool of (CPU and GPU-based) components, many alternatives with the same functionality may result. For example, a vision system may have two alternatives, where one alternative contains only components with CPU functionally, while the other alternative contains only components with GPU computations. Regarding the hardware, the platform encloses, besides the traditional CPU, the GPU that has different characteristics such as the available GPU memory.

The aim of our work is to alleviate the allocation challenge by mitigating the increased amount of (software and hardware) information. In the context of applications with multiple alternatives, and CPU–GPU hardware, the allocator does not need to take in consideration all the system information. For instance, the information that describes the component communication from inside the alternatives may be neglected. An implicit constraint considered in this work is, due to the closely connected nature of the GPU to the CPU, the allocator needs to deploy a (entire) variant that has GPU computations, onto a CPU–GPU processing node. Enforcing this requirement improves the overall system performance.

As a solution, we propose a two-layer architecture to decrease the allocation effort. Both layers describe a same system that has GPU computation; the difference resides in the level of information that characterize each layer. The first layer, seen as regular description of the architecture, encloses all information (e.g., component communication links) of all alternatives. The second layer compacts different alternatives with the same functionality into single components with multiple variants. Each of the variants of the resulted components, is characterized by a set of properties that reflect the requirements of all components contained by its corresponding alternative.

To abstract certain information, the allocator uses the first layer description and selects a suitable component alternative. Once the allocation scheme is computed, the selected alternative is unfolded with the corresponding structure from the second layer. The core idea of a two-layer allocation method is to decrease the information load and constraints that may increase the overhead of the allocator. Another benefit of our solution is the increased scalability characteristic, where our allocator may handle more complex systems (e.g., characterized by a high number of components) due to the decrease of the information and constraints load.

In this work, we use in the evaluation section an already existing (constraint-based) allocator that we constructed in a previous work [3]. Our approach is independent of what allocator is used or how it is implemented. By proposing the two-layer allocation design, we improve the scalability of the allocator used for heterogeneous CPU–GPU systems, and decrease its burden.

The remaining of the paper is organized as follows. Section 2 introduces the context details of this work. Furthermore, the section presents a running example, and, based on it, a way to develop component-based systems for heterogeneous embedded systems. The overview of our method is described in Section 3 while Section 4 evaluates our method using the introduced running example. Related work is covered by Section 5 followed by conclusions and future work in Section 6.

2. Component-Based Design for Heterogeneous CPU–GPU Architectures

The growing complexity and size of embedded systems emphasizes the use of appropriate development methods that can cope with these issues and scale well. Component-Based Development (CBD) is a promising approach in this regard which promotes building a system out of already existing components, as opposed to building it from scratch. In other words, CBD enables *reusability* in software development by building a system as an assembly of components [6,7] selected from a repository of verified existing components. Considering the constraints of embedded systems in terms of available resources, how software components are allocated onto the hardware platform can play an important role in the performance of the system and optimal use of the resources. This is, however, not a trivial task and as the number of software components as well as processing units and computing nodes on the hardware platform increases, different combinatorial allocation schemes need to be evaluated in order to determine an appropriate one. For this reason, having an automated solution for evaluation of allocation schemes is necessary. One aspect that adds to the challenge of allocating software components onto hardware platforms is the move towards the use of heterogeneous hardware architectures such as multi-core CPU and GPU. Use of GPUs along with CPUs is particularly interesting as it can provide increased computation power and diversity due to the parallel processing capability of GPUs. This brings along additional constraints that need to be taken into account for allocation of software components to hardware. For instance, a GPU cannot be used independently of a CPU, and it is the CPU that triggers all GPU specific operations such as data transfer between the main memory unit (i.e., RAM) and the GPU memory system. Therefore, there is constant and high communication between these two processing units.

From the perspective of the processing unit, software components can be categorized as: (i) those that require only CPU for their functionality (ii) components that use GPU (in addition to CPU) to fulfill their functionality. For the rest of the paper, these types of components will be referred to as GPU components. A specific functionality may be implemented as any of these component types. Therefore, in the component repository both types of components may exist as different implementation versions of a specific function. For instance, for an image processing component we may have three different implementation versions in the repository; one that uses CPU only (type i), and two others as GPU-based implementations that require GPU as well (type ii). Each of these three versions can have different properties and characteristics in terms of resource usage and utilization. For example, the two GPU-based component implementations can have different resource usage properties with respect to GPU memory and GPU computation threads.

In the next paragraphs, we introduce a case study that will be used to present our solution. The case study is an underwater robot [8] that autonomously navigates under water and executes various missions such as finding buoys. The robot is a typical embedded system that contains sensors (e.g., cameras), an embedded board with an incorporated GPU, and actuators (e.g., thrusters). An interesting part of the robot is the vision part which is described in Figure 1. The vision (sub-)system is developed with a state-of-practice component model i.e., the Rubus component model [9]. This particular component model follows a *pipe-and-filter* interaction style, where each component computes its received data and sends it to the next connected component(s). The vision system consists of six components as follows. The first two components (i.e., *Camera1* and *Camera2*) receive raw data from two camera sensors, convert it into readable color (i.e., RGB) frames and forward it to the *MergeAndEnhance* component. After the two frames are merged into a single RGB frame and its noise is reduced, the *ConvertGrayscale* component converts it into a grayscale format and sends it to the *EdgeDetection* component. This component converts the frame into a black-and-white frame, where the white lines delimit the objects from the frame. Finally, the *ObjectDetection* component detects a particular object, such a buoys.

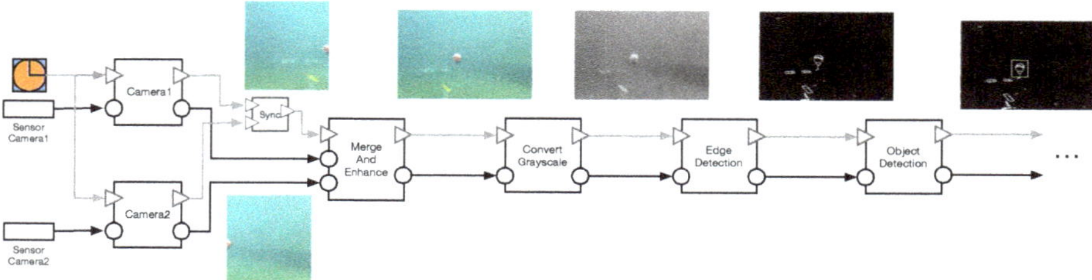

Figure 1. The vision system of an underwater robot.

Due to the fact that several components (i.e., *MergeAndEnhance*, *ConvertGrayscale*, *EdgeDetection* and *ObjectDetection*) have functionalities (i.e., image processing) that can be executed on the GPU, the vision system may have different alternatives. Figure 2 presents possible alternatives of the vision system. Assuming that we have a repository with ten components where there are components with the same behavior but constructed to be executed on different processing units (i.e., either on CPU or GPU). For instance, there is the *EdgeDetection(GPU)* component that is constructed to be executed on the GPU and there is *EdgeDetection(CPU)* that has the same behaviour but it requires to be executed on the CPU. The total amount of alternatives that can be constructed by using the repository components is six, as illustrated in the figure. The first alternative, where all components are executed on the CPU, has a low performance but also zero-demand on the GPU. This alternative can be selected to be used in a system that is not equipped with a GPU or in a system that possesses a GPU but it is used by a different part of the system. The last alternative, containing four components that need to be executed on the GPU, has the highest performance compared to the rest of the alternatives but it also has high GPU requirements (e.g., GPU memory and computation threads usage). The other four possible alternatives contain variations of components with different requirements on the CPU and GPU.

In this context where there are several alternatives and each one contains different components versions characterized by distinct characteristics, the information load on the allocator is much, influencing, in a negative way, the allocation efficiency. Furthermore, because of the tight connection between CPU and GPU, the components executed on a GPU are desired to be placed, alongside with their connected CPU-based components, on the same CPU–GPU chip. In the opposite case, placing a GPU-component on a different CPU–GPU chip than its connected CPU-component brings additional communication overhead which negativly influences the system performance. In general, the allocation complexity is directly influenced by the number of considered alternatives and by the number of the components and their versions included in the alternatives.

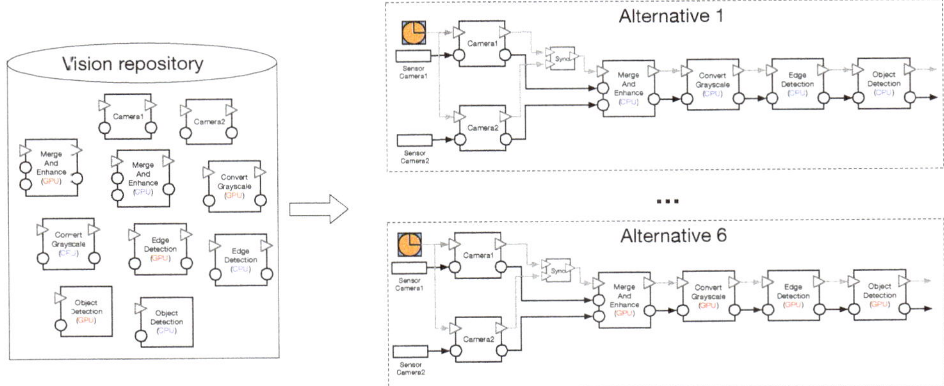

Figure 2. The vision system alternatives. GPU, graphics processing units; CPU, central processing unit.

3. Solution Overview

To diminish the burden of the allocation process, we introduce a two-layer allocation method. The layers correspond to a two-layer architecture view of the system, where the bottom layer describes a detailed system (i.e., composition of different component versions), while the top layer abstracts the complexity of the detailed system, by compacting the resulting alternatives into units (i.e., multi-variant units) that behave as regular components. The resulted components are characterized by different alternatives, where an alternative contains the properties that correspond to all its enclosed components.

Figure 3 depicts the vision system alternatives compacted into a multi-variant component where each alternative is characterized by set of properties. These properties are derived from the properties of the enclosed components. For instance, the alternative 6 where there are four GPU-based components, is characterized by a GPU memory property that describes the GPU memory usage of the enclosed components. This property is derived by e.g., summing the GPU memory usage of all four GPU-based components. As the GPU threads are highly reusable between GPU-based components, the alternative property that describes the number of GPU threads usage is the highest value of threads usage among the four GPU-based components. Furthermore, some attributes may be abstracted away. For example, due to the fact that connected (CPU- and GPU-based) components are enclosed together into single units and implicitly they will be allocated on the same CPU–GPU chip, the bandwidth property required for data transfer between two connected components, is abstracted away.

Figure 4 illustrates the overview description of our solution, that contains six stages as follows:

1. The first stage refers to the component pool from which the system developer constructs the application. The components from the repository may be provided by a 3rd-party or developed in-house. The repository contains regular (CPU-based) components but also components with GPU capability. For instance, there are two component versions (i.e., *C2 GPU* and *C2 CPU*) with the same behaviour but different (hardware) requirements.

2. Using the available components from the repository, the system architect composes them in different alternative systems, as described in the second stage. For example, while the first alternative uses *C1 GPU*, the second system alternative contains *C1 CPU*. All the system alternatives have the same behavior and different requirements. For example, while the first alternative requires GPU memory for two components, the second alternative has only one component with GPU requirement.

3. The third stage compacts all the component variants into multi-variant components. For example, the several alternatives that contain three components are grouped into a unit with different alternatives. These alternatives have the same functionality but with different (CPU and GPU) requirements. This is a simplified system architecture in which we abstracted away some information.

4. The information from the previous stage is forwarded to an allocator which we assume it already exists. Our approach is independent of how the allocator itself is implemented and based on which solver. Other information is fed to the allocator such as the description of the platform or different system constraints. In this stage, the allocator computes component-to-hardware allocation schemes where each component is mapped to a single processing unit.

5. The fifth stage describes the system architecture based on the result computed by the allocator. The architecture contains only (single-variant) components, that is each component has a single set of requirements and is allocated to a particular processing unit.

6. The last stage contains a fully detailed system architecture where the (single-variant) components from the previous stage are unfolded (when it is possible). The figure illustrates the detailed system architecture where the alternative selected by the allocator is unfolded into three connected components, i.e., *C1 GPU, C2 GPU* and *C3 CPU*.

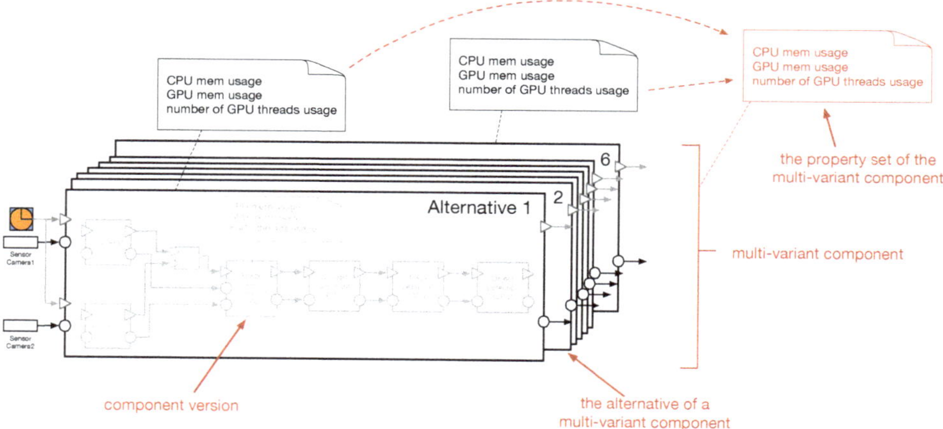

Figure 3. A multi-variant vision component.

In general, using a simplified version of the system architecture decreases the amount of information forwarded to the allocator, and hence, the allocation complexity. The simplified version is obtained by applying a high-level layer on top of the detailed architecture layer. Instead of describing each component by a set of requirements as it is done in the detailed layer, connected components are grouped into units with the same functionality, that behave as regular components, and are described by a condensed number of properties as it is achieved in the high-lever layer. Furthermore, some desired allocation constraints may be automatically taken care of and provided as a by-product of our method. For example, instead of specifically requesting that connected CPU–GPU components to be allocated together onto the same CPU–GPU chip units (e.g., to improve the system performance), our method implicitly introduces this request by compressing the components into single (component-like) units.

We construct our solution around an existing allocator. The (mathematically defined) model of the allocator is based on constraints and optimization goals. The constraints assure that the allocator does not use more than existing hardware resources such as the memory required by all the components placed on a platform does not exceed the available physical memory. The optimization goals allow the user to determine essential features of the system such as performance. The actual allocation is done by using a constraint solver, the details of which we have described in a previous paper [3].

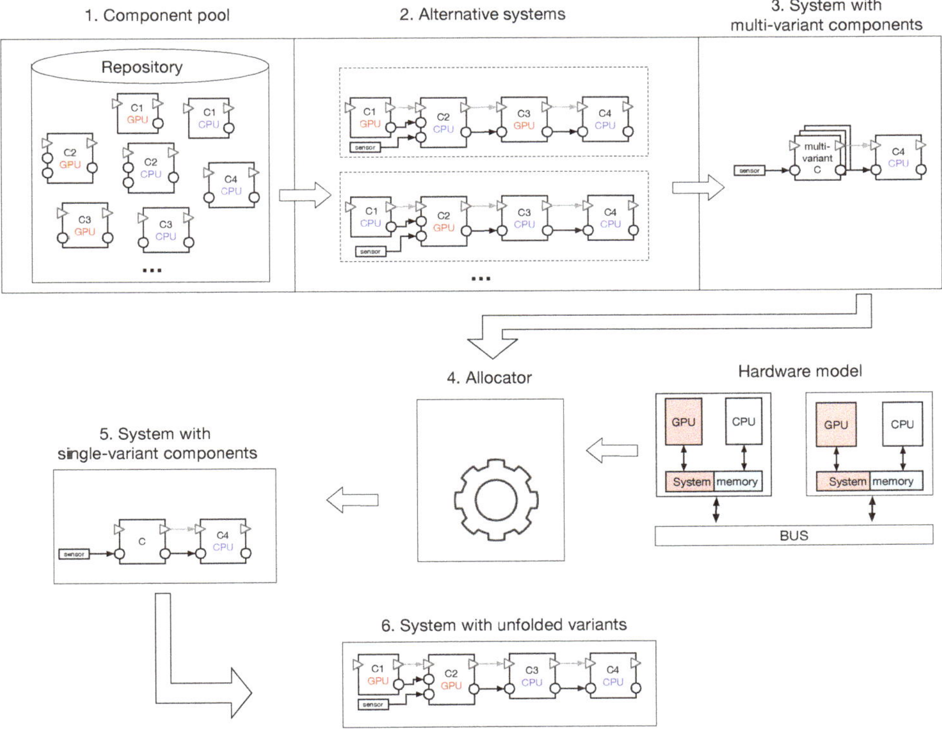

Figure 4. The overview description of our solution.

4. Evaluation

In this section, we analyze our solution in two parts. In the first part, we look into the feasibility aspect where the proposed solution is implemented using an existing system. The second part analyzes the scalability aspect of the solution.

4.1. Feasibility

To analyze the feasibility of our solution, we apply our allocation method on the underwater robot, partially introduced in Section 2. The robot contains two connected embedded controller boards: (i) a System-on-Chip board that contains a CPU–GPU chipset, and (ii) a regular board with a one-core ARM CPU. The boards communicate via a Controller Area Network (CAN) bus, and are connected to various sensors (e.g., cameras, pressure sensor) and actuators (e.g., thrusters). We characterize each board by a set of properties as follows:

- *availMem* represents the available memory of the board, and is measured in megabyte (MB).
- *availCpu* represents the available load of the CPU and its value is compared to a reference unit (e.g., 1 Cpu load unit is a particular amount of work over a period of time).

- *availGpu* represents capacity of a GPU. As a metric we use the amount of threads a GPU has. Although it is not an accurate description, we use this measurement unit to characterize, at a high-level, the GPU power.

A high-level model of the underwater robot is presented in the right-hand side of Figure 5. We notice that for the board that has only a (one-core) CPU, the GPU-related property (i.e., *availGpu*) is set to zero.

The robot is equipped with two vision systems, the front system using two from cameras and the bottom system using a single bottom camera. For both vision systems we constructed different CPU- and GPU-based components as follows. The front vision functionality merges two RGB frames, converts the merged frame into the grayscale format, applies an edge-detection filter and detects the objects from the frame (for more details see Section 2). To construct the functionality, we developed: (i) six CPU-based components (i.e., Camera1, Camera2, MergeAndEnhance CPU, ConvertGrayscale CPU, EdgeDetection CPU and ObjectDetection GPU), and (ii) four GPU-based components (i.e., MergeAndEnhance GPU, ConvertGrayscale GPU, EdgeDetection GPU and ObjectDetection GPU).

The bottom vision has a similar functionality. Due to the fact that there is only one camera, there is no need to merge frames. Therefore, to construct the bottom vision functionality, we reused four of the components developed for the front vision. In this case, there are: (i) four CPU-based components (i.e., Camera1, ConvertGrayscale CPU, EdgeDetection CPU and ObjectDetection GPU), and (ii) three GPU-based components (i.e., ConvertGrayscale GPU, EdgeDetection GPU and ObjectDetection GPU).

Besides the vision systems, five more CPU-based components are also needed. The VisionManager component takes vision decisions based on the data received from the front and bottom vision systems. DecisionCenter is the brain of the system which controls, based on the information received from VisionManager, the system settings (e.g., water pressure) and selects between the robot missions (e.g., find red buoys). The robot thrusters are managed by the MovementNavigation component that maneuvers the underwater robot using the data received from the DecisionCenter component.

Each component is characterized by the following properties:

- *reqMem* characterizes the memory usage requirement of a component and is measured in MB.
- *reqCpu* presents the CPU usage requirement of a component.
- *reqGpu* describes the component GPU usage requirement and is measured in number of threads.
- *Exec* is related to the performance of the component and describes the execution time expressed in milliseconds.

These components, constructed by the component developer, are placed into a Component repository which is illustrated in the upper part of Figure 5. The system developer uses the available components and constructs the system architecture. The (front and bottom) vision systems have multiple alternatives as illustrated in Figure 5. Each alternative has a distinct set of properties and when all alternatives are combined into a single multi-variant component, the properties are described as a sequence of values. Each value of the sequence represents the resource usage of the corresponding variant. For example, the FrontVision multi-variant component requires, for its first alternative (i.e., all CPU-based components), 6 MB of memory, 0.6 CPU load, 0 GPU threads and has an execution time of 22 ms.

Our solution is constructed around an existing allocator. In the following paragraphs, we introduce the formal model of the allocator. It contains three parts, i.e., the input, the constraints and the optimization function, as follows.

1. The input part describes the software components and the platform. Let be C a set of n software components, and four functions $reqMem : C \rightarrow \mathbb{Q}^+$, $reqCpu : C \rightarrow \mathbb{Q}^+$, $reqGpu : C \rightarrow \mathbb{Q}^+$ and $Exec : C \rightarrow \mathbb{Q}^+$, where:

$$
\begin{aligned}
reqMem(c) &= \text{the required memory of component } c \\
reqCpu(c) &= \text{the CPU workload required by component } c \\
reqGpu(c) &= \text{the GPU threads required by component } c \\
Exec(c) &= \text{the execution time of } c
\end{aligned}
$$

   The platform is characterized by a set H of k computation nodes (i.e., either CPU or CPU–GPU based), and three functions $useMem : H \rightarrow \mathbb{Q}^+$, $useCpu : H \rightarrow \mathbb{Q}^+$ and $useGpu : H \rightarrow \mathbb{Q}^+$, where:

$$
\begin{aligned}
useMem(h) &= \text{the usable memory on node } h \\
useCpu(h) &= \text{the CPU capacity on node } h \\
useGpu(h) &= \text{the available number of GPU threads on node } h
\end{aligned}
$$

2. The constraints are defined in order to ensure a feasible allocation. Given the allocation function $allocation : C \rightarrow H$, we define the following constraints:

   - The usable memory of a node h should not be exceeded by the summed required memory of components placed on h.

$$
\forall h \in H \left(\sum_{c \in \{c | c \in C \wedge allocation(c) = h\}} reqMem(c) \leq useMem(h) \right)
$$

   - The usable CPU workload of a node h should not be exceeded by the summed required workload of components placed on h.

$$
\forall h \in H \left(\sum_{c \in \{c | c \in C \wedge allocation(c) = h\}} reqCpu(c) \leq useCpu(h) \right)
$$

   - The total amount of GPU threads of a node h should not be exceeded by the summed number of threads required by the components placed on h.

$$
\forall h \in H \left(\sum_{c \in \{c | c \in C \wedge allocation(c) = h\}} reqGpu(c) \leq useGpu(h) \right)
$$

3. The optimization function:

$$
P(allocation) = \sum_{c \in \{c | c \in C \wedge allocation(c) = h\}} Exec(c)
$$

   provides the best performance of the allocation:

$$
\text{minimize } (P)
$$

The system properties and the constraints and optimization goal are fed to the allocator that computes allocation schemes. The allocation model alongside with all its required information (i.e., system properties, constraints and optimization goals) are translated into the IBM-CPLEX solver. The advantage of employing a mathematical solver is that the computed solution is optimal.

The front vision is considered the main vision of the robot, while the bottom vision is seen as a secondary system being activated when e.g., the main vision does not detect any objects. Therefore, the front vision priority is higher in accessing the GPU resources. The allocator computes allocation schemes as presented in Figure 5, where both of the vision systems are allocated onto the H_1, and the rest of the system is allocated onto H_2. With a higher priority, the front vision accesses more GPU resources (i.e., threads) than the bottom system.

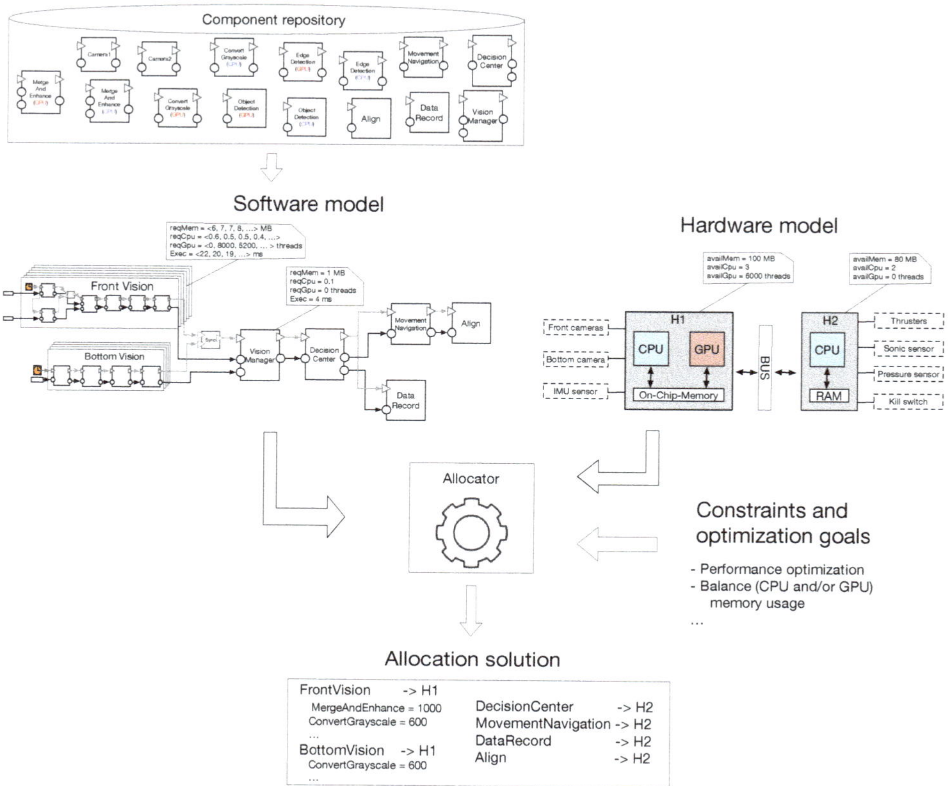

Figure 5. The allocation process of the underwater robot.

The optimal result computed by the CPLEX solver is (partially) described in Figure 6 where the selected front vision alternative contains four GPU-based components and the bottom vision alternative contains only one GPU-based component. The system description corresponds to the detailed architecture view, where the selected single-variant components are unfolded.

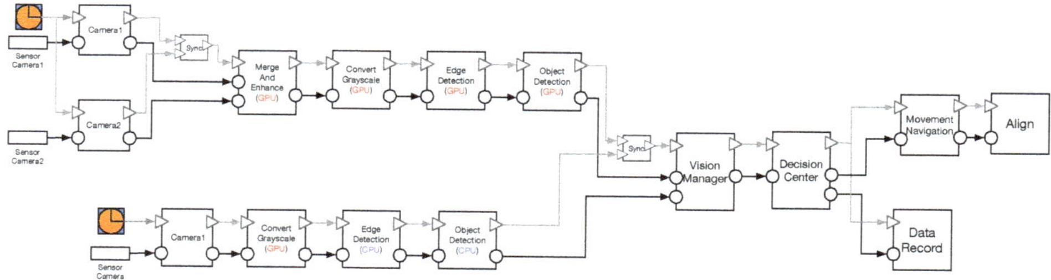

Figure 6. The allocation scheme result with unfolded variants.

The computed solution is a feasible system considering the available hardware resources and the configured optimization goals.

4.2. Scalability

For the scalability analysis we designed a test system composed of $n + 1$ chained components. Each of the first n components has two versions with the same functionality, i.e., one version with CPU-based functionality and the other with GPU-based functionality. For this part of the evaluation, we compare two versions of the system, where one version (referred as the naïve system) contains a chain of $n + 1$ components, and the other version contains a multi-variant component.

The multi-variant component is constructed from the different alternatives that result from composing different component versions. For simplicity, we consider only a two-variant component, where one alternative contains all n components with CPU-based functionality, while the other alternative contains all n components with GPU-based functionality, as illustrated by Figure 7. Each individual component is characterized by memory usage, CPU usage, GPU thread usage and execution time. The multi-variant component is characterized by a set of properties derived from the enclosed individual component properties, where each property is described as a sequence of the values (i.e., one for the CPU- and the other for the GPU-based components). For example, one value of the memory sequence property that characterizes the CPU-based component variant, is computed by summing the memory usage requirements of all its n CPU-based components.

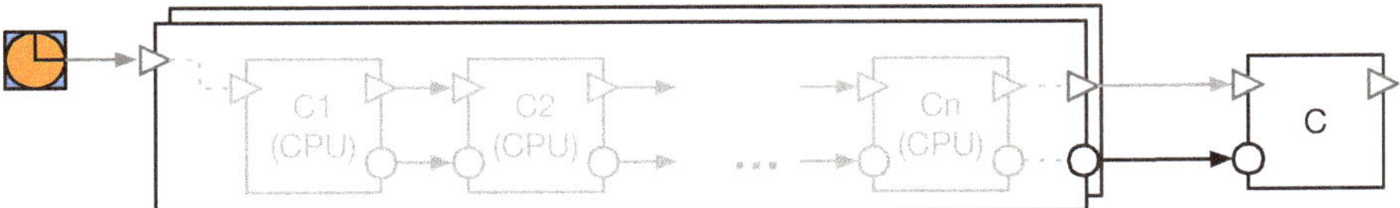

Figure 7. A system composed of chained components.

To calculate the allocation computation time, we have constructed three systems; the first system contains 31 components (i.e., $n = 30$), the second system contains 41 components (i.e., $n = 40$) and the last system has 51 components (i.e., $n = 50$). We randomly provided values for component properties (i.e., for each CPU- and GPU-based version). For instance, the component memory usage is randomly assigned a value between 1 and 100. Regarding the hardware, we assume that we have six connected boards, where only three of them have GPUs. Similarly, we characterized the boards resources in a random manner. For example, the available memory of a board is randomly assign a value between 100 and 2500.

Using the implemented CPLEX allocator from the previous part, we compute allocation schemes for our test systems. As optimization goal, we set the allocator to provide the best performance (i.e., execution time).

The scalability results are presented by Table 1. Using a machine with a 2.6 GHz i7 CPU and 16 GB of memory, we ran the allocation 1000 times for three systems, i.e., a naïve system that contains CPU-based/GPU-based components, and a system that contains the two-variant component. The results show that the allocator uses less time to compute results for the system with the two-variant component. In other words, for the CPU- and GPU-based systems, where there are $n + 1$ components and each has its own set of properties, the allocator analyzes a higher number of properties than for the two-variant system where the two-variant component has one set of properties. Furthermore, the computation time for the CPU-based system is relatively the same as for the GPU-based system due to the fact that the analyzed number of properties are the same for the two systems. By proposing the two-layer allocation design, we show in this part of the evaluation, how the scalability of an allocator for heterogenous CPU–GPU systems is improved.

Table 1. The time used to compute allocation schemes.

	Average Allocation Time (ms)		
n	Naïve *		Two-Variant **
	CPU-Based	GPU-Based	
30	18.2	18.3	15.4
40	29.1	29.0	24.6
50	49.5	49.7	39.8

* A system with (CPU-based/GPU-based) $n + 1$ components; ** A system with a two-variant component; GPU, graphics processing units; CPU, central processing unit.

5. Related work

We introduce in [3] the initial idea of the two-layer allocation method. We extended the initial work by describing the solution using an existing component model and presenting the overview using existing components. Furthermore, in the current paper we introduced an existing system and applied the solution on it to analyze the feasibility aspects. Moreover, the evaluation section analyzes the scalability aspects, which were not covered by the previous work.

Software-to-hardware allocation and optimization of the allocation mechanism have been the topic of many research works in the literature. A systematic literature review on the software architecture optimization methods is provided in [10]. The authors in this work analyzed 188 papers and identified 30 papers related to the optimization of component-based systems. Of this set of papers, only 13% (i.e., 4 papers) use exact optimization algorithms (vs. approximate algorithms). While exact optimization algorithms can provide optimal solutions, their applications poses several challenges when it comes to adopting them, such as difficulty of formally defining the allocation model, search-space, and the usually non-linearity of the object functions (and thus being computationally expensive). The approach we presented in this paper, enabled us to formally define our optimization model which in turn allows to use exact optimization algorithms and methods. Moreover, one of the important characteristics of our approach is that, as we demonstrated, it simplifies the search-space and therefore complexity for allocation optimization.

Consideration of quality attributes and satisfaction of non-functional requirements play an important role in designing embedded systems due to resource constraints of these systems. While our proposed approach can address different quality attributes (such as memory, processing capacity and number of threads, etc.) and is generic in this regard, there are some research in the literature that target specific quality attributes. For instance, in [11], a detailed optimization model and framework for energy consumption in component-based distributed systems in Java is provided. The main goal in this work has been to help system architects make informed decisions such that the energy consumption is reduced in a designed system. An interesting aspect discussed in this work is the energy consumption of the communication of components that reside in different Java Virtual Machines, on the same host. From this perspective, the communication aspect is implicitly address in our optimization model where connected CPU- and GPU-based components are tried to be allocated on the same node. Furthermore, in our optimization model, the energy is treated as any other property in a generic fashion. [12] is another example of works that address energy usage in heterogeneous multiprocessor embedded systems. In this work, an optimization model using integer linear programming is introduced that minimizes the system energy usage when the end-to-end time constraints are given. Moreover, the CPLEX solver is used to compute allocation solutions. It is shown that, for a system with more than 30 components, the solver computes solutions in up to couple of minutes. The solution introduced in our work aims at decreasing the allocator burden and we show that, for a system with 31 components, the allocation computation time is reduced.

Wang et al. in [13] introduce a method to allocate the software components in a design model to a given platform while meeting multiple platform resource constraints. In the method, different types of resources are considered and weights are used to define the importance of each in the allocation process. The components that require more resources get higher priority getting allocated first. In contract, in our approach all components have same and equal allocation priority with respect to their resource requirements. Also using the flexible component concept, we increase the flexibility of the allocation regarding the component resource requirements. Weight parameters are used in our approach to define the importance of properties in the allocation process.

For the systems in the automotive domain [14] proposes optimization of software allocation and deployment to hardware nodes (i.e., ECUs) as a as a bi-objective problem using an evolutionary algorithm. It considers the reliability of data communications between components as one and the communications overhead as a second objective.

6. Conclusions

In this paper, we proposed a method to reduce the burden of the allocator and ease the software-to-hardware allocation in the design of component-based embedded systems. Our solution works by introducing a two-layer architecture design for heterogeneous CPU–GPU embedded systems, where detailed component information is abstracted using the two layers to ease allocation decisions. The work is independent of the employed allocator, i.e., how the allocator itself is implemented and based on which solver.

To show the feasibility of the approach, we applied it on a underwater robot which we participated in its development. Additionally, to also demonstrate and evaluate the scalability of the approach, we analyzed it on three test systems consisting of 31, 41, and 51 components respectively. We compared the average allocation time for two versions of each of these test systems: (i) containing all components, (ii) a two-variant component model of the system, based on the two-layer allocation concept. The results show that the allocator does its computations faster and requires less time for the two-variant component version. Although CPLEX solver was used in this work, the proposed solution can be implemented in any mixed-integer non-linear solver. However, the usage of a different solver may influence the allocation computation time.

In terms of quality attributes and non-functional properties, the proposed two-layer allocation solution is generic and be be applied in allocation optimization based on any set of properties. From this perspective, it is property-agnostic. Deriving variant properties (i.e., aggregated from its constituting components), however, can be less trivial for certain non-functional properties such as energy. In our case study in this paper, we characterized the system through simple properties such as static memory or GPU thread usage. Then to derive the multi-variant properties, we simply used addition operation for these properties. As a future direction, we plan to investigate how energy usage can be derived for variants, and thus, enable its inclusion and evaluation as part of our proposed solution. Another extension of this work is to extend the scope of heterogeneity of the approach to include other processing units such as DSPs and FPGAs as well.

Author Contributions: Conceptualization, writing—review & editing, G.C. and M.S.

Conflicts of Interest: The authors declare no conflict of interest.

References

1. Svogor, I.; Crnkovic, I.; Vrcek, N. An Extended Model for Multi-Criteria Software Component Allocation on a Heterogeneous Embedded Platform. *J. Comput. Inf. Technol.* **2013**, *21*, 211–222.
2. Baruah, S.K. Task partitioning upon heterogeneous multiprocessor platforms. In Proceedings of the 10th IEEE Real-Time and Embedded Technology and Applications Symposium, Toronto, ON, Canada, 25–28 May 2004; pp. 536–543. [CrossRef]
3. Campeanu, G.; Carlson, J.; Sentilles, S. Allocation Optimization for Component-Based Embedded Systems with GPUs. In Proceedings of the 2018 44th Euromicro Conference on Software Engineering and Advanced Applications (SEAA), Prague, Czech Republic, 29–31 August 2018; pp. 101–110.

4. AUTOSAR—Technical Overview. Available online: http://www.autosar.org (accessed on 27 July 2018).

5. John, K.H.; Tiegelkamp, M. *IEC 61131-3: Programming Industrial Automation Systems: Concepts and Programming Languages, Requirements for Programming Systems, Decision-Making Aids*; Springer Science & Business Media: Berlin, Germany, 2010.

6. Torngren, M.; Chen, D.; Crnkovic, I. Component-based vs. model-based development: A comparison in the context of vehicular embedded systems. In Proceedings of the 31st EUROMICRO Conference on Software Engineering and Advanced Applications, Porto, Portugal, 30 August–3 September 2005; pp. 432–440. [CrossRef]

7. Saadatmand, M.; Leveque, T. Modeling Security Aspects in Distributed Real-Time Component-Based Embedded Systems. In Proceedings of the 2012 Ninth International Conference on Information Technology—New Generations, Las Vegas, NV, USA, 16–18 April 2012; pp. 437–444. [CrossRef]

8. Ahlberg, C.; Asplund, L.; Campeanu, G.; Ciccozzi, F.; Ekstrand, F.; Ekstrom, M.; Feljan, J.; Gustavsson, A.; Sentilles, S.; Svogor, I.; et al. The Black Pearl: An autonomous underwater vehicle. In Proceedings of the AUVSI Foundation and ONR's 16th International RoboSub Competition, San Diego, CA, USA, 22–28 July 2013.

9. Hänninen, K.; Mäki-Turja, J.; Nolin, M.; Lindberg, M.; Lundbäck, J.; Lundbäck, K.L. The Rubus component model for resource constrained real-time systems. In Proceedings of the International Symposium on Industrial Embedded Systems, Le Grande Motte, France, 11–13 June 2008; pp. 177–183.

10. Aleti, A.; Buhnova, B.; Grunske, L.; Koziolek, A.; Meedeniya, I. Software Architecture Optimization Methods: A Systematic Literature Review. *IEEE Trans. Softw. Eng.* **2013**, *39*, 658–683. [CrossRef]

11. Seo, C.; Malek, S.; Medvidovic, N. An Energy Consumption Framework for Distributed Java-based Systems. In Proceedings of the Twenty-Second IEEE/ACM International Conference on Automated Software Engineering, Atlanta, GA, USA, 5–9 November 2007; ACM: New York, NY, USA, 2007; pp. 421–424. [CrossRef]

12. Goraczko, M.; Liu, J.; Lymberopoulos, D.; Matic, S.; Priyantha, B.; Zhao, F. Energy-optimal software partitioning in heterogeneous multiprocessor embedded systems. In Proceedings of the 2008 45th ACM/IEEE Design Automation Conference, Anaheim, CA, USA, 8–13 June 2008; pp. 191–196. [CrossRef]

13. Wang, S.; Merrick, J.R.; Shin, K.G. Component allocation with multiple resource constraints for large embedded real-time software design. In Proceedings of the 10th IEEE Real-Time and Embedded Technology and Applications Symposium, Toronto, ON, Canada, 25–28 May 2004; pp. 219–226. [CrossRef]

14. Moser, I.; Mostaghim, S. The automotive deployment problem: A practical application for constrained multiobjective evolutionary optimisation. In Proceedings of the IEEE Congress on Evolutionary Computation, Barcelona, Spain, 18–23 July 2010; pp. 1–8. [CrossRef]

Article

Developing Self-Similar Hybrid Control Architecture Based on SGAM-Based Methodology for Distributed Microgrids

Pragya Kirti Gupta [1],* and Markus Duchon [2]

[1]   Chair of Software and Systems Engineering,Department of Informatics, Technical Universität München, Boltzmannstraße 3, 85748 Garching bei München, Germany
[2]   Fortiss GmbH, Forschungsinstitut des Freistaats Bayern für Softwareintensive Systeme und Services, Guericke Str. 25, 80805 München, Germany; duchon@fortiss.org
*   Correspondence: pragya.gupta@tum.de; Tel.: +498-936-035-2226

Received: 1 August 2018; Accepted: 18 October 2018; Published: 23 October 2018

Abstract: Cyber-Physical Systems (CPS) are the complex systems that control and coordinate physical infrastructures, which may be geographically apart, via the use of Information and Communication Technology (ICT). One such application of CPS is smart microgrids. Microgrids comprise both power consuming and power producing infrastructure and are capable of operating in grid connected and disconnected modes. Due to the presence of heterogeneous smart devices communicating over multiple communication protocols in a distributed environment, a system architecture is required. The objective of this paper is to approach the microgrid architecture from the software and systems' design perspective. The architecture should be flexible to support various multiple communication protocols and is able to integrate various hardware technologies. It should also be modular and scalable to support various functionalities such as island mode operations, energy efficient operations, energy trading, predictive maintenance, etc. These requirements are the basis for designing the software architecture for the smart microgrids that should be able to manage not only electrical but all energy related systems. In this work, we propose a distributed, hybrid control architecture suited for microgrid environments, where entities are geographically distant and need to operate in a cohesive manner. The proposed system architecture supports various design philosophies such as component-based design, hierarchical composition of components, peer-to-peer design, distributed decision-making and controlling as well as plug-and-play during runtime. A unique capability of the proposed system architecture is the self-similarity of the components for the distributed microgrids. The benefit of the approach is that it supports these design philosophies at all the levels in the hierarchy in contrast to a typical centralized architectures where decisions are taken only at the global level. The proposed architecture is applied to a real system of 13 residential buildings in a low-voltage distribution network. The required implementation and deployment details for monitoring and controlling 13 residential buildings are also discussed in this work.

Keywords: microgrid; distributed design; self-similar architecture; plug-n-play; distributed control; distribution network; field test

1. Introduction

Smart energy systems have been viewed as a typical example of Cyber-Physical Systems (CPS) [1], where the physical and organization processes are carried out, both at local and global scales. Across the world, the electrical grid is going through numerous changes in the transmission and distribution systems. Towards the realization of smart grids, the changes in the transmission grid (high-voltage energy network) are less drastic in comparison to the enhancements in the distribution networks below 10 kV. Some major breakthroughs in the distribution network are the decentralized power generation, increased use of renewable energy generation and integration of storage units [2,3]. These changes mostly affect network districts and ring and mesh networks. Despite the challenges involved in using renewable energy sources in the distribution system, the motivation to use clean energy sources to reduce greenhouse gases is pushing the technological development and regulatory (legal) boundaries. As a result, this has led to the development of advanced technologies [4,5] and novel business techniques. Improvements in the hardware infrastructure include better energy storage systems, smart sensors and advanced materials that can withstand the operational wear and tear. These smart devices are capable of monitoring the real-time information of the environment (e.g., weather data) and can transfer such information over the communication network. The advancements in the communication networks include the advanced communication protocols, the communication channels (Power Line Communication, Ethernet, mobile networks, etc.) and the communication quality features (such as latencies, signal strength, etc.). These advancements in the communication technology facilitate a reliable and robust interaction between the distributed entities.

According to the IEEE Standard for the Specification of Microgrid Controllers (MC) [6], a microgrid is a group of interconnected loads and distributed energy resources with clearly defined electrical boundaries that acts as a single controllable entity with respect to the grid and can connect and disconnect from the grid to enable it to operate in both grid-connected or island modes. A typical microgrid has several consumption devices (also known as loads), as well as the Distributed Energy Generation (DER) units such as solar panels, wind turbines, biogas plants, and combined heat and power plants. DERs are often connected with battery backup systems to make up for the intermittent behaviour of renewable energy generation. During the grid-disconnected mode, also known as *island* mode, the microgrid uses locally available energy sources [7]. Since the mode of operation depends on several factors such as the energy demand, supply from the renewables, operational constraints of the hardware, etc., a robust MC is required to monitor and control the microgrid.

For harnessing the full potential of renewable energy resources and for supporting stable power in a microgrid, the use of the latest hardware and advanced communication technologies is not sufficient. A Distribution Management System (DMS) [6], which is a collection of applications designed to monitor and control a distribution network efficiently and reliably, is required to the handle advanced hardware and the communication technologies. The design of the DMS involves several innovation areas, such as data processing and data analysis techniques, decision-making using mathematical or statistical models, real-time execution of the control signals, etc. Moreover, several microgrids operating in the distribution network may have smart and non-smart infrastructure and may also vary in size and scale. Therefore, the design of the DMS must be based on flexible, scalable, modular and secure software architecture to support and meet the objectives of the microgrid.

Several approaches for designing the DMS architecture are present in the literature [8–10]. They mostly focus on interoperability, flexibility, distributed controlling and on the scalability aspects in microgrids. Chandy et al. [11] proposed a System-of-Systems (SoS) architecture for smart grids. Their architecture focus on the modularity, flexibility and scalability aspects. Most of the literature in the area of the design of DMS architectures either include all the aforementioned aspects but restrict themselves only to the simulation environments or are developed in the real environment while focusing only on a few of the aspects of the microgrid. Practical problems like dealing with weak communication infrastructure and ability to integrate heterogeneous and inexpensive hardware become critical in a rural setting.

Florea et al. [12] proposed a fractal architecture for power grids, using the concept of holons. Holons are the autonomous entities with fixed functionalities and have the ability to interact with other holons. The functionality of each layer in the fractal architecture is the primary, intermediary and tertiary controlling. Holons at the lowest layer interacts with hardware. Holons at the intermediary level act as supplementary generator controllers and holons at the highest level perform smart grid core optimization controls. They focused on two characteristics of fractal systems, namely, self-similarity and self-organization. A centralized control hub holon is simulated and tested to emulate the dialogue with other control holons.

In the direction of fractal design research, there are also ongoing research projects such as Retière et.al. [13], who presented the methodology and initial results of the ongoing project *Fractal Grid* in the urban areas of Jura and Haute-Saone in France. Their work focused only on the criteria and constraints to be considered for creating the fractals for high and low voltage electrical networks. Initial survey results on the grid management and planning approaches have been discussed. The spatial organization of the electrical grid was considered so the needs of the users located in the urban areas were better addressed and voltage drop was taken as a measure of power quality.

In this work, we address the design challenges for the DMS or Microgrid Controller (MC) [6] for distributed microgrids in rural and remote locations. Based on our prior experience of designing the architecture of an MC for single nodes or smart homes [14], this paper proposes a hierarchical architecture for a distributed microgrid. The proposed architecture is flexible, scalable, modular, capable of handling heterogeneous hardware infrastructure and supports numerous microgrid objectives such as grid balancing, fault management, etc. Although the security and data privacy features are essential for microgrids, these aspects are not directly focused in this work. Architecture is capable of supporting advanced features such as security and data privacy, data analytics, energy efficiency, etc. A basic level of security (role-based authorization and password-based authentication) is supported by the architecture. In addition, data encryption techniques are also used to allow secure data transfer over the communication channel between distributed entities. While focusing on decision-making and intelligent controlling, we propose a self-similar and hybrid control architecture that is embedded throughout the layers of the hierarchy. As proposed by Schaetz [1], the architecture supports the three different dimensions of CPS i.e., self, cross and live dimensions. *Self-dimension* relates to the local operations required for autonomous operations of a microgrid, such as data handling at different layers and island mode operation in a smart home. *Cross-dimension* refers to the coordination required among microgrids when connected with the main grid, such as the area level islands, which have multiple smart homes and other loads along with the low-medium generation units. *Live-dimension* deals with the changes during runtime e.g., switching from grid connected to island mode operations. The proposed architecture is implemented and deployed on a 13-node low-voltage radial distribution system at Amrita University in Kerala, India. The cost of setting up microgrids is usually high and therefore in this work a major consideration was setting up low-cost microgrid in rural areas of India to digitize the electrical infrastructure. This work is carried out as part of the Indo-European research initiative.

This paper is organized in five major sections: Section 2 discusses the existing approaches for MC architecture in both centralized and distributed microgrids. In Section 3, detailed requirements and main design challenges for distributed microgrids are enumerated. Requirements are linked to standards and existing design principles to highlight the applicability and shortcomings of any one single design philosophy. This results in the need for combining different design styles leading to a hybrid control architecture. The proposed self-similar, hybrid control architecture is described in detail in Section 4 where every constituent and fulfillment of requirements are explained. Section 5 provides the implementation details of the proposed architecture and deployment details in real environment. The paper is concluded in Section 7 highlighting contributions and advantages of the proposed architecture. Finally, in Section 6, the lessons learnt, further challenges and future work of the proposed architecture are described.

2. Review on the Architectural Concepts for Microgrid Environments

A massive transformation in the electrical grid is rapidly increasing the possibility of new technologies and concepts. The larger vision and the purpose of the smart grids are relevant to understand the business objectives of the smart grids. In this section, we highlight the need of the architecture concepts that fulfill the business objectives of the microgrid, without depending heavily on the standards and methodology. Farhangi [2] highlights that the system architecture is a key ingredient for smart grids. He emphasizes that the fundamental attribute of future grids is the highly distributed Software Controller (SC), which is capable of monitoring and controlling distributed entities. The SC that is responsible for distributed control should be scalable to handle large amounts of data and offer a wide variety of functionalities. It should be flexible enough to accommodate the latest and future technologies. Amin [15] asserts that overall smartness in the distribution grid can be introduced by adding *smartness* in each independent entity in the grid. For example, intelligent algorithms implemented independently at device level, energy efficiency included in smart homes and solutions for demand-response at grid operator level, etc. improves the smartness of the overall power grid. He further suggests that this approach not only ensures application of wide variety of solutions but also makes it possible to apply solutions tailored for specific energy needs. Erlinghagen et al. [16] observe that Information and Communication Technology are transforming the electricity sector in Europe. As a result, there is a rise in ICT firms that are coming up with innovative solutions for smart grids. This study highlights that the MC must also support business requirements suitable for microgrids.

In order to fulfill the objectives, the design requirements are essential. Towards creating architecture of MCs, the survey by Martin et al. [17] is helpful, as they provide a clear understanding of microgrids, their requirements, compositions and attributes. These requirements are considered in the proposed design approach as well. Rohjans et al. [18] have systematically worked on the requirements for the SC architectures. They have listed out the essential non-functional requirements (NFR) for SC. The selection of NFRs is according to Sommerville [19].

Chandy et al. [11] highlight the System of System (SOS) architecture approach, which includes architectural styles such as silo, enterprise service bus, adapter principle and Open standard service mechanism, also known as Service-Oriented Architecture (SOA). Although SOA is helpful for distributed applications, the orchestration of services for distributed microgrids is crucial and a real challenge for architecture of MC, which is considered in the proposed architecture. Various software architecture design principles [19] such as Service-Oriented, Client-Server, peer-to-peer, Model-View-Controller and pipes-filters are frequently used in the existing design approaches of MCs, but the usage of one or more design principles depends on the specific microgrid setting. Zhabelova et al. [20] propose multi-agent smart grid architecture based on the communication protocol for intelligent electronic devices (IEC 61850/61499), with a special objective of Fault Location Isolation and Supply Restoration (FLISR), supported by design. The lack of support to other business requirements is missing in their work, such as data visualization, data analytics, predictive maintenance, etc. Towards the system architecture, the most known architecture style for smart grids is Supervisory Control And Data Acquisition (SCADA) [21]. Since SCADA supports only centralized decision-making and controlling, recent efforts have been made to include the distributed decision-making. These efforts include retrieve information from several SCADA systems as servers by a single OPC-UA client using web services. Open Platform Communication-Unified Architecture (OPC-UA) is widely known protocol in the industrial automation community, it is gaining relevance in smart grid research as well. There are also approaches that use the SCADA system with additional control systems. As mentioned before, another system architecture relevant for the SC is holonic architecture style [22,23], originally used for the manufacturing domain. *Holons* are the autonomous, cooperative and recursive entities that interact with other *holons* in a holarchic manner. Fractal approach is also proposed for designing smart grid system architecture. Towards the fractal design approaches, emphasis is given on both *self-similarity* and *self-orgnization* as described by [12,13]. Taking the fractal concept further, Bytschkow et al. [24] focused on the compositionality of self-similar components. *Goal-oriented* designing is also another popular approach used by many researchers. Marzband et al. [25] propose Transactive Energy (TE) framework, which comprises many home microgrids. These home microgrids are the individual smart homes or home energy management systems that interact with each other to create coalitions. The basis of creating coalition is the optimal use of installed resources by considering demand fluctuations, energy production, etc. From the design perspective, their work focuses more on the *self-organizational* aspect.

Not all architectural approaches lend themselves as suitable when applied to microgrids. Various European research projects focusing on the design and architecture of microgrid show the relevance and the insufficiency of the existing solutions. Girbau-Llistuella et al. [26] present the experience and approach used in setting up smart grid pilot network in the rural distribution network in Spain. Towards the distributed controlling, they have created two new agents, communicating with the SCADA system at the Distribution System Operator (DSO) side. In addition, there are many other research projects in Europe, especially focusing on the design of architecture for smart grids [27–30].

3. General Requirements on the Software Architecture for Distributed Microgrids

In this section, we describe the requirements taken into consideration for the architecture of MCs for the distributed microgrid. As per the functions described in the Standard for the Specification of Microgrid Controllers [6], a microgrid controller must fulfill the following:

- Operation in grid-connected and islanded modes.
- Automatic transition from grid-connected to islanded mode by providing a managed transition to islanded mode for microgrid loads during abnormal bulk power system conditions and planned interruptions of the system.
- Resynchronization and reconnection from islanded mode to grid-connected mode.
- Energy management to optimize, both real and reactive power generation and consumption.
- Ancillary services provision, support of the grid, and participation in the energy market and/or utility system operation, as applicable.

Based on the above requirements, it can be inferred that, in a distributed microgrid, an MC should be capable of monitoring and controlling the distributed entities, which could be real or virtual. Real entities correspond to the physical devices, whereas virtual entities mimic the behavior of physical entities.

From these functional features and, as per the survey conducted by Fang et al. [31] on the future of smart grids, we extract the functional and non-functional requirements for a Microgrid Control System (MCS) [6]. These are listed in Table 1 and described below.

Table 1. Functional and non-functional requirements for microgrids according to the surveys and standards for microgrid controllers. [6,31]

Functional Requirements	Non-Functional Requirements
Distributed control for island mode operations	Modularity
Centralized control for grid-connected operations	Flexibility
Hierarchical structure for coordination and synchronization	Scalability
Plug-n-play to support both features and devices	

1. Distributed control for island mode operations: Several microgrids should be able to operate in an independent manner, when operating in an island mode. These microgrids could communicate with each other without creating any adverse impact on the infrastructure. During the island mode operation, any individual microgrid should be able to balance itself to meet the required power demand at all times. Therefore, microgrids must be monitored and be allowed to run independently.
2. Centralized control for grid-connected operations: In case several microgrids are connected with each other in a grid connected mode, these individual microgrids should be able to communicate and share their resources in a collaborative way. Any microgrid can offer to publish certain services, while others requiring these services can subscribe to them in a collaborative manner.

In order to access the services, it is essential to know whether the services are internally available (within a microgrid) or external to the microgrid. The internal services within a microgrid perform local actions. For example, intelligent decisions within a smart home. In contrast, several smart homes within a microgrid may offer external services of energy sharing, which can be executed by the node of common coupling.

3. Hierarchical structure for coordination and synchronization: When several microgrids combine to form a larger microgrid, synchronization and coordination is required. They can again disjoin to make several distributed microgrids operating in islanded mode. To make this possible, functionalities of microgrids should be organized into a structure. Control functionalities and global controlling should be organized as a hierarchy allowing physical devices to be controlled for both local and global objectives. This can enable a command to be propagated through the hierarchy to all the connected microgrids, e.g., to change to island mode. Additionally, information from individual microgrids can be aggregated and analyzed on higher levels to make a larger grid, supporting a more global decision.

4. Plug-n-play: Also known as hot-plugging, this is another crucial requirement for a microgrid. It provides the flexibility to add, remove and modify microgrids or even hardware infrastructure at any time. The microgrid should be able to alter functionalities during island and grid-connected modes. Such plug-n-play features make the microgrid more intelligent and real-time responsive.

In order to highlight the need for the architecture of MC that can incorporate all the above explained requirements of microgrids, we present an association of each microgrid requirement with existing software architecture styles. This association helps in understanding the compatible and incompatible requirements between software design and the microgrid. Thus, the resulting enumeration, shown in table 2, forms the foundation to identify appropriate architecture design patterns suited for the application in distributed microgrids.

Table 2. Compatibility of functional requirement with the architecture style.

Requirements	Architectural Style
Distributed control for island mode operations	Peer-to-peer
Centralized control for grid-connected operations	Client-server
Hierarchical structure for coordination and synchronization	Layered
Plug-n-play to support both features and devices	Component-based or Service Oriented

The detailed description of the suitable architectural style corresponding to the functional requirement is as follows:

1. For the distributed operation of various entities, a *peer-to-peer* architecture style [19] is better suited, where all peers are able to communicate directly without interfering in the operation of other peers. All the individual microgrids have equal privileges and have direct communication with each other. MCs of individual microgrids are capable of sharing information and are able to switch modes form grid-connected to islands and vice versa.

2. A *client–server* architecture style [19] is suited where a server hosts certain services and multiple clients are able to use these services. The server publishes a service and waits for clients to make requests. When several microgrids switch back to the grid connected mode, the MC for the newly formed larger grid must be able to control the imbalances. Therefore, MC acts as master and the MCs of all the individual microgrids act as slaves.

3. The requirement for hierarchical structure of the functionalities is supported by the *Layered* architecture style [19], where multi-layered design provides abstraction to both data and services. Analysis and planning are done on the higher abstraction layers, as it requires larger response time. Data requirements also depend on the functionalities. Higher layers often have more

planning and analytical functionalities resulting in large response times. Lower layers closer to the physical devices are often responsible for the local actions and have shorter reaction times. Most common multi-layer designs have a presentation layer for visualization, a business layer for post processing of data and a device layer for interacting with the hardware in the field.

4. To support the plug-and-play requirement, the *component-based architecture* design suits well, as the functionalities are developed as loosely coupled components. Complex functionalities can be created as composite components. For instance, the smart switching between grid-connected and island mode operations can be applied only when there is a backup energy generation or storage resource available.

There are more requirements which could be addressed in the design of microgrids, but they are either too specific or only slightly influence the architecture design. *Component-based architecture* and *layered design* are taken into consideration to fulfill the modularity, flexibility and scalability requirements of the target microgrid. We propose a self-similar, layered and hybrid-control architecture for MC by combining different design styles to support distributed microgrids. This is described in Section 4.

4. Design of a Hybrid, Self-Similar but Distributed Architecture

MCs controlling microgrids both in island and grid-connected mode are required to perform certain functions, as described in the previous section. Consider a scenario, where multiple grids connect with each other to create a larger microgrid. In such a situation, all the MCs of individual microgrids must perform some common core functions such as avoidance of harmonic and voltage fluctuation, etc. Along with these core functionalities, there are local functionalities only relevant within the microgrid. The MC for the resulting larger microgrid now has a new function. Towards such scenarios, we propose a self-similar-, distributed-, hybrid control- software architecture for MCs.

Self-similarity coupled with the *hierarchical* design helps better cooperation and coordination of microgrids. All three perspectives of *self-similarity* are embedded in the design of the building blocks namely, structural, functional and behavioral. *Structural similarity* refers to the replication of the static structure or core functionalities for each individual MC. This also leads to a more robust design consisting of many low cost devices which can replace/take over each other's functionality on demand. The reuse of core functions in each MC enforces structural self-similarity. *Functional similarity* refers to the similarity in the dynamic processes. Due to the replication of core functions in each microgrid, certain dynamic processes also remain the same. For example, health check mechanisms, notifications, access control and data storage are the dynamic processes that can be reused. *Behavioral similarity* refers to the commonalities in the system dynamics. Behavioral similarity is introduced by the hierarchical structure, where each building block has either at least a parent or a child. For instance, behavior of operating in island mode as a single child in hierarchy with a single parent or having several children microgrids, in a grid connected mode.

4.1. Proposed Architecture

For each MC, the proposed architecture can be extended with components to support the objectives of the zone dimension as given by the Smart Grid Architecture Model (SGAM) reference architecture [32]. The interoperability dimension is reflected in the layered design of our architecture, which will be detailed in Section 5.2. In addition, the functionality can be adapted to different SGAM domains like Prosumer, DER, Distribution and Transmission System. In this work, the focus is mainly on the distribution system. This concept has been applied on the different layers in the distributed microgrid.

Considering the non-functional and architectural requirements of the MC, a schematic of the proposed distributed MCs is shown in Figure 1. The distributed microgrid is organized in a hierarchical structure. The structure is composed of layers where each layer denotes the extent of the control i.e., local, regional, global, etc. The global MC provides overall management of the microgrid. It is responsible for the overall network management and imposes global functionalities on the complete microgrid system. A global MC is composed of several intermediary MCs. The intermediary MC is composed of several smaller MCs. MCs operate independently, have their own decision-making capabilities and can communicate with other MCs in the higher or other layers.

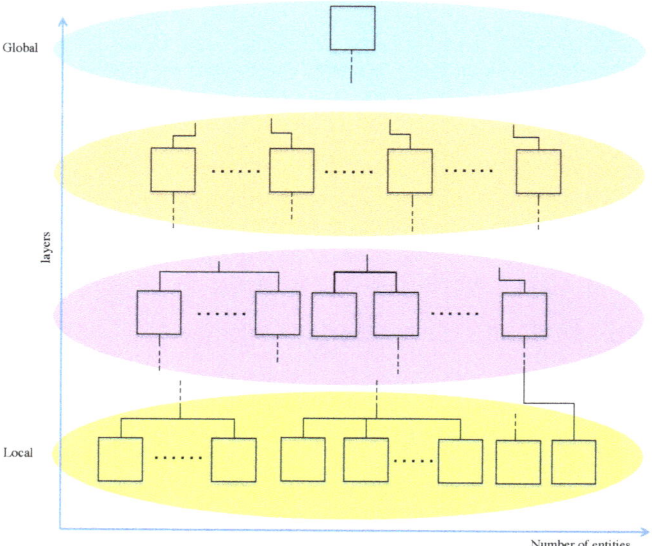

Figure 1. Generic layered architecture for a distributed smart microgrid.

The local layer is shown in the bottom of Figure 1. At the local layer are MCs, which are directly communicating with independent DERs and consumption units of a microgrid. Apart from these, sensors and actuators are also connected to the MCs at the local layer. The local MC controls all the devices connected locally within the microgrid. The physical devices provide data to the MC. The data from the physical devices is of two types: either measure of power quality (voltage, current, active-reactive power, etc.) or measure of the environmental factors (such as temperature, brightness, occupancy, etc.). Based on the data, the MC can make decisions to control the actuators. Apart from the capability to control physical devices, the MCs can also perform data analytics based optimization of the connected resources. For example, the decision to operate in an island mode with the help of Photo Voltaic (PV) coupled with battery backup system or to operate in a grid connected mode is a decision that can be taken by the local MC. The local MC operates in a distributed manner and satisfies the idea of edge computing [33] where data processing is closer to the data source, thereby resulting in faster decision-making of resource utilization.

Several local microgrids combine to form intermediary microgrids. These are shown in the middle layers of Figure 1. For example, at the intermediate layer could be streets or districts, which comprise individual buildings. Management of intermediary microgrids is done by the intermediary MCs. The intermediary MC is a collection of the local MCs. It provides a communication channel between individual MCs. For example, the intermediate street MC could balance the power supply between smart buildings and normal buildings. The intermediary MC can also take decisions for all the connected local MCs. Therefore, intermediary MCs play an important role in organizing and clustering towards grid balancing.

The functionalities of an intermediary MCs could include device and data aggregation, which is similar to the aggregation functionalities of a Head-End System (HES) [34]. Device aggregation is the mechanism of combining multiple devices to create a single virtual device. Data aggregation implies combining multiple independent devices to create a single virtual entity, which provides an aggregated single value computed on the data from multiple independent devices. Data aggregation helps in the extraction of knowledge from the raw data provided by devices. In addition to data processing, intermediary MCs provide the capability to generate control actions from automated decisions for predefined objectives. For instance, power theft can be detected by the intermediary MC either by comparing the metered value at the supply point with the aggregated data from the consumers or by aggregating the consumption data coming from the meters installed outside the building. Another kind of information that can be generated by intermediary MC are the Key Performance Indicators (KPI), regarding the performance of a cluster. The intermediary MCs can also be formed by aggregating several other intermediary MCs. For instance, a city is made of blocks, which are made of streets and finally individual buildings. Both blocks and streets could form intermediary MCs.

A single global MC provides overall management of the grid. The is shown in the top layer of Figure 1. The global MC is the final aggregation of the entire available intermediary MC. It is responsible for the overall microgrid management. It monitors and controls all the nodes in the lower layers. For example, at the global level, data processing techniques can be used to determine the performance and health status of the complete microgrid. Statistical data from the intermediary MC such as KPIs can be used to make decisions that affect the entire microgrid. The global MC generates control signals for the microgrid and sends them to intermediary MCs. Intermediary MCs receive control signals and forward these signals to the local MCs. Local MCs convert control signals to appropriate device specific format and send them to the physical devices for final execution of the control command.

In the proposed architecture, self-similarity of the MC enables it to be used for various objectives. For example, any MC at the lower layer can take the role of an intermediary MC (at the higher layer) in case of MC failure at the higher layer. The flexibility of the design allows addition or removal of functionalities. Diverse hardware can be integrated and the architecture is not bound to any particular hardware. The support to scalability ensures the inclusion of both layers and MCs. In other words, the same architecture can be applied to smart buildings, districts, regions or state levels. Modularity is incorporated in the MC by design. Each MC is an independent entity and can be added or removed from the microgrid without affecting the operation of the microgrid. In the case of an island mode, the MC is capable of managing its energy generation and consumption devices.

5. Implementation of Architecture in the Real Environment

In this section, we present the implementation details of the proposed architecture. Following the concepts described so far, a distributed management system was developed and deployed in a real world environment, which consists of a 13 node distribution grid. Each node is connected to a residential district, representing its own microgrid environment. Some districts where equipped with their own energy storage and generators. The architecture is designed for this distribution system and is done in an Indo-European research effort [35] towards integrating solutions from individual technologies starting with an electrical network over a robust communication network and finally a distributed software controller enabling management and control capabilities. This section is divided into three parts: first, the physical and the environmental conditions of the field demonstrator are described. Second, the implementation details of a single MC for a microgrid in the demonstrator are described. In the third part, we describe a hierarchical architecture of the complete demonstration site consisting of self-replicating MCs. Additionally, we explain in detail the similarities and differences of each MC, when placed at various hierarchical levels of the proposed architecture.

5.1. Field Demonstrator

The real world demonstrator site is located amidst the residential buildings in the campus of Amrita University, Kerala, India. The demonstrator on which the ICT architecture is applied is grid-connected with 3-phase power supply of 240 V at 50 Hz.

Figure 2 illustrates the systematic view of the distribution system, with the 13-nodes labeled as buildings. Each floor of the building is considered as a single load point or a power consumption entity and therefore represented as independent box in Figure 2. These buildings are connected with over-head and underground electrical lines. The over-head electrical lines supply power to the buildings, whereas underground lines (depicted as dashed lines) are tie-lines, which are activated only when one or more over-head line(s) are broken resulting in a change of the network topology to restore power supply. Overhead and tie-lines are connected to the electrical poles. At each pole, a smart measuring and a control box is mounted to monitor and control the power consumption of each building. Node 1 is connected to the general power grid (S) on one end and on the other end with the substations in the field demonstrator.

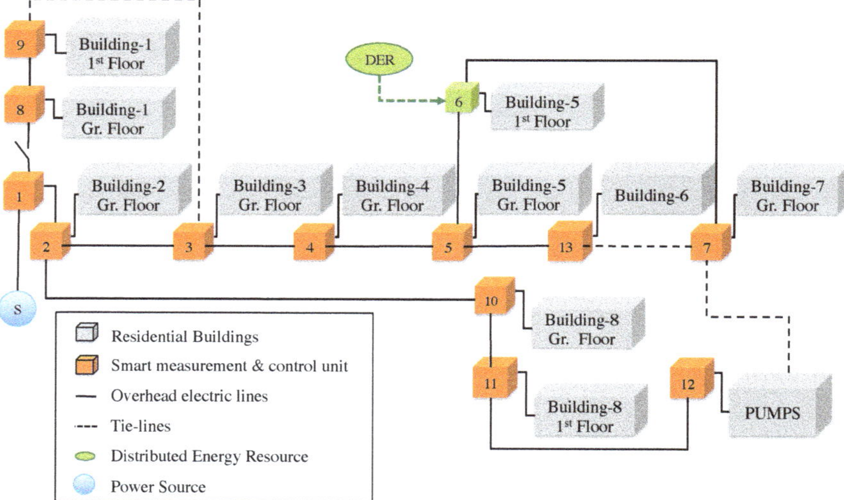

Figure 2. 13-node distribution system at Amrita University Campus in India.

Out of thirteen nodes, eleven consumption entities are residential buildings. A water pump represented as node 12 is connected with node 7. These consumption entities are called as *consumer* nodes. Building-5 is equipped with rooftop solar panels, labeled as DER in Figure 2. This node is powered by both solar power and power from the grid. It is called a *prosumer* node as it can both produce and consume power. Other nodes do not have alternative power generation units and rely only on the conventional power supply from the grid.

5.2. Architecture of the Microgrid Controller (MC)

The design of the MC is based on the requirements and architecture principles discussed in Sections 3 and 4.1. Figure 3 shows the architecture of a single MC. The architecture of MC is based on our previous work [14] for setting up an intelligent node in the office environment.

Components are organized as low-level, core, high-level and external components. *Hierarchical structure* is established by implementing device interfaces in the low-level layer also known as a device layer, in which physical devices are integrated. The low-level functional components (or device interfaces) interact with hardware including virtual peripherals. For instance, in the 13-node distribution system, smart meters are installed at each building. Voltage, current, active power, reactive power, etc. are measured by smart meters, which are integrated as *actuatorclients* software components in the MC. The smart meters are also connected with the switches, which can be controlled by the actuator client component. The feedback of the change in the switch status is read using I2C serial protocol to ensure that the control command generated translates into a real hardware control signal.

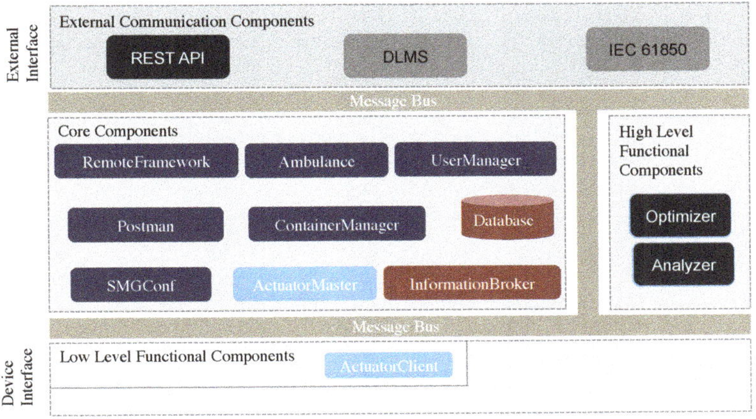

Figure 3. Component-based architecture of the Microgrid Controller (MC).

The core-component layer has components that carry out data handling, data organizing and data persistency functionalities. These core-components must be implemented for each MC to carry out specific core objectives. By the use of core-components, replication is introduced towards self-similarity. Specific objectives of each individual MC are implemented in the high-level components. Examples of each individual specific objective could be forecasting consumption or generation using data analytics, optimization of energy, etc. These high-level components are optional and do not affect the operation of core components and hence can be added, removed or modified anytime, providing plug-n-play capability to the MC. For coordination among MCs, each MC can communicate with other using external components. The external interfaces provide communication to the entities which use the data generated by the individual MC. Few examples are interaction with other MCs, offering information conforming to standards like IEC 61850 or Device Language Message Specification (DLMS), information sharing with energy markets or enabling services for network operators like demand side management.

We provide a quick overview of each core component shown in Figure 3. Core components are mandatory for the operation of single instance of the basic unit:

- InformationBroker provides the data logging/persistency functionality. This component provides several interfaces to common backend storage technologies such as MySQL, PostgreSQL, MongoDB, etc.
- ContainerManager models the real world entities/physical infrastructure by defining the unit's relationship in the existing physical setup. In the previous work [14], we described the smart office building structure (floor, rooms, devices), where a *building* is the root container, which has several *floor* child containers. Each floor has several rooms and rooms contain several devices as children.
- UserManager provides the access control functionality for the operation of the unit by defining roles and authorization mechanisms for the authorized/restricted access.
- Ambulance acts like a watchdog for monitoring health of each component and making the user aware by sending notifications using the postman component. This component initiates the fault detection and handling techniques internal to the software application.
- Postman provides interface to various notification technologies such as alarms, emails, tweets and SMS, etc.
- Remoteframework Library connects with the message bus, acting as a communication channel between the components to interact with each other.

- SMGConf acts as a configuration library containing information of the addresses and ports for databases and device specifications.

The proposed architecture enables the replication of MC for a single microgrid and provides the ability of distributed decision-making among the participating microgrids. Therefore, *self-similarity* is embedded in the architecture of each MC.

5.3. Implementation of a Microgrid Controller (MC) Following the Proposed Hybrid Architecture

The proposed hybrid architecture (Figure 1) adopted to the 13-node system (in Section 5.1) is shown in Figure 4. The 13-nodes are organized as a set of so called Intelligent nodes at the local control layer, three aggregators at the area control layer and a single control station at the global control layer. All the intelligent nodes have physical devices (smart meters) connected with them, which are installed, inside the smart measuring box. Each area is monitored and controlled by a unique aggregator. Note that the number of independent areas is proportional to the number of aggregators. This information is processed and forwarded to the control station. As highlighted before, the aggregators are also capable of executing commands or accessing information depending on their configuration.

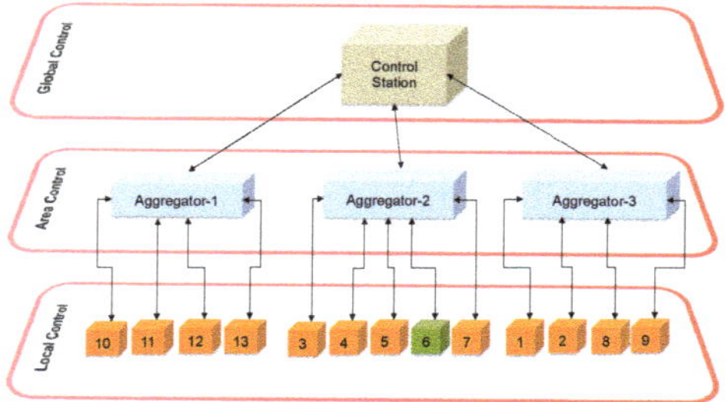

Figure 4. Proposed self-similar hybrid architecture implemented in the 13-node radial distribution system.

Aggregator-2 receives consumption (Voltage, Current) from the Intelligent nodes 3, 4, 5, 6 and 7 to generate KPIs. These KPIs are the indicators of the power quality supplied at the intelligent nodes summarized as average, maximum or minimum daily consumption of each intelligent node.

Figure 5 shows the 13-node distribution system from the ICT perspective. In the cyber infrastructure, the 13 nodes are shown, using the Raspberry Pi deployment platform. A Java-based application based on the MC design as shown in Figure 3 is deployed on each Raspberry Pi. There is a wired connection between every Raspberry Pi and smart meter. The communication between nodes and aggregators (represented as AGG-1, AGG-2, AGG-3) is currently using REpresentational State Transfer (REST)-based web interfaces [36]. These interfaces are flexible and, in the future, these can be replaced by appropriate Standards like DLMS, IEC 61850, etc. depending on the site requirements. The three aggregators communicate with a single control station.

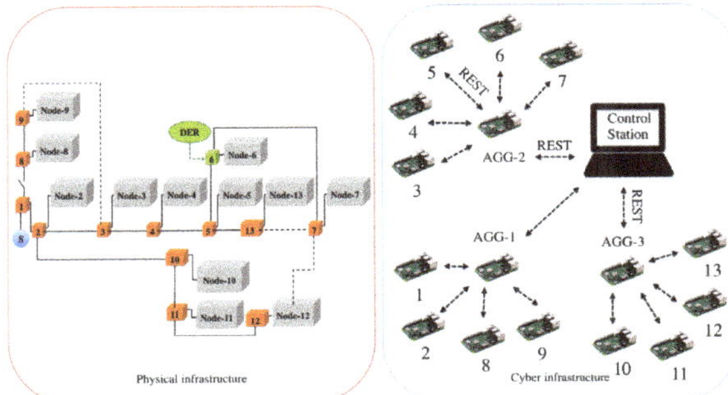

Figure 5. The implementation overview with respect to physical 13-node distribution system and cyber system.

Figure 6 provides a combined view of the internal architecture of a distributed microgrid, MCs placed in different layers and the communication channel across layers. Tracing the communication across the three layers between intelligent node-6, aggregator-2 and the control station, Figure 6 emphasizes the self-similarity of the proposed hybrid architecture. For the sake of simplicity, the detailed view of core components have not been shown as they are the same as in Figure 3.

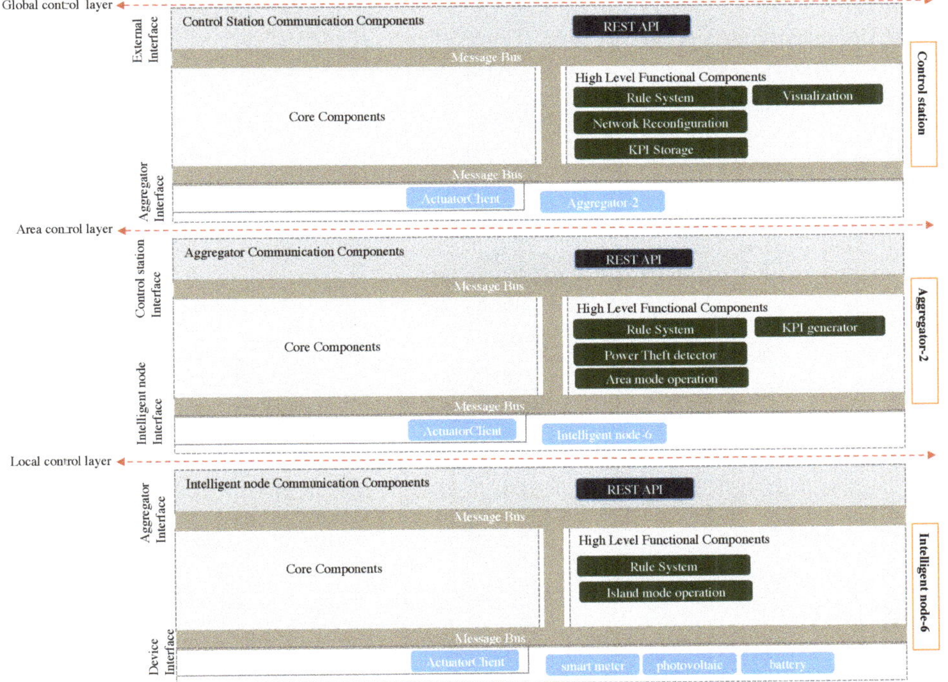

Figure 6. Detailed view of the Microgrid Controller (MC) implementation at each layer of the proposed architecture.

At the bottom of Figure 6, local control layer shows components of node-6. The physical devices connected with Intelligent node-6 are: photovoltaic, battery and smart meter(s). Devices are connected as actuatorclients, which implement appropriate communication protocols and logic to convert the device-specific information into a homogeneous format. The high level functional components for node-6 are a rulesystem and an island mode operation component. As the name indicates, the rulesystem provides the capability to make smart decisions based on the available information. Constraints are developed as set of rules in the rule system. Island mode operation means the actions carried out to use either solar power or the stored energy from the battery. For communication with aggregator-2, REST calls are made to send the device data and to receive the control signals either from aggregator-2 or from the control station. A REST interface at each layer provides the flexibility of external communication such as visualization of raw data at node level and KPI data at the aggregator level.

The raw data from node-6 is sent to the aggregator. The high level components of the aggregator are a rulesystem, a power theft detector, area mode operation and a KPI generator. The functionality of the rule system is the same as at node-6, i.e., smart decision-making. Power theft detection is based on the consumption values from all the connected Intelligent nodes (3, 4, 5, 6 and 7) and then checking the power consumption values with pre-defined threshold values. KPI shows the quality of power supplied provided to Intelligent nodes. These quality indicators are Voltage fluctuations, harmonic distortions, daily peak consumption, etc. The Area node operation enables power sharing from photovoltaic at node-6 with the adjacent nodes. For communicating with the control station, REST calls are made to send the generated KPIs and to receive control signals from the control station.

The global control layer is shown as the top layer in Figure 6. The KPI data and the regular backup data from the aggregator are received by the control station and saved in the local database. The high level functional components include global decisions concerning the complete network such as network reconfiguration, KPI storage, visualization and rule system. Similarly, post processing is done on the stored KPI.

Finally, Figure 7 shows the complete implementation of MCs in the 13-node system. This 13-node system is in operation since March 2018 and data is being generated and stored at regular intervals on the control station server, where post processing can be done for statistical analysis or to support energy related services.

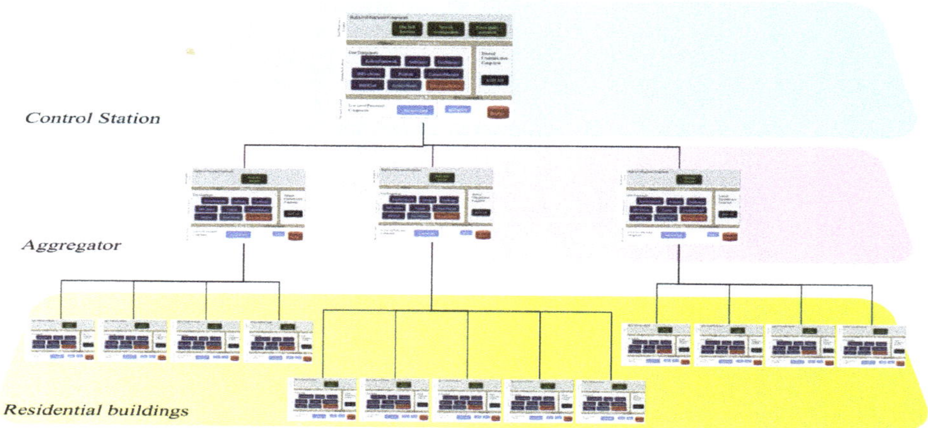

Figure 7. Complete view of the Microgrid Controller (MC) implementation at each layer in the 13-node distribution system.

6. Discussion

As mentioned above, in this work, the architecture concept to support low cost equipment to fulfill the complex tasks in a distributed microgrid is established. A robust underlining architecture enables the infrastructure which supports relevant services of microgrid. The architecture of intelligent nodes, aggregators and control station is related to the general trend of edge computing. Shifting more processing and decision processes closer to the embedded devices results in increased scalability. Another benefit of the decision-making at the intelligent nodes level results in lower upstream bandwidth requirement for the communication network. In situations where the communication networks are unstable, like at the implemented site, decision-making at the intelligent node is clearly advantageous. Unlike a classical hierarchical network, where the control signals are generated only at the highest layer, the proposed architecture offers decision-making and controlling at all the layers. Self-similarity enables on-the-fly change of roles. In case of a failure at a higher-level component, e.g., failure of an aggregator, one of the intelligent nodes can take over the functionality until the aggregator is restored or replaced. Moreover, the proposed self-similar building blocks support additional energy services for data preprocessing and decision-making. The proposed architecture has been implemented in a rural setting in India, where communication network and weather conditions were major constraints. The stability and signal strength of the communication network was a challenge. Appropriate communication networks were selected based on the detailed investigation of impact of weather conditions and other infrastructure constraints. This is out of the scope of the paper and not described here. Therefore, use of only a single communication protocol was not feasible in our work. A wired connection exists between the smart sensors and the local MCs, while a wireless communication channel is present between intelligent nodes at the local layer and the aggregator MCs.

Further investigation is required for the seamless operation of microgrids. An apparent limitation is the communication network connection issue; if the aggregators are not very well interconnected, then intermittent communication network failures may cause partial or total information loss of dependent building blocks. Therefore, interdependency among the self-similar blocks must be taken into account at the design time to cope up with the communication network glitches. Another consideration is maintenance of such systems. Each residential owner should have the capability to maintain and modify the local MC, but with the increase in the system complexity and higher modularity, maintenance and ownership of application and information becomes an additional challenge. It becomes increasingly difficult for multiple owners of the same application to maintain the complete system. These aspects are worth investigating from the maintainability perspective. In this work, the ownership and maintenance of the application is done by a single entity i.e., Distribution Station Operator. Lastly, the integration of self-similar blocks using heterogeneous services requires highly skilled developers and might require considerable time and effort. These limitations were out of scope of this research work and require experts from different areas of research.

The architecture concept for the demonstrator allows flexibility for future extensions such as the integration of additional buildings or floors. To scale the system, the existing intelligent nodes can be replicated and connected to the existing or new aggregators. The integration of an additional hierarchical layer of sub-control stations managing a subset of the whole network is feasible. Aside from the extension by additional adding nodes, the functionalities of the respective subsystems can be extended too. The fine-grained monitoring and control capabilities of the deployed system allows implementation of additional analytical and visualization services to increase energy awareness, replacement of components and dynamically changing the role of any MC.

As the next step in this work, data analysis and optimization techniques will be carried out to assess the performance of the system. Based on the data availability, various intelligent algorithms like network reconfiguration, frequent switching from island mode to grid-connected mode and vice versa will be tested. Graceful degradation strategies will be implemented in order to maintain a grid-disconnected condition for as long as possible. The system is also planned to equip smaller subnetworks with PV and battery backup systems to operate in an island mode. Here, each island could be responsible for the operation of a subnetwork and for interacting with the neighboring subnetworks.

7. Conclusions

In this paper, one of the application domains of Cyber Physical System is discussed. A distributed, hybrid control architecture is proposed for smart micro grids. In the proposed architecture, various functional and non-functional requirements are satisfied and a direct relationship is established between the distributed microgrids and design concepts of software architecture. The distinct contribution of this work is the use of fundamental software architecture concepts for a complex microgrid application domain. This work demonstrates the need of software engineering concepts towards realization of CPS. Peer-to-peer architecture is observed to be relevant for distributed control for island mode operations. Similarly, client–server architecture for centralized control, layered architecture for hierarchical structure and component or service-oriented architecture for plug and play of features and devices are chosen. The distributed microgrid is organized in a hierarchical structure, composed of layers. The layers could be local, regional, global, etc. Architecture of each individual MC is also presented. MC architecture is replicated for every node, at every layer of the hierarchy. The MCs in local layers communicate and interact directly with the DERs and consumption devices of the microgrid via smart sensors and actuators. Multiple local MCs can be aggregated to form an intermediate MC, which can take decisions for all connected local MCs. The intermediate MC further acts as a communication channel between local MCs and the global MC. The global MC is the final aggregation of all individual MCs and provides overall management to the entire microgrid. It also can take decisions that affect the entire microgrid. The proposed architecture has self-similarity embedded at every layer of the hierarchy. Components in the self-similar building blocks could be classified as low-level, core, high-level and external components. Low-level components interact with the physical hardware. The core components are the same in every MC irrespective of the layer in which it is present. The core components carry out data handling, data organizing, and data persistency functionalities. Layer specific objectives are included in the high-level components. The high-level components are optional and do not affect the working of the core components. Interfacing with other MCs is done using external components. By using a common structure for every MC, replication is introduced towards self-similarity. This enables flexibility, modularity and scalability required for the local and global scale operations. Towards the evaluation of the architecture, it is implemented in an actual 13-node radial distribution system at Amrita University Campus, India. The details of the implementation and the challenges faced are discussed in this work. Data from the implementation site is being collected since March 2018. The implemented architecture satisfies both the functional and non-functional requirements and thereby demonstrates the relevance of the proposed architecture for distributed microgrids.

Author Contributions: P.K.G. proposed the idea and performed the experiment. P.K.G. and M.D. prepared the manuscript. M.D. provided supervision and direction to the work.

Funding: This work was funded by the Federal Ministry of Education and Research (BMBF) and the Technical University of Munich (TUM) in the framework of the Open Access Publishing Program.

Acknowledgments: This work is carried out as part of the Stabiliz-Energy project (November 2014 until March 2018), a joint research work between India and Europe (Finland, Germany and Portugal). We especially thanks our project partners, Amrita Center for Wireless Networks and Applications for providing details of the radial distribution system used as a case study in this work. We acknowledge the valuable suggestions and constant support from all the partners in this work.

Conflicts of Interest: The authors declare no conflict of interest.

References

1. Schätz, B. Platforms for cyber-physical systems-fractal operating system and integrated development environment for the physical world. In Proceedings of the 3rd International Workshop on Emerging Ideas and Trends in Engineering of Cyber-Physical Systems (EITEC), Vienna, Austria, 11–14 April 2016; pp. 1–4.
2. Farhangi, H. The path of the smart grid. *IEEE Power Energy Mag.* **2010**, *8*, 18–28. [CrossRef]

3. Amin, S.M.; Giacomoni, A.M. Smart grid, safe grid. *IEEE Power Energy Mag.* **2012**, *10*, 33–40. [CrossRef]

4. Tavakoli, M.; Shokridehaki, F.; Marzband, M.; Godina, R.; Pouresmaeil, E. A two stage hierarchical control approach for the optimal energy management in commercial building microgrids based on local wind power and PEVs. *Sustain. Cities Soc.* **2018**, *41*, 332–340

5. Hussain, S.S.; Ustun, T.S.; Nsonga, P.; Ali, I. IEEE 1609 WAVE and IEC 61850 standard communication based integrated EV charging management in smart grids. *IEEE Trans. Veh. Technol.* **2018**, *67*, 7690–7697. [CrossRef]

6. IEEE Power Electronics Society. *2030.7-2017 IEEE Standard for the Specification of Microgrid Controllers*; IEEE: Piscataway, NJ, USA, 2017.

7. Tavakoli, M.; Shokridehaki, F.; Akorede, M.F.; Marzband, M.; Vechiu, I.; Pouresmaeil, E. CVaR-based energy management scheme for optimal resilience and operational cost in commercial building microgrids. *Int. J. Electr. Power Energy Syst.* **2018**, *100*, 1–9. [CrossRef]

8. Tebekaemi, E.; Wijesekera, D. A communications model for decentralized autonomous control of the power grid. In Proceedings of the International Conference on Communications (ICC), Kansas City, MO, USA, 20–24 May 2018; pp. 1–6.

9. Obaidat, M.S.; Reddy, N.R.; Venkata, K.P.; Saritha, V. Context-aware middleware architectural framework for intelligent smart grid data management. In Proceedings of the International Conference on Computer, Information and Telecommunication Systems (CITS), Colmar, France, 11–13 July 2018; pp. 1–5.

10. Diaconescu, A.; Frey, S.; Müller-Schloer, C.; Pitt, J.; Tomforde, S. Goal-oriented holonics for complex system (self-) integration: Concepts and case studies. In Proceedings of the 10th International Conference on Self-Adaptive and Self-Organizing Systems (SASO), Augsburg, Germany, 12–16 September 2016; pp. 100–109.

11. Chandy, K.M.; Gooding, J.; McDonald, J. Smart grid system-of-systems architectures. *Asset. Sce. Com.* **2010**. Available online: https://www.gridwiseac.org/pdfs/forum_papers10/gooding_gi10.pdf (accessed on 22 October 2018).

12. Florea, G.; Chenaru, O.; Popescu, D.; Dobrescu, R. A fractal model for power smart grids. In Proceedings of the 20th International Conference on Control Systems and Computer Science (CSCS), Bucharest, Romania, 27–29 May 2015; pp. 572–577.

13. Retière, N.; Muratore, G.; Kariniotakis, G.; Michiorri, A.; Frankhauser, P.; Caputo, J.G.; Sidqi, Y.; Girard, R.; Poirson, A. Fractal grid—Towards the future smart grid. In Proceedings of the 24th International Conference on Electricity Distribution CIRED, Glasgow, Scotland, 12–15 June 2017; p. 1236.

14. Duchon, M.; Gupta, P.K.; Koss, D.; Bytschkow, D.; Schätz, B.; Wilzbach, S. Advancement of a sensor aided smart grid node architecture. In Proceedings of the 6th International Conference on Cyber-Enabled Distributed Computing and Knowledge Discovery, Shanghai, China, 10–12 October 2014; pp. 349–356.

15. Amin, S.M.; Wollenberg, B.F. Toward a smart grid: Power delivery for the 21st century. *IEEE Power Energy Mag.* **2005**, *3*, 34–41. [CrossRef]

16. Erlinghagen, S.; Markard, J. Smart grids and the transformation of the electricity sector: ICT firms as potential catalysts for sectoral change. *Energy Policy* **2012**, *51*, 895–906. [CrossRef]

17. Martin-Martínez, F.; Sánchez-Miralles, A.; Rivier, M. A literature review of Microgrids: A functional layer based classification. *Renew. Sustain. Energy Rev.* **2016**, *62*, 1133–1153. [CrossRef]

18. Rohjans, S.; Dänekas, C.; Uslar, M. Requirements for smart grid ICT-architectures. In Proceedings of the International Conference and Exhibition on Innovative Smart Grid Technologies (ISGT Europe), Berlin, Germany, 14–17 October 2012; pp. 1–8.

19. Sommerville, I. *Software Engineering*, 10th ed.; Pearson: London, UK, 2015.

20. Zhabelova, G.; Vyatkin, V. Multiagent smart grid automation architecture based on IEC 61850/61499 intelligent logical nodes. *IEEE Trans. Ind. Electr.* **2012**, *59*, 2351–2362. [CrossRef]

21. Daneels, A.; Salter, W. What is SCADA? In Proceedings of the International Conference on Accelerator and Large Experimental Physics Control Systems, Trieste, Italy, 4–8 October 1999.

22. Negeri, E.; Baken, N.; Popov, M. Holonic architecture of the smart grid. *Smart Grid Renew. Energy* **2013**, *4*, 202. [CrossRef]

23. Negeri, E.; Baken, N. Architecting the smart grid as a holarchy. In Proceedings of the 1st International Conference on Smart Grids and Green IT Systems, Porto, Portugal, 19–20 April 2012; SciTePress: Setubal, Portugal, 2012.

24. Bytschkow, D. Towards composition principles and fractal architectures in the context of smart grids. *it-Inf. Technol.* **2016**, *58*, 3–14. [CrossRef]

25. Marzband, M.; Azarinejadian, F.; Savaghebi, M.; Pouresmaeil, E.; Guerrero, J.M.; Lightbody, G. Smart transactive energy framework in grid-connected multiple home microgrids under independent and coalition operations. *Renew. Energy* **2018**, *126*, 95–106. [CrossRef]

26. Girbau-Llistuella, F.; Díaz-González, F.; Sumper, A.; Gallart-Fernández, R.; Heredero-Peris, D. Smart grid architecture for rural distribution networks: Application to a Spanish pilot network. *Energies* **2018**, *11*, 844. [CrossRef]

27. Giordano, V.; Gangale, F.; Fulli, G.; Jiménez, M.S.; Onyeji, I.; Colta, A.; Papaioannou, I.; Mengolini, A.; Alecu, C.; Ojala, T.; et al. Smart grid projects in Europe. Available online: https://ses.jrc.ec.europa.eu/sites/ses/files/documents/smart_grid_projects_in_europe.pdf (accessed on 22 October 2018).

28. Zaballos, A.; Vallejo, A.; Selga, J.M. Heterogeneous communication architecture for the smart grid. *IEEE Netw.* **2011**, *25*, 30–37. [CrossRef]

29. Rohjans, S. A standard-compliant ICT-architecture for semantic data service integration in smart grids. In Proceedings of the Innovative Smart Grid Technologies (ISGT), Bangalore, India, 10–13 November 2013; pp. 1–6.

30. Kok, K.; Karnouskos, S.; Nestle, D.; Dimeas, A.; Weidlich, A.; Warmer, C.; Strauss, P.; Buchholz, B.; Drenkard, S.; Hatziargyriou, N.; et al. Smart houses for a smart grid. In Proceedings of the 20th International Conference and Exhibition on Electricity Distribution-Part 1, Prague, Czech Republic, 8–11 June 2009; pp. 1–4.

31. Fang, X.; Misra, S.; Xue, G.; Yang, D. Smart grid—The new and improved power grid: A survey. *Commun. Surv. Tutor. IEEE* **2012**, *14*, 944–980. [CrossRef]

32. CENELEC. *ETSI, "Smart Grid Reference Architecture," CEN/Cenelec/ETSI Smart Grid Coordination Group Std.*; Technical report; CENELEC: Brussels, Belgium, November 2012.

33. Gaber, M.M.; Gomes, J.B.; Stahl, F. Pocket data mining. In *Big Data on Small Devices. Series: Studies in Big Data*; NSTA: Arlington VA, USA, 2014.

34. Sarfi, R.; Green, B.D.; Simmins, J. *AMI Network (AMI Head-End to/from Smart Meters)*; Technical Report; Onramp Wireless: San Diego, CA, USA, 2012.

35. Gupta, P.K.; Martins, R.; Chakarbarti, S.; Remanidevi, A.D.; Mäki, K.; Krishna, G.; Schätz, B.; Ramesh, M.V.; Singh, S. Improving reliability and quality of supply (QoS) in smart distribution network. In Proceedings of the National Power Systems Conference (NPSC), Bhubaneswar, India, 19–21 December 2016; pp. 1–6.

36. Fielding, R. Representational state transfer. *Architectural Styles and the Design of Network-based Software Architecture*; University of California: Irvine, CA, USA, 2000; pp. 76–85.

MDPI

St. Alban-Anlage 66

4052 Basel

Switzerland

Tel. +41 61 683 77 34

Fax +41 61 302 89 18

www.mdpi.com

Designs Editorial Office

E-mail: designs@mdpi.com

www.mdpi.com/journal/designs